Dolphin Communication and Cognition

Dolphin Communication and Cognition

Past, Present, and Future

edited by Denise L. Herzing and Christine M. Johnson

The MIT Press
Cambridge, Massachusetts
London, England

This book was set in ITC Stone Serif by Toppan Best-set Premedia Limited.

Library of Congress Cataloging-in-Publication Data is available.

ISBN: 978-0-262-02967-4 (hardcover)
ISBN: 978-0-262-54962-2 (paperback)

To all the dolphins that showed us their world or interacted in ours. Thank you for sharing the details of your sounds, signals, and societies.—DLH

To Ken, for inspiring so much of the work in this book, and for showing such faith in us to do it.—CMJ

Contents

Preface

When we were students, our favorite place in the library stacks was QL737—because that's where the dolphin books could be found. Even when the field was in its infancy, the early works (Norris, 1966; Caldwell & Caldwell, 1972), including some excellent popular books (e.g., Norris, 1974; McIntyre, 1974; Cousteau & Diole, 1975; Pryor, 1975), were full of fascinating descriptions of dolphin behavior and remain important sources to this day. As field and laboratory sites began to develop, new compilations appeared, providing fresh insights into the range of dolphins' cognitive abilities (Herman, 1980; Schusterman, Thomas, & Wood, 1986; Leatherwood & Reeves, 1990; Roitblat, Herman, & Nachtigall, 1993), as well as their behavior in the wild (Norris, Würsig, Wells, & Würsig, 1994; Pryor & Norris, 1991; Mann, Connor, Tyack, & Whitehead, 1999). Further books on their extraordinary echolocation abilities (Au, 1993) and other sensory systems (Thomas & Kastelein, 1990) helped to round out our understanding of the constraints on, and resources for, dolphin communication and cognition.

Since then, although a number of fine books have included updates on dolphin cognition (e.g., Bearzi & Stanford, 2008) and field behavior (e.g., Würsig & Würsig, 2010; Herzing, 2011; Whitehead & Rendell, 2014), some time has passed since an edited volume, with a broad scope and up-to-date details, has been available. This book is designed to fill that niche. Thanks to advances in technology for monitoring, recording, and analyzing behavior, and a sustained interest in exploring and modeling the dolphin "mind," an impressive amount of data has been collected over the past twenty years, despite obvious practical difficulties. It is our hope that this book can serve as a comprehensive, contemporary reference for anyone interested in conducting or evaluating dolphin research.

The book's organization reflects the array of information that we believe is essential for an informed approach to the study of dolphin communication and cognition. Because those processes are embodied, we begin by reviewing information about the dolphin brain, the anatomy of its unique sound production and reception systems, and the full complement of its sensory abilities. Part 1, especially, reflects the dolphin's evolutionary history of aquatic adaptation and so highlights the species-specific constraints on communication and cognition.

Part 2 focuses on communication. It examines the variability in dolphin social ecology, reviews what we know of dolphins' complex vocalizations, and demonstrates the multimodality of their communication systems. These approaches all emphasize diversity, the importance of context, and the demands for multifaceted models if we are to develop ecologically valid approaches to future research.

Part 3 looks at the research on cognition, both in the laboratory and in the field. The extensive experimental work enables us to compare dolphin performance with similar studies in other species, including humans. And when we consider cognition in terms of the set of problems and resources faced by a developing dolphin, it becomes apparent how critical ontogeny is to understanding cognitive function.

The book's final part, on the future of dolphin research, offers alternative approaches, ethical considerations, and an outline of provocative questions that remain in the field. It aims to provide a set of methodological and conceptual tools that, together with the data and issues presented in the earlier chapters, can help prepare us to better conduct the upcoming decades of research. In what follows, a brief summary of each chapter will give a sense of each contributor's approach to this exciting enterprise.

In chapter 1, Lori Marino discusses the evolutionary factors that may have helped shape the dolphin's large brain, and reviews the comparative neuroanatomy of contemporary species. This highlights the hypertrophy of auditory structures and their integration with the other senses, the reproportioning of the limbic and paralimbic systems, and the corresponding enlargement and rearrangement of the cerebral cortex. While we still have much to learn about the dolphin brain, especially concerning connectivity and functionality, research to date suggests that it may be a multitiered system, specializing in elaborate vocalization and social regulation.

In chapter 2, Ted Cranford, Mats Amundin, and Petr Krysl focus on the anatomy of the sound production and reception systems, which have undergone significant elaboration since the animals' return to the sea. They discuss current issues in the field such as the number of "click" sources in dolphin phonation, the potential for internal beam steering, and the possibility of a continuity between whistles and burst pulse sounds. They characterize the integrated structures of the sound reception system as an internal "acoustic antenna" and show how techniques for the study of its operation—including jawphones, head scans, and computational modeling—affect our understanding of these systems.

In chapter 3, Wolf Hanke and Nicola Erdsack review what is known about multiple sensory modalities of dolphins, and how their range, acuity, specificity, and so on, are constrained by both anatomy and environment. These include vision, hearing and echolocation, mechanoreception, chemoreception, and, in some species, magnetoreception and electrical senses. If we aim to study how dolphins discriminate, evaluate, and respond to their social and physical worlds, we need to understand all these sources of information.

In chapter 4, Bernd Würsig and Heidi Pearson review what we know about dolphin societies in the wild. By examining, in some detail, the best-known examples of a few key species, they show the role that differences in habitat can play in group size, social structure, diel activity patterns, foraging strategies, and so on. They also discuss how life history parameters interact with ecotype to influence learning opportunities. The authors also remind us how important such details are in the formation of conservation plans for dolphin societies in the wild.

In chapter 5, Marc Lammers and Julie Oswald focus on the details of acoustic analysis. They review the challenges involved in categorizing and analyzing dolphin vocalizations, including whistles, clicks, and those mysterious burst pulse sounds. The many challenges of both recording and analyzing signals become clear in the authors' discussions of amplitude modulation, harmonics, and graded signals. With a focus on emerging technology and a hint of the potential functions of these sounds, we are left with a sense of hope that someday we may better understand the acoustic world of these aquatic beings.

In chapter 6, Denise Herzing takes us through the chronological progression of studies that correlate sound with behavior. As she points out, before the 1990s most of the studies on sound linked to behavior were quite anecdotal, if extensive. As the 1990s ushered in new opportunities for developing not only field sites in the wild but advanced technology to aid researchers in signal recording, new perspectives on communication in the wild emerged. This review of the multimodal world of dolphins gives us new perspective but also reminds us that we have much work to do to understand the complexities of dolphin communication.

In chapter 7, Adam Pack reviews the details of experiments conducted in the lab on a wide array of cognitive abilities, including perception, attention, memory, abstract reasoning, and decision making. He also provides an overview of the work involving "language" training in dolphins, in which the animals responded to sign sequences or engaged with underwater keyboards, and describes more recent research on imitation and social and self-awareness. Taken together, this body of work reveals an animal capable of remarkable sophistication and flexibility and provides a host of points for comparison with research on other species.

In chapter 8, Stan Kuczaj and Kelley Winship examine cognition from a developmental perspective. This includes learning basic skills like swimming and echolocation, mastering the postures and vocalizations involved in communication, and acquiring the foraging and social skills that result in adult proficiency and dolphin cultural diversity. This chapter examines how knowledge is acquired about and from others, and discusses how factors like personality, age, and gender affect what is learned and from whom. It also discusses the roles that synchrony, imitation, and teaching can play in dolphin social learning, and how the study of play can provide exciting insights into cognitive development.

In chapter 9, Christine Johnson describes a new approach—"cognitive ecology"— for studying dolphin social cognition through the systematic and detailed observation of naturally occurring behavior. By treating such behavior as the embodied "media of information flow," this system-based approach provides a metric for characterizing and comparing cognitive complexity across species. By considering how it might be applied to complex, multiparty engagements in dolphins—such as collaboration, social tool use, social attention, and information brokering—this chapter illustrates how tracking information flow, situated within a social marketplace, could allow us to investigate real-world social cognition in interacting dolphins.

In chapter 10, Thomas White presents a primer for thinking about the ethical issues surrounding work with cetaceans. We, the editors, feel strongly that research should not proceed in this field without our first giving serious thought to the ethical implications of our work. This chapter provides critical theoretical tools for such an ethical analysis. For example, it defines ethics, proposes criteria for "moral standing," and outlines traditional standards for evaluating pertinent qualities such as suffering, flourishing, and intrinsic worth. It also proposes possible changes in human practices that could be made to address these issues.

In keeping with our theme of past, present, and future, in chapter 11 we took the opportunity to summarize the future thoughts that all our contributors had suggested. In addition, we had the incredible opportunity to ask the experts—those who had spent decades putting in their "10,000 hours" on dolphin research—what three questions they would most like to see answered in the future. Their responses make for a fascinating discursion through the myriad ways of researching dolphin communication and cognition. It is always a treat to be led by a true native into diverse territories and to have subtle and vital issues pointed out. Together, these founders present an enticing future for research with these animals.

Finally, we would like to express our gratitude for the excellent contributions made by all our authors. The comprehensive coverage of broad and deep literatures, the fine quality of the writing, and the cogency of the concepts they have presented are all we could have hoped for and more. We feel so lucky, and proud, to have had a hand in bringing this book about. We hope that it will prove a useful and inspiring tool for future research with these amazing animals.

DLH and CMJ

References

Au, W. L. L. (1993). *The sonar of dolphins*. New York: Springer.

Bearzi, M., & Stanford, C. (2008). *Beautiful minds: The parallel lives of great apes and dolphins*. Cambridge, MA: Harvard University Press.

Caldwell, D. C., & Caldwell, M. C. (1972). *The world of the bottlenose dolphin*. Philadelphia, PA: J. B. Lippincott.

Cousteau, J.-Y., & Diole, P. (1975). *Dolphins*. Garden City, NY: Doubleday.

de Waal, F. B. M., & Tyack, P. L. (2003). *Animal social complexity: Intelligence, culture, and individualized societies*. Cambridge, MA: Harvard University Press.

Herman, L. M. (1980). *Cetacean behavior: Mechanism and functions*. New York: John Wiley and Sons.

Herzing, D. L. (2011). *Dolphin diaries*. New York: St. Martin's Press.

Leatherwood, S., & Reeves, R. R. (1990). *The bottlenose dolphin*. New York: Academic Press.

Mann, J., Connor, R. C., Tyack, P. L., & Whitehead, H. (1999). *Cetacean societies: Field studies of dolphins and whales*. Chicago: University of Chicago Press.

McIntyre, J. (1974). *Mind in the waters*. New York: Scribner.

Norris, K. S. (1966). *Whales, dolphins, and porpoises*. Berkeley: University of California Press.

Norris, K. S. (1974). *The porpoise watcher: A naturalist's experiences with porpoises and whales*. New York: W. W. Norton.

Norris, K. S., Würsig, B., Wells, R. S., & Würsig, M. (1994). *The Hawaiian spinner dolphin*. Berkeley: University of California Press.

Pryor, K. (1975). *Lads before the wind: Adventures in porpoise training*. New York: Harper & Row.

Pryor, K., & Norris, K. S. (1991). *Dolphin societies: Discoveries and puzzles*. Berkeley: University of California Press.

Roitblat, H. L., Herman, L. M., & Nachtigall, P. E. (1993). *Language and communication: Comparative perspectives*. Hillsdale, NJ: Erlbaum.

Schusterman, R. J., Thomas, J. A., & Wood, F. G. (1986). *Dolphin cognition and behavior: A comparative approach*. Hillsdale, NJ: Erlbaum.

Thomas, J. A., & Kastelein, R. A. (1990). *Sensory abilities of cetaceans: Laboratory and field evidence*. New York: Plenum Press.

Whitehead, H., & Rendell, L. (2014). *The cultural lives of whales and dolphins*. Chicago: University of Chicago Press.

Würsig, B., & Würsig, M. (2010). *The dusky dolphin*. San Diego, CA: Academic Press.

I Anatomy and Senses

1 The Brain: Evolution, Structure, and Function

Lori Marino

Introduction

In the past three decades, new research has uncovered the large, complex cetacean brain that forms the neurobiological basis for the considerable cognitive abilities of dolphins and whales. Furthermore, the evolutionary history of cetacean brains gives us a powerful example of the way that brain structure-function relationships can follow a complicated pattern of divergence and convergence at different levels. The study of cetacean brains has revealed that the human brain is not the only brain that has undergone significant increases in size and complexity. Cetacean brains have, as well, but have done so by taking a very different neuroanatomical path to complexity, and although not the only one, they may be the most compelling example of an alternate route to intelligence on par with that of humans.

The Evolution of Large Brains in Cetaceans

The most recent terrestrial ancestor of cetaceans, *Pakicetus* sp., looked somewhat like a medium-sized dog and stood on the hoofed feet of its ungulate ancestors (Gingerich & Uhen, 1998). *Pakicetus* was a member of the first cetacean suborder, Archaeoceti, which originated about 52.5 million years ago (Ma) (Bajpai & Gingerich, 1998; Gingerich & Uhen, 1998) and lasted until the end of the Oligocene epoch (23 Ma) (Fordyce, 2004; Uhen, 2010). *Pakicetus* possessed an average brain in absolute and relative size (Marino, 2009). Archaeocete brains changed relatively little over the first 15 million years of adaptation to a fully aquatic lifestyle (Marino, McShea, & Uhen, 2004), refuting a long-standing belief that cetacean brains became large and complex as a direct result of adaptation to an aquatic existence (for further discussion, see Tartarelli & Bisconti, 2006). All the evidence taken together suggests that, despite the obvious differences in behavior driven by a terrestrial versus aquatic lifestyle, brain size and morphology indicate there was not a fundamental shift in general level of intelligence and basic cognitive profile between *Pakicetus* and the fully aquatic archaeocetes.

Dramatic environmental changes at the Eocene-Oligocene boundary (around 35 Ma) coincided with the emergence of Neoceti ("new cetaceans"), which includes the two modern suborders Odontoceti and Mysticeti (Fordyce, 2003; Uhen, 2010; Geisler, McGowen, Yang, & Gatesy, 2011). And with the Neoceti came the earliest detectable changes in cetacean brain size and structure (Marino, McShea, & Uhen, 2004).

Because most of the data on cetacean brain evolution come from odontocetes, extrapolations to mysticetes should be made conservatively, bearing in mind that both suborders adapted early on to two very different behavioral niches, which then led to differences in modern brain structure. But modern mysticetes and odontocetes also share several characteristics of brain evolution, including hemisphere elaboration, level of cortical folding, and neocortical architecture, which indicate they are on a par with each other in terms of complexity (Hof & Van Der Gucht, 2007; Oelschlager & Oelschlager, 2009). Recent results suggest the difference in relative brain size between odontocetes and mysticetes is largely due to a higher rate of body mass evolution in mysticetes, along with decreases in body mass at the origin of odontocetes, resulting in a complete grade shift between the two (Montgomery et al., 2013).

Because brain and body size are positively correlated (i.e., brains, like other organs, are larger in larger bodies), brain size is often expressed as an Encephalization Quotient, or EQ (Jerison, 1973). EQ is a value that represents how large or small the average brain of a given species is compared with other species of the same body weight. Thus EQ accounts for brain-body allometry. An EQ of 1 means brain size is average for species of that body size, less than 1 means the brain is smaller than expected for its body size, and greater than 1 larger than expected. For example, modern human EQ is 7.0; our brains are seven times larger than one would expect for a species of our body size.

The first major change in odontocete relative brain size was due mainly to a significant decrease in body size along with a more moderate increase in absolute brain size. This combination of changes drove the EQ levels of these early Neoceti groups significantly beyond the range of the archaeocetes. Recent findings suggest that cetacean brain and body mass evolved under strong directional trends of increase through time, but decreases in EQ were widespread as well (Montgomery et al., 2013).

The average EQ for archaeocetes was about 0.5, below average, while that of the early odontocetes soared to 2.0, above average, with, importantly, no overlap in EQ across the two groups (Marino, McShea, & Uhen, 2004). Modern cetacean brains are among the largest of all mammals both in absolute mass and in relation to body size. The largest brain on earth today, that of the adult sperm whale (*Physeter macrocephalus*), at 8,000

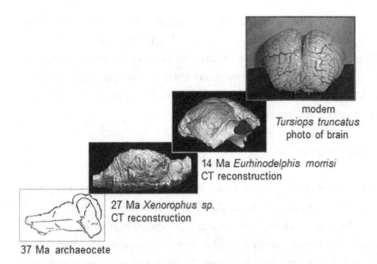

modern
Tursiops truncatus
photo of brain

14 Ma *Eurhinodelphis morrisi*
CT reconstruction

27 Ma *Xenorophus sp.*
CT reconstruction

37 Ma archaeocete

Figure 1.1
Evolution of cetacean brains from archaeocetes to modern species.

grams (Marino, 2009) is about 60 percent larger than the modern elephant brain and six times larger than the human brain. All modern odontocetes possess above-average encephalization levels compared with other mammals. Numerous odontocete species possess EQs in the range of 4 to 5; that is, they possess brains four to five times larger than one would expect for their body weights. There is strong evidence for a directional trend in increasing brain size and in decreasing body mass, but little evidence for a directional trend in increasing EQ per se (Montgomery et al., 2013). However, lineage-specific selection led to some cetaceans, such as the Delphinoidea (porpoises, oceanic dolphins, and some of the toothed whales), becoming very highly encephalized, with some possessing EQs in a range second only to those of modern humans (Marino, 1998). Moreover, cetacean, or at least odontocete, brain evolution appears to have been in stasis for the past 15 million years (Marino, McShea, & Uhen, 2004).

As mentioned earlier, archaeocete brain morphology was almost indistinguishable from that of the Pakicetids and many other predatory terrestrial species of the time. However, after the initial change in relative size, Neoceti cetacean brains underwent a number of structural and organizational modifications with important functional consequences. Along with their modest size, archaeocete brains were characterized by small, elongated cerebral hemispheres ending rostrally in large olfactory peduncles and bulbs (Edinger, 1955). The signature changes occurring in odontocetes were the substantial hyperproliferation and reorganization of the cerebral hemispheres, enlargement of auditory processing areas, development of a unique neocortical architecture, and reduction of olfactory structures with a reproportioning of associated limbic and paralimbic regions (Ridgway, 1990; Oelschlager & Oelschlager, 2009). These changes, along with many others, placed odontocetes on an unusual evolutionary trajectory that would result in brains that are among the most massive, complex, and sophisticated in the animal kingdom and engendering numerous hypotheses and debates about the ultimate drivers of cetacean brain evolution.

Drivers of Cetacean Brain Evolution

Archaeocetes underwent several changes in body structure and physiology during their period of adaptation to a fully aquatic lifestyle. Skull elongation and the migration of the nares to the top of the head, called "telescoping," produced significant changes in the architecture of the cranium over time. Archaeocetes also began the pattern of change in dentition from heterodonty to, ultimately, homodonty in odontocetes and baleen in mysticetes. In addition, changes in jaw and ear bone structure indicate that early odontocetes developed increased underwater directional hearing abilities. But no evidence exists for echolocation in archaeocetes (for a review, see Uhen, 2007). Finally, as mentioned, fossil cranial morphology indicates they still possessed olfactory abilities. What we do know is that the emergence of the Neoceti from the archaeocetes around 35 million years ago signaled a host of significant transformations that were probably accompanied by dramatic changes in behavioral ecology. Discussion and debate about the ultimate causes of the initial major shift in cetacean brain evolution that occurred with the Neoceti have been ongoing for decades. Hypotheses proposed to explain the large brains of modern cetaceans in ways independent of increased cognition (Manger, 2006) have been refuted (Marino et al., 2008). Rather, the scientific consensus is that large brain size (and large relative brain size) is evidence of, or necessary for, the behavioral complexity and cognitive abilities observed in cetaceans (Simmonds, 2006; Marino et al., 2008; Marino, 2009).

One hypothesis as to why the brain-body relationship shifted in odontocetes is that selection associated with the evolution of echolocation drove increases in EQ (Marino, McShea, & Uhen, 2004; Montgomery et al., 2013). The few Oligocene odontocetes whose inner ears have been studied in detail have ears resembling those of modern odontocetes (Fleischer, 1976; Luo & Eastman, 1995), suggesting they were specialized for hearing high-frequency sounds. However, some uncertainty remains about whether the earliest odontocetes had inner ears specialized for the high-frequency hearing specifically used in echolocation (Geisler et al., 2011). Furthermore, it should be noted that although high-frequency hearing is a prerequisite for echolocation, it does not, by itself, provide information about how echolocation is used, that is, if it shifted from a basic sensory system for navigation and hunting to one used in more complex ways in, for instance, social interactions. This idea is supported by the fact that burst pulse sounds, used in social interaction, fall on a continuum with high-frequency echolocation sounds. Echolocation was likely embedded within a suite of characteristics that signaled a change in behavior in early odontocetes.

At the time of the major shift in brain-body allometry in early cetaceans came a cooling in temperate to polar latitudes, triggering changes in biodiversity and productivity in oceanic food chains (Salamy & Zachos, 1999). These changes must have led, in turn, to changes in behavioral ecology (e.g., foraging opportunities and predation risk) (Lindberg & Pyenson, 2007) that drove further cognitive and social changes. In odontocetes, the moderate increase in brain size was matched by the evolution of a novel echolocation-related nasofacial muscle complex (Fordyce, 2002, 2003) and concurrent changes in ear structures associated with echolocation (Fleischer, 1976). Heterodonty was replaced entirely by long rows of pronglike, uniform teeth adapted for grabbing and gulping whole prey items. Taken together, the decreased body size, homodonty, changes in skull and ear bone morphology, and brain enlargement all lead to evidence for a new echolocating niche for the early odontocetes. These early odontocetes were— in a phrase—fast and sleek and possessed new and enhanced sensory-perceptual abilities, leading to the sophisticated levels of intelligence in modern species (Marino et al., 2008; Marino, 2011).

The environmental changes that occurred about 35 million years ago spurred a different set of physical and behavioral adaptations in mysticetes that led to the development of baleen, filter feeding, and large body size. Explanations for why two different suborders emerged so early on in cetacean evolution await further data and analysis. Nevertheless, some modern mysticetes evince similar cognitive capacities to some odontocetes (Whitehead, 2011).

Modern Cetacean Brain Structure and Organization

Modern cetacean brains are not only very different from those of their ancestors but also highly unusual compared with other mammals. Speaking broadly, two different levels of evolutionary change can occur: (1) direct selection, or adaptation to specific selection pressures; and (2) indirect selection, or accommodation of other structures to those directly selected adaptations, including allometrically driven changes. Some of the changes that occurred in cetacean brain evolution were likely direct adaptations to specific pressures in the physical and social environment. A notable example is the increase in size of auditory brain structures. That is, the emerging function of increased sensitivity and range in auditory processing most likely came under direct selection pressure. Second, changes in auditory brain structures and functions necessitated modifications in other components of the brain to accommodate those specific changes. For example, other sensory functions may have been modified to accommodate the burgeoning auditory capacity, or changes may have occurred in more general functions, like sensory integrative processes, which represent broader cognitive modifications. For instance, in the course of human language development, we see a number of changes in brain structures and architecture to accommodate the new capacity (Sakai, 2005). Brain size and structure also drive "universal" allometric changes in brains as well, that is, changes throughout the brain that allow for the maintenance of critical neurobiological parameters, such as connectivity or speed of neural transmission. For instance, when brains enlarge over evolutionary time, they do not simply become larger versions of the smaller form. When brains enlarge, they do so by changing their general architecture to maintain connectivity over longer distances (through increased axon diameter and specialized cellular morphologies that aid in "long-distance" neural transmission) and by decreasing distance between processing areas by increasing modularization and cortical adjacency. Therefore brains cannot enlarge without also reorganizing, thus creating new areas and new features. It is important to keep in mind that the modern cetacean brain, like any other brain, is undoubtedly a product of all such processes.

Development of Auditory Structures

Counter to popular belief, with the exception of river dolphins (who are almost blind), cetaceans have acute vision in and out of water (Mass & Supin, 2009; see also Hanke & Erdsack, chap. 3, this vol.). But the dominant sense is hearing. Cetaceans are highly reliant on audition, and the diameter of their vestibulocochlear nerve (the auditory nerve), with its impressively large and numerous well-myelinated fibers, provides evidence of this fact (Oelschlager & Oelschlager, 2009). Likewise, all brain structures associated with hearing that receive input from the auditory nerve, that is, subcortical auditory regions as well as cortical projection zones, are massive in cetaceans and, particularly, in odontocetes (Ridgway, 1990; Oelschlager & Oelschlager, 2009). For example, the volume of the cochlear nuclei (areas in the brain stem that receive input from the auditory nerve) in the harbor porpoise (*Phocoena phocoena*), La Plata river dolphin (*Pontoporia blainvillei*), and common dolphin (*Delphinus delphis*) is six to ten times greater than in the human brain; in beluga whales (*Delphinapterus leucas*), the same areas are thirty-two times larger than in the human brain (Oelschlager & Oelschlager, 2009). These numbers represent large absolute and relative volumes. Likewise, the primary and secondary auditory projection zones on the cerebral surface are extensive (Supin et al., 1978). Generally, auditory structures are relatively larger in odontocetes than in mysticetes because of the odontocete ability to echolocate, but both groups possess massive hearing structures (Oelschlager & Oelschlager, 2009). Electrophysiological studies of bottlenose dolphin evoked potentials reveal an exquisite sensitivity both to rapid, high-frequency sounds in some parts of the brain and to longer-latency, lower frequencies in others (Bullock & Ridgway, 1972), suggesting a "dual analysis system," one specialized for high-frequency clicks, and the other for the lower-frequency sounds used in communication (Ridgway, 2000). Recent studies have continued to support earlier studies by elucidating the extremely complex acoustic processing functions of the dolphin brain (Ridgway & Au, 1999).

It would be a mistake, despite the specialized acoustic systems of the dolphin brain, to view them as merely fine-tuned sensory capacities. One of the most intriguing and informative aspects of such sophisticated auditory processing in cetaceans is the evidence that they have integrated this perceptual system into the general cognitive domain in a way that may be unprecedented in the animal kingdom. These complex acoustic abilities in dolphins have become amalgamated into a brain that has high-level cognitive capacities. Research has shown that bottlenose dolphins are capable of cross-modal integration, that is, recognizing through echolocation complex objects that they have only seen and vice versa (Pack & Herman, 1995; Herman, Pack, & Hoffmann-Kuhnt, 1998; Harley, Putnam, & Roitblat, 2003). These findings indicate that audition and vision have been so well integrated that echolocation involves the capacity for higher-order mental representation. Lende and Welker (1972) found an area in the dolphin cortex that responds to both acoustic and tactile input, suggesting further that dolphin cortical regions may be functionally well integrated for higher-level representation.

Thus, from an evolutionary point of view, echolocation in odontocetes may have evolved initially as a sensory device to provide information about the environment and objects in it. In this scenario, the large brains of odontocetes were initially used, at least partly, for processing this entirely new sensory modality. But experimental evidence and behavioral field observations suggest echolocation plays a role in more highly integrated cognitive processing and communication in modern odontocetes (see Jerison, 1986; Ridgway, 1986). For instance, echolocation sounds are often produced simultaneously or in conjunction with other sounds that are communicative in nature. In addition, many authors have noted that sonar provides an efficient way to keep track of a moving social group over distances (Wood & Evans, 1980). Plus, research indicates that such intense sounds also have a tactile impact (Kolchin & Bel'kovich, 1973), and the animals use this feature, particularly during courtship and aggression (Herzing, 1996, 2004). Therefore the initial function (which was likely navigation and hunting) may have evolved into an increasingly complex cognitive-behavioral feature with a more global impact on odontocete lifestyle.

Reproportioning of Limbic and Paralimbic Structures

Olfaction in cetaceans has been severely reduced in mysticetes and entirely lost in odontocetes. Fetal odontocetes possess small olfactory structures (Buhl & Oelschlager, 1988; Marino, Murphy, Gozal, & Johnson, 2001) that typically regress completely by birth. Infrequently, a short olfactory peduncle is found in adult sperm whales and northern bottlenose whales (*Hyperoodon ampullatus*) (Oelschlager & Oelschlager, 2009). Olfaction in adult mysticetes is vastly reduced; unlike odontocetes, as adults mysticetes maintain small olfactory bulbs, a thin olfactory peduncle, and an olfactory tubercle (Oelschlager & Oelschlager, 2009).

The limbic system is an evolutionarily old and highly conserved set of structures that supports a variety of functions, including memory, emotional processing, and olfaction. It includes the hippocampus, amygdala, mammillary bodies, and the various structures of the limbic cortex. Limbic structures most intimately connected with olfaction, that is, the hippocampus, fornix, and mammillary bodies, have been greatly reduced in cetaceans (Jacobs, McFarland, & Morgane, 1979; Morgane, Jacobs, & McFarland, 1980).

In contrast to these reductions, the cetacean limbic system features a remarkably large amygdala (Schwerdtfeger, Oelschlager, & Stefan, 1984) and extremely well-developed cortical limbic lobe and entorhinal region above the corpus callosum (Oelschlager & Oelschlager, 2009), creating a highly elaborated paralimbic region. Dense connections exist between paralimbic structures and the amygdala in mammals in general. Broadly, the paralimbic region connects limbic and "higher-order" cortical structures, and the degree of elaboration may be related to the complexity of how emotions are integrated with higher-order thought processes. The reduced hippocampus in cetaceans is likely due to a number of factors, which include loss or reduction of olfaction and the lack of reliance on spatial memory in an open three-dimensional environment (the hippocampus is involved in spatial memory in many species, such as caching birds, which rely on remembering places). That cetaceans possess a reduced hippocampus and well-developed paralimbic region adjacent to the limbic system suggests, but does not prove, that there may have been a transfer of memory functions from the olfactory-based hippocampus to higher-order cortical regions. This hypothesis is consistent with the fact that cetacean learning, memory, and spatial navigation abilities (all hippocampal functions) are sophisticated and robust (Marino et al., 2008).

Enlargement and Rearrangement of Cerebral Hemispheres

The cetacean forebrain represents a radical departure from the lobular architecture of many mammal brains, including that of humans. Instead of lobes, cetacean brains are organized into three concentric tiers of tissue that include the limbic, paralimbic, and supralimbic regions. In addition to its unusual organization, the cetacean forebrain is among the most highly convoluted of all mammalian forebrains, revealing that its neocortical surface area and volume increased substantially throughout cetacean evolutionary history. One measure of convolution is the "gyrification index," which compares neocortical surface area to total brain weight. The index for modern humans is approximately 1.75. Known gyrification indices for odontocetes range from 2.4 to 2.7, substantially exceeding that of modern humans (Ridgway & Brownson, 1984). The gyrification index is positively correlated with brain mass across the mammals, and cetacean (as well as human) brains appear to be consistent with this pattern.

The pattern of elaboration of the neocortex (the evolutionarily newest region of the forebrain) in cetaceans has resulted in a highly unusual configuration of sensorimotor processing regions as well. The map of sensory projection regions (the cortical regions that receive sensory information) in the cetacean brain stands in striking contrast to that of other large-brained mammals. In primates, for instance, the visual and auditory projection regions are located in the occipital and temporal lobes, respectively (Macaluso & Driver, 2005). This means that visual information is first processed in the cortex in the back of the brain (occipital region) and auditory information on the side of the brain (temporal region). An expanse of nonprojection or association cortex intervenes between these two regions. Therefore, in primates, for instance, to be integrated, visual and auditory information must be sent from primary and secondary projection zones to this intervening cortex. In cetaceans, by contrast, the visual and auditory projection zones are located in the parietal region atop the hemispheres and are immediately adjacent to each other. They are not separated by association cortex (Ladygina, Mass, & Supin, 1978; Supin et al., 1978). This pattern of cortical adjacency in cetacean brains is unique among large-brained mammals but likely associated with their abilities for highly integrated auditory and visual processing.

Cetaceans possess a vast expanse of nonprojection or association cortex for even higher-order cognitive information processing, which lies outside the visual-auditory region in the remaining temporal and occipital regions. This idiosyncratic pattern of visual-auditory adjacency could allow for highly developed cross-modal sensory processing abilities in cetaceans (Pack & Herman, 1995; Herman, Pack, & Hoffmann-Kuhnt, 1998; Harley, Putnam, & Roitblat, 2003), and the vast expanse of association cortex suggests complex higher-order mental functions.

The Cetacean Neocortex Revisited

Early neuroanatomical studies of cetacean neocortex appeared to show that cortical regions are homogeneous and fairly simple in architecture (Kesarev, 1971; Gaskin, 1982). These data led to a view labeled the "initial brain hypothesis," which engendered much confusion and debate about how a relatively "primitive" brain could be the basis of the considerable cognitive and behavioral complexity observed in cetaceans (Glezer, Jacobs, & Morgane, 1988). But more recent studies, using more sophisticated histological methods, reveal a very different picture, providing evidence for substantial neocortical complexity in both odontocetes and mysticetes (Hof, Chanis, & Marino, 2005; Hof & Van Der Gucht, 2007). The cellular architecture of various regions of the cetacean neocortex is characterized by a wide variety of organizational features associated with complex brains. These include cell clusters in the insular cortex (Manger, Sum, Szymanski, Ridgway, & Krubitzer, 1998), a variety of clearly defined cytoarchitectural patterns (Hof & Van Der Gucht, 2007), well-defined laminar patterns in frontal regions (Hof, Chanis, & Marino, 2005), highly distinct vertically organized modules in the insular cortex (Hof, Chanis, & Marino, 2005), and a rich diversity of cytoarchitectonic fields in all the lobes (Hof, Chanis, & Marino, 2005). In a comprehensive histological analysis of cellular organization in cetacean neocortex, Hof, Chanis, and Marino (2005) concluded: "In spite of a pervasive notion that neocortical structure is rather uniform throughout the cortical mantle with minor local variations, it is in fact quite complex, with a degree of regional parcellation at least comparable to that of large-brained terrestrial mammals, such as anthropoid primates, carnivores and ungulates" (p. 1144). Studies continue to support this more informed view of cetacean brains as comparable in complexity to other mammal, and possibly even human, brains.

Whereas there appears to be a high degree of organizational complexity throughout the cetacean neocortex, some regions are especially elaborated. The cingulate and insular cortices (both part of the paralimbic system) in odontocetes and mysticetes are extremely well developed (Jacobs, McFarland, & Morgane, 1979; Hof & Van Der

Gucht, 2007), and the expansion of these areas in cetaceans is consistent with high-level cognitive functions such as attention, judgment, and social awareness (Allman, Watson, Tetrault, & Hakeen, 2005). Moreover, recent studies show that the anterior cingulate and insular cortices in larger cetaceans contain a type of projection neuron, known as a spindle cell or von Economo neuron (Hof & Van Der Gucht, 2007). Von Economo neurons are highly specialized projection neurons thought to be involved in maintaining transmission between distant brain regions in large brains. Because of their location, they may also be part of neural networks subserving aspects of social cognition (Allman et al., 2005). These cells have been found in humans and great apes (Allman et al., 2005), as well as elephants (Hakeem et al., 2008). It is important to note that although spindle cells are thought to play a role in adaptive intelligent behavior, there is no consensus on their specific purpose. However, the presence of these neurons in cetaceans is, at the least, not inconsistent with the complex cognitive abilities found in this group.

Despite general similarities in level of cellular and organizational complexity, there are striking differences in the specific connectivity patterns of cetacean brains and primate, including human, brains. Specifically, cetacean neocortex is characterized by five layers instead of the six typical of primates and many other mammals. Cetacean neocortex possesses an extremely thick layer 1, in combination with the absence of a granular layer 4. In primates, a granular layer 4 is the primary input layer for fibers ascending from the brain stem to the cortex, and this layer is also the source of important connections within the neocortex. However, since granular layer 4 is absent in cetaceans, information comes into the cetacean neocortex through a different pathway than in most other mammals (Glezer, Jacobs, & Morgane, 1988). The prevailing view is that the thick layer 1 is the primary layer receiving incoming fibers in cetaceans. The strikingly different connectivity patterns of primate and cetacean brains are another compelling example of two distinctly different evolutionary trajectories taken toward neurological and behavioral complexity.

Future Directions

Future directions in our study of cetacean brains and intelligence require taking into account the current evidence for sophisticated cognitive, emotional, and social abilities in cetaceans. We still have much to do in the domain of understanding patterns of brain evolution in cetaceans within a phylogenetic context, as the phylogeny of the order and subtaxonomic groups is far from resolved. Paleoneurological methods using fossil material continue to elucidate evolutionary patterns and relationships among cetaceans. And comparative phylogenetic methods using postmortem brains of modern specimens (from individuals who have died naturally), using increasingly sophisticated histological and imaging methods, make possible more profound analyses of cetacean neuroanatomy and comparisons with other taxonomic groups. In combination, studies of paleontological and modern material will undoubtedly yield greater insights into cetacean brain evolution.

In an applied domain, conservation measures will need to take into account the known psychological complexity of these animals on an individual level. As such, there needs to be more integration of the basic science on who cetaceans are and conservation efforts and a greater emphasis on individuals in efforts to protect wild cetaceans.

Finally, beyond all these suggestions, the main point is that our scientific understanding and current knowledge of cetacean brains and cognitive abilities demand that we develop a new ethic of respect and coexistence with them, requiring changes in human behavior as we move into the future.

References

Allman, J. M., Watson, K. K., Tetrault, N. A., & Hakeen, A. Y. (2005). Intuition and autism: A possible role for von Economo neurons. *Trends in Cognitive Sciences*, *9*, 367–373.

Bajpai, S., & Gingerich, P. D. (1998). A new Eocene archaeocete (Mammalia, Cetacea) from India and the time of origin of whales. *Proceedings of the National Academy of Sciences of the United States of America*, *95*, 15464–15468.

Buhl, E. H., & Oelschlager, H. A. (1988). Morphogenesis of the brain in the harbour porpoise. *Journal of Comparative Neurology*, *277*, 109–125.

Bullock, T. H., & Ridgway, S. H. (1972). Evoked potentials in the central auditory system of alert porpoises to their own and artificial sounds. *Journal of Neurobiology*, *3*, 79–99.

Connor, R. C., Wells, R., Mann, J., & Read, A. (2000). The bottlenosed dolphin: Social relationships in a fission-fusion society. In J. Mann, R. C. Connor, P. Tyack, & H. Whitehead (Eds.), *Cetacean societies: Field studies of whales and dolphins* (pp. 91–126). Chicago: University of Chicago Press.

Edinger, T. (1955). Hearing and smell in cetacean history. *Monatsschrift für Psychiatrie und Neurologie*, *129*, 37–58.

Fleischer, G. (1976). Hearing in extinct cetaceans as determined by cochlear structure. *Journal of Paleontology*, *50*, 133–152.

Fordyce, R. E. (2002). *Simocetus rayi* (Odontoceti: Simocetidae) (new species, new genus, new family), a bizarre new archaic Oligocene dolphin from the eastern North Pacific. *Smithsonian Contributions to Paleobiology*, *93*, 185–222.

Fordyce, R. E. (2003). Cetacean evolution and Eocene-Oligocene oceans revisited. In D. R. Prothero, L. C. Ivany, & E. Nesbitt (Eds.), *From greenhouse to icehouse: The marine Eocene-Oligocene transition* (pp. 154–170). New York: Columbia University Press.

Fordyce, R. E. (2004). The transition from Archaeoceti to Neoceti: Oligocene archaeocetes in the southwest Pacific. *Journal of Vertebrate Paleontology, 24,* 59A.

Gaskin, D. E. (1982). *The ecology of whales and dolphins.* London: Heinemann.

Geisler, J. H., McGowen, M. R., Yang, G., & Gatesy, J. (2011). A supermatrix analysis of genomic, morphological, and paleontological data from crown Cetacea. *BMC Evolutionary Biology, 11,* 112. doi:10.1186/1471-2148-11-112.

Gingerich, P. D., & Uhen, M. D. (1998). Likelihood estimation of the time of origin of Cetacea and the time of divergence of Cetacea and Artiodactyla. *Paleo-electronica, 2,* 1–47.

Glezer, I., Jacobs, M., & Morgane, P. (1988). Implications of the "initial brain" concept for brain evolution in Cetacea. *Behavioral and Brain Sciences, 11,* 75–116.

Hakeem, A. Y., Sherwood, C. C., Bonar, C. J., Butti, C., Hof, P. R., & Allman, J. M. (2008). Von Economo neurons in the elephant brain. *Anatomical Record, 292,* 242–248.

Harley, H. E., Putnam, E. A., & Roitblat, H. L. (2003). Bottlenose dolphins perceive object features through echolocation. *Nature, 424,* 667–669.

Herman, L. M., Pack, A. A., & Hoffmann-Kuhnt, M. (1998). Seeing through sound: Dolphins perceive the spatial structure of objects through echolocation. *Journal of Comparative Psychology, 112,* 292–305.

Herzing, D. L. (1996). Vocalizations and associated underwater behavior of free-ranging Atlantic spotted dolphins, *Stenella frontalis,* and bottlenose dolphins, *Tursiops truncatus. Aquatic Mammals, 22,* 61–80.

Herzing, D. L. (2004). Social and nonsocial uses of echolocation in free-ranging *Stenella frontalis* and *Tursiops truncatus.* In J. A. Thomas, C. F. Moss, & M. Vater (Eds.), *Echolocation in bats and dolphins* (pp. 404–410). Chicago: University of Chicago Press.

Hof, P., Chanis, R., & Marino, L. (2005). Cortical complexity in cetacean brains. *Anatomical Record, 287A,* 1142–1152.

Hof, P. R., & Van Der Gucht, E. (2007). The structure of the cerebral cortex of the humpback whale, *Megaptera novaeangliae* (Cetacea, Mysticeti, Balaenopteridae). *Anatomical Record, 290,* 1–31.

Jacobs, M. S., McFarland, W. L., & Morgane, P. J. (1979). The anatomy of the brain of the bottlenose dolphin (Tursiops truncatus). Rhinic lobe (rhinencephalon): The archicortex. *Brain Research Bulletin, 4,* 1–108.

Jerison, H. J. (1973). *Evolution of the brain and intelligence.* New York: Academic Press.

Jerison, H. J. (1986). The perceptual world of dolphins. In R. J. Schusterman, J. A. Thomas, & F. G. Wood (Eds.), *Dolphin cognition and behavior: A comparative approach* (pp. 141–166). Mahwah, NJ: Erlbaum.

Kesarev, V. S. (1971). The inferior brain of the dolphin. *Soviet Science Review, 2,* 52–58.

Kolchin, S. P., & Bel'kovich, V. M. (1973). Tactile sensitivity in Delphinus delphis. *Zoologicheskiy zhurnal, 52,* 620–622.

Ladygina, T. F., Mass, A. M., & Supin, A. I. (1978). Multiple sensory projections in the dolphin cerebral cortex. *Zhurnal Vysshei Nervnoi Deiatelnosti Imeni I P Pavlova, 28,* 1047–1054.

Lende, R. A., & Welker, W. (1972). An unusual sensory area in the cerebral neocortex of the bottlenose dolphin, *Tursiops truncatus. Brain Research, 45*(2), 555–560.

Lindberg, D. R., & Pyenson, N. D. (2007). Things that go bump in the night: Evolutionary interactions between cephalopods and cetaceans in the Tertiary. *Lethaia, 40,* 335–343.

Luo, Z., & Eastman, E. R. (1995). Petrosal and inner ear of a squalodontoid whale: Implications for evolution of hearing in odontocetes. *Journal of Vertebrate Paleontology, 15,* 431–442.

Macaluso, E., & Driver, J. (2005). Multisensory spatial interactions: A window onto functional integration in the human brain. *Trends in Neurosciences, 28*(5), 264–271.

Manger, P. R. (2006). An examination of cetacean brain structure with a novel hypothesis correlating thermogenesis to the evolution of a big brain. *Biological Reviews of the Cambridge Philosophical Society, 81,* 293–338.

Manger, P. R., Sum, M., Szymanski, M., Ridgway, S., & Krubitzer, L. (1998). Modular subdivisions of dolphin insular cortex: Does evolutionary history repeat itself? *Journal of Cognitive Neuroscience, 10,* 153–166.

Marino, L. (1998). A comparison of encephalization between odontocete cetaceans and anthropoid primates. *Brain, Behavior, and Evolution, 51,* 230–238.

Marino, L. (2009). Brain size evolution. In W. F. Perrin, B. Würsig, & H. Thewissen (Eds.), *Encyclopedia of marine mammals* (pp. 149–152). San Diego: Academic Press.

Marino, L. (2011). Brain structure and intelligence in cetaceans. In P. Brakes & M. P. Simmonds (Eds.), *Whales and dolphins: Cognition, culture, conservation and human perceptions* (pp. 115–128). London: Earthscan.

Marino, L., Butti, C., Connor, R. C., Fordyce, R. E., Herman, L. M., Hof, P. R., et al. (2008). A claim in search of evidence: Reply to Manger's thermogenesis hypothesis of cetacean brain structure. *Biological Reviews of the Cambridge Philosophical Society, 83,* 417–440.

Marino, L., McShea, D., & Uhen, M. D. (2004). The origin and evolution of large brains in toothed whales. *Anatomical Record, 281A,* 1247–1255.

Marino, L., Murphy, T. L., Gozal, L., & Johnson, J. I. (2001). Magnetic resonance imaging and three-dimensional reconstructions of the brain of the fetal common dolphin, *Delphinus delphis. Anatomy and Embryology, 203,* 393–402.

Marino, L., Sudheimer, K., Murphy, T. L., Davis, K. K., Pabst, D. A., McLellan, W., et al. (2001). Anatomy and three-dimensional reconstructions of the bottlenose dolphin (*Tursiops truncatus*) brain from magnetic resonance images. *Anatomical Record, 264*, 397–414.

Mass, A. M., & Supin, A. Y. (2009). Vision. In W. F. Perrin, B. Würsig, & J. G. M. Thewissen (Eds.), *Encyclopedia of marine mammals* (pp. 1200–1211). San Diego: Academic Press.

Montgomery, S. H., Geisler, J., McGowan, M. R., Fox, C., Marino, L., & Gatesy, J. (2013). The evolutionary history of cetacean brain and body size. *Evolution: International Journal of Organic Evolution, 67*(11), 3339–3353.

Morgane, P. J., Jacobs, M. S., & McFarland, W. L. (1980). The anatomy of the brain of the bottle-nosed dolphin (*Tursiops truncatus*): Surface configuration of the telencephalon of the bottlenosed dolphin with comparative anatomical observations in four other cetacean species. *Brain Research Bulletin, 5 (Suppl. 3)*, 1–107.

Oelschlager, H. A., & Oelschlager, J. S. (2009). Brains. In W. F. Perrin, B. Würsig, & H. Thewissen (Eds.), *Encyclopedia of marine mammals* (pp. 134–158). San Diego: Academic Press.

Pack, A. A., & Herman, L. M. (1995). Sensory integration in the bottlenosed dolphin: Immediate recognition of complex shapes across the senses of echolocation and vision. *Journal of the Acoustical Society of America, 98*, 722–733.

Ridgway, S. H. (1986). Physiological observations on dolphin brains. In R. J. Schusterman, J. A. Thomas, & F. G. Wood (Eds.), *Dolphin cognition and behavior: A comparative approach* (pp. 31–59). Mahwah, NJ: Erlbaum.

Ridgway, S. H. (1990). The central nervous system of the bottlenose dolphin. In S. Leatherwood & R. R. Reeves (Eds.), *The bottlenose dolphin* (pp. 69–97). New York: Academic Press.

Ridgway, S. H. (2000). The auditory central nervous system of dolphins. In W. Au, A. Popper, & R. Fay (Eds.), *Hearing by whales and dolphins* (pp. 273–293). New York: Springer.

Ridgway, S. H., & Au, W. W. (1999). *Hearing and echolocation: Dolphin. Encyclopedia of neuroscience* (2nd ed., pp. 858–862). Berlin: Springer.

Ridgway, S. H., & Brownson, R. H. (1984). Relative brain sizes and cortical surface areas of odontocetes. *Acta Zoologica Fennica, 172*, 149–152.

Sakai, K. (2005). Language acquisition and brain development. *Science, 310*(5749), 815–819.

Salamy, K. A., & Zachos, J. C. (1999). Latest Eocene–Early Oligocene climate change and Southern Ocean fertility: Inferences from sediment accumulation and stable isotope data. *Palaeogeography, Palaeoclimatology, Palaeoecology, 145*(1–3), 61–77.

Schwerdtfeger, W. K., Oelschlager, H. A., & Stephan, H. (1984). Quantitative neuroanatomy of the brain of the La Plata dolphin, *Pontoporia blainvillei. Anatomy and Embryology, 170*, 11–19.

Simmonds, M. P. (2006). Into the brains of whales. *Applied Animal Behaviour Science, 100*, 103–116.

Supin, A. Y., Mukhametov, L. M., Ladygina, T. F., Popov, V. V., Mass, A. M., & Poliakova, I. G. (1978). *Electrophysiological studies of the dolphin's brain* (pp. 29–85). Moscow: Izdatel'ato Nauka.

Tartarelli, G., & Bisconti, M. (2006). Trajectories and constraints in brain evolution in primates and cetaceans. *Human Evolution, 21,* 275–287.

Uhen, M. (2007). Evolution of marine mammals: Back to the sea after 300 million years. *Anatomical Record, 290,* 514–522.

Uhen, M. (2010). The origin(s) of whales. *Annual Review of Earth and Planetary Sciences, 38,* 189–219.

Whitehead, H. (2011). The cultures of whales and dolphins. In P. Brakes & M. P. Simmonds (Eds.), *Whales and dolphins: Cognition, culture, conservation, and human perceptions* (pp. 149–165). London: Earthscan.

Wood, F. G., & Evans, W. E. (1980). Adaptiveness and ecology of echolocation in toothed whales. In R. Busnel & F. Fish (Eds.), *Animal sonar systems* (pp. 381–426). New York: Plenum.

2 Sound Production and Sound Reception in Delphinoids

Ted W. Cranford, Mats Amundin, and Petr Krysl

Introduction

The biosonar system in dolphins and other odontocetes has two biomechanical or vibroacoustic components, plus the analytical components in the central nervous system (CNS). The vibroacoustic systems that make up the biosonar apparatus in dolphins can be divided into the output side (sound generation and beam formation) and the input side (sound reception and hearing). The sound generation and beam formation components are located primarily within the enlarged forehead containing mostly the nasal anatomy. The sound reception components are found on the ventral aspects of the head and are relatively more complex, structurally, materially, and geometrically.

Ironically, the less complex output side has been the least studied and continues to be incompletely understood, while the sound reception and hearing side is understood to a greater degree despite its complexity, although many intriguing conundrums persist.

The biosonar sense is well suited to the aquatic environment because, unlike light, sound penetrates equally well into all depths and environmental light levels where odontocetes live. The biomechanical properties of biosonar could be characterized as tactile or *touch at a distance*, providing a perceptual channel for investigating the environment. This forms the basis for communication, prey capture, and social facilitation through enlarged dedicated portions of a highly organized central nervous system.

Sound Production

A wealth of bioacoustic data from laboratory studies on the parameters of delphinid click sounds has been produced (see, e.g., Au, 1993; Thomas, Moss, & Vater, 2003). Traditionally, these studies have necessarily been characterized by strict experimental control, often requiring the animal to be stationed, with its head fixed on a bite plate or in a hoop during sonar detection and discrimination tasks. This undoubtedly under-estimates the full potential and dynamics of dolphin biosonar.

Researchers have gathered new data by using multielement arrays, making it possible to acoustically track phonating free-swimming wild dolphins and back-calculate their click source levels, and to measure other click parameters that describe the sonar beam in some detail. Such new field data have been obtained from a wide range of species (see Benoit-Bird & Au, 2003, 2009; Kyhn et al., 2013; Rasmussen & Miller, 2002; Rasmussen, Wahlberg, & Miller, 2004; Villadsgaard, Wahlberg, & Tougaard, 2007; Wahlberg et al., 2011).

In general, higher click source levels (up to 60 dB) have been observed in the wild (Au, 1993; Evans, 1973; Kastelein, Au, Rippe, & Schooneman, 1999; Villadsgaard et al., 2007; Wahlberg et al., 2011), though source levels as high as 230 dB p–p *re* 1μPa at 1m have been observed in captive bottlenose dolphins in connection with long-distance target detections in open waters (Au, Floyd, Penner, & Murchison, 1974; Au & Snyder, 1980). A review of the different types of sounds and their use in different contexts can be found in Herzing and Santos (2004).

After several decades of controversy about the origin and production of sounds in delphinids (see review by Cranford & Amundin, 2004), there is now a consensus that click sounds are produced by the phonic lips in all odontocetes (Cranford & Amundin, 2004; Cranford, Amundin, & Norris, 1996). The process is pneumatically driven (Dormer, 1979; Evans & Prescott, 1962) by air pressure built up in the bony nasal passages (Amundin & Andersen, 1983; Cranford et al., 2011; Ridgway & Carder, 1988; Ridgway et al., 1980; Ridgway et al., 2001).

How Many Delphinoid Click Sources Are There?

The nonphyseterid odontocetes (all toothed whales except sperm whales, Physeteridae) have bilateral sets of phonic lips (figs. 2.1, 2.2), one associated with each nasal passage (Cranford et al., 1996). In nearly all delphinoids, the bilateral sets of phonic lips are different in size, the right side larger than the left. The phonic lips on the right side are also normally aligned perpendicular to and near the midline, although that geometry differs with the species in question. Despite this bilateral structure, there is a current debate in the literature regarding the existence of a single versus multiple click generators in odontocetes.

Lammers and Castellote (2009), using one fixed on-axis hydrophone and another hydrophone that was moved in 15° increments up to ±90° around the head of a beluga whale (*Delphinapterus leucas*), observed double pulses on both sides of the head, with maximum delays at 90° off-axis and a gradual decrease in this delay until the pulses

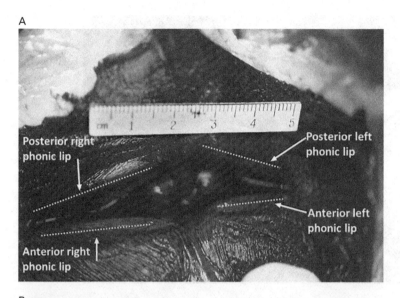

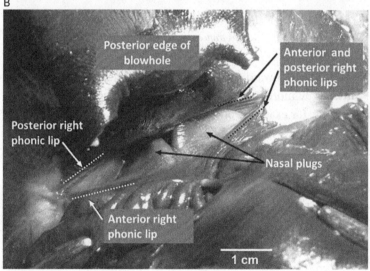

Figure 2.1
Dorsal view of left and right phonic lips in (A) a Risso's dolphin (*Grampus griseus*) and (B) a harbor porpoise (*Phocoena phocoena*; right). Photos by T. Cranford and M. Amundin.

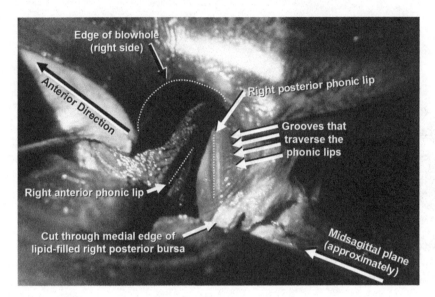

Figure 2.2
Midsagittal section (approximate) through blowhole, exposing right phonic lips and right posterior dorsal bursae in a bottlenose dolphin (*Tursiops truncatus*). Photo by T. Cranford.

apparently merged from ±45° and toward the main beam axis. The center frequencies
of the double pulses differed by more than 25 kHz. The long interpulse intervals (IPI)
ruled out internal reflections, for example, off of air sacs or skull bones from a single
source of sounds. Lammers and Castellote concluded that the animal most likely produced clicks simultaneously or nearly simultaneously by both phonic lip pairs, which
then merged on their way out through the melon to form a single click on axis.

If we assume two sources, a direct path through tissue, and the sound speed in
water, the measured IPIs would place the two pairs of phonic lips three to five times
farther apart than expected in beluga whales. Lammers and Castellote suggested that
the long IPIs might be explained by sounds that first pass through the air-filled vestibular sac before reentering the tissues and eventually the water. However, it is difficult to understand why the lateral hydrophone did not pick up sound traveling by a
direct path from the phonic lips through the tissues between the vestibular sacs and
the maxillary bone, which would not have any such delay. Later, in a paper together
with Madsen et al. (Madsen, Lammers, Wisniewska, & Beedholm, 2013), Lammers put
forward another explanation for the long delays, assuming only one active source:
the first pulse arrived directly from the right phonic lip pair while the second pulse
was radiated from the front of the melon to ensonify the hydrophones off axis at ±90°

(cf. Au, Kastelein, Benoit-Bird, Cranford, & McKenna, 2006; Au et al., 2010). According to Madsen et al. (2013), this hypothesized sound path would be consistent with the dimensions of the melon in the white whale. Clearly, the single/dual source issue remains unresolved.

Madsen, Wisniewska, and Beedholm (2010) conducted a study on harbor porpoises (*Phocoena phocoena*), using a set of contact hydrophones placed on the surface of the head, one lateral to each pair of phonic lips, and one at the center of the sloping forehead near the melon. Since the hydrophone on the right side received clicks before the one on the left in all three animals, the researchers proposed that *all* odontocetes click *only* by using the phonic lips on the right side. But this explanation may not make sense from an evolutionary perspective: if the left phonic lips and bursae are not being used to produce clicks, then it would seem that selective pressure would be relaxed and allow random architectural flaws to creep in, but we see no obvious evidence of this (Cranford et al., 1996; see also Cranford, Trijoulet, Smith, & Krysl, 2014). Harbor porpoise phonic lips are nearly bilaterally symmetrical (like other porpoise species and certain other delphinoids that also generate high-frequency, narrow-band sonar clicks), an exception to the delphinid "rule" that phonic lips are considerably larger on the right side than the left.

Later, Madsen et al. (2013) conducted a similar study using contact hydrophones placed on the head of one bottlenose dolphin (*Tursiops truncatus*) and one false killer whale (*Pseudorca crassidens*). They presented these animals with the simple task of detecting steel targets placed 2.6 meters away, behind a visually opaque screen. An additional three bottlenose dolphins were stationed in a hoop and asked to echolocate ad libitum at various metallic objects placed ~20 m in front of them. Once again, observing the time-of-arrival differences between the contact hydrophones, the authors suggested that not only the tested species but *all* odontocetes click *only* with the right side's phonic lips.

This conclusion disregards the fact that producing clicks primarily on the right side but also, albeit less frequently, on the left side has been reported in the literature multiple times over the years (Cranford et al., 2011; Evans, 1973; Lilly, 1978; Mackay & Liaw, 1981; Norris, Harvey, Burzell, & Kartha, 1972). More recently, Cranford et al. (2011) used multiple endoscopes and pressure catheters placed *inside* the dolphin's nasal apparatus to directly observe the activity at either or both bilateral phonic lips. They also measured the air pressure in the bony nasal passages during the production of clicks in their bottlenose dolphins. Their recordings showed that even though the right side was used most of the time, the left side was also used, but much less often, and one of the bottlenose dolphins evidently used both sets of phonic lips to click simultaneously.

Cranford et al. (2011) also showed that the air reservoirs in the two bony nares are pressurized simultaneously and equally, no matter whether the animal is using the phonic lips on one side or the other. This pattern never varied over hundreds of observations. This means that both phonic lips are powered and potentially ready to be actuated by muscular action, as demonstrated by Ridgway et al. (1980) and Amundin and Andersen (1983).

In addition, Cranford et al. (2011) presented a sonogram (provided by William E. Evans) of clicks recorded from a stationary and solitary bottlenose dolphin in a tank. The beginning of the sonogram shows two sets of clicks with different spectral characteristics, but their rates are changing in different directions, one getting faster and the other getting slower. By the end of the spectrogram, both sets of clicks merge together as the click repetition rate rises rapidly. The parsimonious interpretation here is that there are two click trains from two independent sources whose click rates change gradually until they merge so that clicks from both sources occur simultaneously.

One potential advantage of two click sources would be to generate extremely high repetition rates by alternating between them (Cranford et al., 2011). In response, Madsen, Jensen, Carder, and Ridgway (2012) claimed that two sources were not necessary for high click rates, basing this on the maximum click rates that they measured of only 600 clicks per second (cps) for the porpoise (Madsen et al., 2010) and 430 cps for the bottlenose dolphin (Madsen et al., 2013). These rates are far from the maximum repetition rates reported for dolphins and porpoises: for example, 2,000 cps in Atlantic spotted dolphins, *Stenella frontalis* (Herzing, 1996), 1,000 cps in the harbor porpoise (Amundin, 1991), or 960 cps of doublet clicks in the tucuxi, *Sotalia fluviatilis*, resulting in a total click rate of 1,920 cps (Norris et al., 1972). It may be that highly trained dolphins could have been biased by the training process or the simplicity of the target detection task in these studies that have only elicited unilateral click production (Madsen et al., 2010, 2013). To date, little is known about the circumstances under which these animals choose the parameters of their sound generation system in the wild.

Can Delphinoids Accomplish Internal Beam Scanning?
Researchers have long suspected that dolphins are capable of sonar beam steering, adjusting the width and direction of their sonar beam. This ability has now been demonstrated in three delphinoid species, the harbor porpoise (Amundin, 1991; Madsen et al., 2010), the false killer whale (Kloepper, Nachtigall, Donahue, & Breese, 2012; Kloepper, Nachtigall, Quintos, & Vlachos, 2012), and the bottlenose dolphin (Moore, Dankiewicz, & Houser, 2008; Starkhammar, Moore, Talmadge, & Houser, 2011).

Amundin (1991) provided the first evidence of possible beam "scanning" behavior in a harbor porpoise. He used a two-hydrophone arrangement, with one hydrophone immediately in front of the rostrum fixed to a bite plate, and another hydrophone on an articulating rod that could be positioned within a transverse plane in front of the rostrum. The porpoise was enticed to click but had no sonar task to solve. If the beam pattern was fixed, there should be constant relative amplitude between the two hydrophones, irrespective of the position of the adjustable hydrophone. However, Amundin discovered that the amplitude difference varied between the hydrophones over a range of 30 dB, always gradually and in a cyclic manner, suggesting that the porpoise was "scanning" its beam, and that this scanning was under the control of the animal. Similar cyclic changes were seen when the second hydrophone was placed in contact with the melon at different locations.

Madsen et al. (2010) placed contact hydrophones on either side of the midline on the melon of a harbor porpoise. Again, the received levels at the two hydrophones did not always follow the pattern for a fixed beam. Instead, the majority of clicks showed this relationship, where the right-side hydrophone recorded the higher levels; but in some click trains, the left hydrophone recorded the highest amplitudes.

A false killer whale (Kloepper, Nachtigall, Donahue, & Breese, 2012) stationed in a hoop was asked to discriminate between hollow aluminum cylinders with different wall thicknesses and from different distances up to 7 m, increasing task difficulty. A sixteen-element array, placed between the animal and the targets, recorded the outgoing sound beam. The whale reportedly increased sonar beam width when discriminating between targets with small differences in wall thickness and with increasing distance to these hard-to-discriminate targets.

Moore et al. (2008) recorded the sonar beam of a bottlenose dolphin with a curved 24-hydrophone array placed 1.2 m in front of an animal stationed on a bite plate. The dolphin detected water-filled stainless steel spheres (target strength –27 dB) and a hollow aluminum cylinder (target strength –17 dB) that were moved both horizontally and vertically at a distance of 9 m. The hydrophone array covered a 60° horizontal and 32.8° vertical sector. Moore and colleagues reported an angular detection threshold for the sphere at 26° to the left and 21° to the right. The threshold for the cylinder was 19° to the left and 13° to the right. They also found that the animal was able to steer the maximum response axis of the beam up to 18° in the horizontal, 12° in the upward vertical, and 4° in the downward vertical direction.

Starkhammar et al. (2011) studied beam dynamics in a bottlenose dolphin using a similar setup as that used by Moore et al. (2008). The dolphin detected a water-filled stainless steel sphere (target strength of –32 dB) placed 9.4 m in front of its head. A curved 29-hydrophone array was placed in front of the rostrum with all elements equi-distant at 90 cm. The target was moved horizontally, and the animal was capable of steering the beam internally to detect the target up to 28° to either side of the center, which was the limit of the array.

Starkhammar et al. (2011) also made amplitude maps of the sonar beam in different frequency ranges and found that between 20 and 70 kHz an energy peak was consis-tently projected downward at 6°. Often a second energy peak appeared, with energy between 30 kHz and 120 kHz, and focused some 3° above the center of the array and slightly to the right. This second energy peak occurred only at higher amplitudes, and never in the beginning of a click train. The two peaks resulted in an energy "valley" in the center of the beam, suggesting that if a target was placed in this valley and moved relative to the beam, it would optimize the changes in the returning echo. This would greatly facilitate target localization. Similar beam formation has been found in Egyp-tian fruit bats (Yovel, Falk, Moss, & Ulanovsky, 2010).

Cranford et al. (2014) used vibroacoustic finite element modeling (FEM), based on CT scans from live and postmortem bottlenose dolphins, to simulate sound generation and sound beam formation. They found that a center beam axis was produced when activating the bursae pair contained within the phonic lips on right side. This bursae pair is located near the midline and perpendicular to the long axis of the skull in bottle-nose dolphins. The left side's pair of bursae is located off the midline and canted some 30° toward the long axis of the head; activating them created a beam axis pointing between 10° and 30° to the right.

The internal mechanism(s) responsible for sonar beam width modulation, steering, and scanning have not yet been conclusively explained. Moore et al. (2008) proposed two mechanisms: (1) manipulation of the air sac inflation/deflation and melon geom-etry, and (2) phase shifting between two click generators. Melon deformation can read-ily be seen in phonating dolphins and porpoises, as noted by Moore and colleagues. The two-source phase shift hypothesis may gain support from a short-time fractional Fourier transform analysis of bottlenose dolphin clicks, reported by Capus et al. (2007). They showed that bimodal clicks contained two separable chirplike components with a higher-frequency component slightly delayed relative to a lower-frequency com-ponent. This finding agrees with Starkhammar et al. (2011), suggesting that the two energy peaks found in the beam might originate from two sources, where the second, higher-frequency peak comes in later in a click train, augmenting the first. Cranford et al. (2014), though they found an energy peak that was offset to the right when activat-ing the left dorsal bursae, also found that extremely small changes in the relative posi-tion of the right side's anterior and posterior bursae resulted in significant adjustments

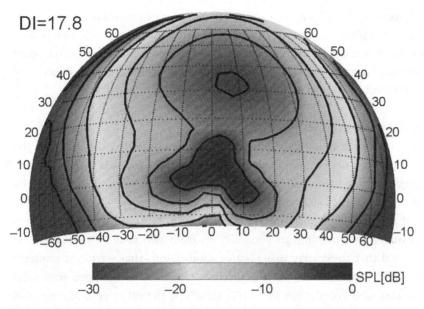

Figure 2.3

In finite element modeling simulations, extremely small changes in the relative position of the right side's anterior and posterior bursae resulted in significant adjustments in the sonar beam. This energy distribution was achieved by moving both bursae forward approximately 6 mm and the posterior bursae down approximately 6 mm. For details see fig. 4 in Cranford et al. (2014).

in the beam focus and axis direction. They tested a large number of such positions and found several that produced double energy peaks (fig. 2.3), similar to that found by Starkhammar et al. (2011).

When a dolphin has locked its sonar on a target, there is often a 30 to 50 ms delay from the reception of an echo until the next click is emitted (Verfuß, Miller, Pilz, & Schnitzler, 2009). To explain such a short delay, which is shorter than the minimum motor neuron response time, Ridgway (2011) suggested that dolphins tune to a click rate rhythm to indirectly achieve this short delay. A similar tuning of the click rates of two sources might allow for fine timescale control of phase and hence beam steering. Another possibility might be a master–slave synchronization between the two sides' phonic lips, where activation of one side triggers the other to achieve changes in their relative timing, and hence adjusting the phase. This could be accomplished by varying the tension on the posterior labium and bursa using the intrinsic musculature of the posterior nasofrontal sac (Cranford & Amundin, 2004; Cranford, 1992) or by moving the two sources closer to, or farther from, each other (Cranford et al., 2014). Since the dolphin cannot get feedback from a point in the beam somewhere in front of its rostrum where such phase variations take effect, it will most likely make these source changes in a cyclic manner, for example, to achieve a side-to-side *scanning* of the beam axis. This notion is supported by the cyclic or gradual amplitude changes observed by Amundin (1991) and Madsen et al. (2010).

Whistles Are Not Really Whistles?

The traditional view of dolphin whistles is that they are resonance phenomena in the air-filled cavities of the upper nasal pathways, as originally suggested by Lilly (1962), or "some edge- or hole-tone, vortex-shedding mechanism related to that used by humans," as suggested by Mackay and Liaw (1981).

The view that whistles are airborne was challenged by Murray, Mercado, and Roit-blat (1998). Analyzing recordings of vocalizations from two false killer whales, the authors suggested that these were tissue-borne vocalizations that were best modeled as lying along a continuum with trains of discrete, exponentially damped sinusoidal pulses at one end, and continuous sinusoidal signals at the other end. Madsen and colleagues (2012) analyzed recordings of a bottlenose dolphin that was given heliox to breathe, and came to a similar conclusion. Since the speed of sound is higher in this gas mixture than in air, air-space resonance should have resulted in an upward frequency shift of the whistles. However, Madsen and his coworkers found that the fundamental frequency of the whistles remained unchanged, and they suggested that the whistles were generated in a way similar to the operation of vocal cords in human speech or the syrinx in birdsong, that is, by vibrating tissues.

Whistling requires higher intranarial air pressure, with greater and presumably faster airflow than the generation of biosonar clicks (Cranford et al., 2011; Ridgway et al., 2001). Jensen, Perez, Johnson, Soto, and Madsen (2011) recorded fewer whistles from pilot whales during deep dives, suggesting that it may be more difficult to generate whistles at high ambient pressure, where the volume of air would be considerably reduced and the air density much higher. Ridgway et al. (2001) made similar observations. They studied two beluga whales, subjected to hearing tests at increasing water depths down to 300 m. The whales were required to whistle in response to hearing the test signal. The researchers found that the whistle fundamental frequency, as well as the harmonics, remained basically unchanged irrespective of depth, but the relative amplitude of harmonics was irregularly increasing with increasing depths, concurrent with a reduction in the amplitude of the fundamental. One of the whales did not whistle at 300 m and instead responded with a burst pulse sound, indicating that it may have had difficulty producing whistles at that depth.

Although whistling species like bottlenose dolphins and beluga whales are capable of diving to at least 300 m (Ridgway, Scronce, & Kanwisher, 1969; Ridgway et al., 2001), where pressure would be an issue for whistling, these species and many other delphinids also occupy shallow waters (Alekseeva, Panova, & Bel'kovich, 2013; Chmelnitsky & Ferguson, 2012; Dudzinski, 1996; Herzing, 1996; Rossbach & Herzing, 1997; Scott, Wells, & Irvine, 1990) or socialize at the surface (Jensen et al., 2011). Here the timbre of the whistles would be rather stable and hence capable of conveying socially important cues, for example, identity (Janik, Sayigh, & Wells, 2006), gender (Miller, Smarra, & Perthuison, 2007), or direction of movement (Lammers, Au, & Herzing, 2003). Lammers and Au (2003) recorded Hawaiian spinner dolphins traveling from their shallow resting areas to offshore feeding grounds. They found that higher-frequency whistle harmonics were clearly directional. This was later corroborated by Branstetter, Moore, Finneran, Tormey, and Aihara (2012), who reported directivity indices of 8.5 to 9.3 for higher whistle harmonics. Lammers and Au (2003) hypothesized that this would facilitate group coordination by offering directional cues. Therefore, in these circumstances, whistle timbre may indeed be used, and even if timbre does change with depth, the higher harmonics would not disappear, continuing to give directional information. Even so, Janik and colleagues showed that only the fundamental contour of signature whistles, which is unaffected by depth, is necessary for individual recognition (Janik, Sayigh, & Wells, 2006). This also allows individual recognition to function over long distances, where higher frequencies would be attenuated to a greater degree.

Sound Reception

Historical Perspective

Some of the earliest work on dolphin sensory systems revealed their keen sense of high-frequency hearing (Kellogg & Kohler, 1952; Kellogg, Kohler, & Morris, 1953; McBride, 1956; Schevill & Lawrence, 1953). By then it was clear that dolphins could hear above 100 kHz, but few studies succeeded in describing or probing the *undisrupted* anatomic geometry, between the skin and the ears, for a systemic or holistic view of the mechanisms involved. Ridgway, Norris, and their colleagues provided welcome exceptions and sparked the discovery of mechanisms that drove much of the research in this field (McCormick, Wever, Ridgway, & Palin, 1980; Norris, 1968, 1975, 1980; Ridgway, 1980; Ridgway et al., 1980, 1981). After more than a half century of studies on dolphin biosonar, a case can be made that we are still in our infancy of understanding how the system works.

The sensory system that produces the perception of hearing combines several structural complexes, including the outer-ear, middle-ear, inner-ear, and CNS components. In this chapter, we concentrate on the outer- and middle-ear components because their functional morphology is not well studied. We do not intend to indicate that the inner-ear and CNS components are unimportant; quite the contrary. In fact, much of the work that has been conducted on these portions of the sensory system has revealed essential functional aspects (Bullock et al., 1968; Bullock & Ridgway, 1972; Ketten, 1992, 1997, 2000; McCormick, Wever, Palin, & Ridgway, 1970; McCormick et al., 1980; Ridgway, 1980, 1983, 1985; Ridgway, McCormick, & Wever, 1974; Wever, McCormick, Palin, & Ridgway, 1971a, 1971b, 1972).

Over the past half century, investigators have learned a great deal about the structure and function of the inner ear. For example, we have a basic understanding of the frequency and intensity ranges of hearing sensitivity (Johnson, 1967), temporal resolution (Mooney, Nachtigall, & Yuen, 2006; Mooney, Li, Ketten, Wang, & Wang, 2011; Mooney, Nachtigall, Taylor, Rasmussen, & Miller, 2009), the effects of depth on hearing (Ridgway et al., 2001), the maximum range of sonar target detection (Au, 1993), the abilities of dolphins to perceive signals in the presence of masking noise (Au et al., 1974) or during complex attention requirements (Penner, 1988), and critical bandwidths and ratios (Au & Moore, 1990), as well as some understanding of the exposure levels that can cause temporary or permanent shifts in hearing thresholds (Houser & Finneran, 2006), frequency discrimination acuity (Herman & Arbeit, 1972), target localization, distance discrimination, detection, and wall thickness discrimination or material composition, and receive beam directivity (Au, 1993). Despite all that we have learned from these studies, almost none of them were conducted under dynamic conditions that might approximate free-swimming dolphin behaviors in their natural environment. There are a few exceptions (Verfuß et al., 2009; Verfuß, Miller, & Schnitzler, 2005), only investigating a handful of species. The crux of the subject—what we understand about *how* echolocation actually works—is still a pertinent question.

In fact, one of the few early classic studies to use a swimming dolphin produced results that should still cause us to remain humble about what we understand about dolphin hearing. This revealing study was designed and conducted by two of the pioneers in dolphin bioacoustics, F. G. Wood and W. E. Evans (1980). A bottlenose dolphin named Scylla was trained to discriminate between two targets during a swimming approach while recordings were made on magnetic tape. The dolphin was instrumented with *seven* hydrophones connected to an umbilicus: four on the forehead and one on the rostrum to collect outgoing signals produced by the dolphin, and two mounted on the eyecups that served as blindfolds to capture the returning echoes. Three more hydrophones and two video cameras were mounted within the round pool. During the initial set of trials, Scylla approached the targets while producing echolocation pulses and made her decision. For the last trial, the researchers removed one of the spherical steel targets. As Scylla approached, they dropped a live fish into the water at the position of the missing target, and she followed it around the tank without making any detectable sounds. Blindfolded, the dolphin repeatedly tracked and captured live fish five different times without making any sounds. Wood and Evans pondered the question of how often these animals can use passive sound reception to identify and track targets in the wild. That is still an intriguing question.

The Internal Acoustic Pinnae: Acoustic Antenna of the Delphinoid Head

The cone shape of the cetacean head, and the fact that the acoustic impedance of water is close to that of the soft tissues of the head, predispose it to function somewhat like an antenna. As such, sounds may enter the head at multiple locations. This new paradigm expands the sound reception concept from the long-standing hypothesis that sounds enter the head only through bilateral acoustic windows, through the lower jaws (Norris, 1964, 1968). Several published studies have provided hints about this new multipathway paradigm (Brill, Moore, Helweg, & Dankiewicz, 2001; Bullock et al., 1968; Ketten, 2000; Møhl, Au, Pawloski, & Nachtigall, 1999; Popov, Supin, Klishin, Tarakanov, & Pletenko, 2008).

Although sounds can enter the head from many directions, which sounds reach the ears appear to be determined by the incident location, trajectory, and frequency spectrum. This implies the existence of an internal acoustic pinna (IAP), a concept that originated with Norris (1964, 1968). The IAP is a collection of structurally and materially integrated tissues and organs within the head, including multiple fat bodies, dense connective tissue planes and discrete structures, bony surfaces of various curvatures and thicknesses, as well as air spaces and several muscles. Based on FEM simulations, we provided the first mechanistic demonstration of the IAP for Cuvier's beaked whale, *Ziphius cavirostris* (Cranford, Krysl, & Hildebrand, 2008; Cranford & Krysl, 2012; Cranford, Krysl, & Amundin, 2010) and recently for the long-beaked common dolphin, *Delphinus capensis* (Krysl & Cranford, 2015). Our FEM simulations show that some sounds that reach the head may not make it to the ears because they are almost completely attenuated or filtered out, while others are amplified by constructive interference or other unknown mechanisms. In addition, different pathways may have different lengths and variable characteristic sound velocities (see Krysl & Cranford, 2015).

It appears as if the ventral aspect of the head contains multiple bilateral pathways that effectively pass sounds through to the ears. The dorsal side of the head apparently has fewer pathways that can pass sound to the ears (Bullock et al., 1968; Møhl et al., 1999; Mooney et al., 2008), and only after considerable attenuation (Cranford et al., 2008). The pathways on the ventral side of the head can be grouped into channels that wrap around and underneath the mandibles and then run posteriorly, where they eventually coalesce into the internal mandibular fat body (iMFB) on their respective side, and then interact with the ears. We have termed these coalescing channels the *gular pathways* because they primarily occur within the gular anatomy between and below the mandibles (Cranford et al., 2008).

Understanding the functional implications of the IAP has buttressed the little-studied notion that sound reception in odontocetes is highly directional and frequency dependent, and generally narrower for higher frequencies (Au & Moore, 1984; Kastelein, Janssen, Verboom, & de Haan, 2005; Krysl & Cranford, 2015). We do not know whether any crossover pathways connect the two sides, but evidence suggests that the IAP treats sounds *unequally* on each side (Brill et al., 2001; Bullock et al., 1968; Møhl et al., 1999), although the two bony ear complexes (TPCs) within individuals are nearly identical (Ary, Cranford, Berta, & Krysl, 2015). This inherent asymmetry in acoustic functional morphology was long ago proposed by Norris (personal communication) to help resolve left–right ambiguity in the sonar problem, particularly when a dolphin is quickly closing the distance to fast-moving prey and decision intervals are constantly becoming shorter.

Main Components of the "Antenna"
Mandibles and Mandibular Fat Bodies
The mandibular structure in odontocetes is atypical for mammals (Barroso, Cranford, & Berta, 2012). The posterolateral wall of each mandible (pan bone) is thin, and the medial wall is absent, resulting in an enlarged hollow mandibular fossa that is filled with a specialized fat body, the iMFB, whose chemical composition and distributional topography are acoustically functional (Koopman, Budge, Ketten, & Iverson, 2006). Similar fat bodies bulge from the outside of the posterolateral wall of each mandible and expand underneath it: the external mandibular fat body (eMFB). Collectively and bilaterally, the iMFBs and eMFBs form portions of the pathways that carry sound to each TPC (fig. 2.4).

The Tympanoperiotic Complexes (TPCs)
The TPC combines bony elements that house the middle and inner ear, along with a few muscles, various fluid-filled channels, and the transduction system that produces input to the CNS. The posterior terminus of each iMFB bifurcates and attaches to its

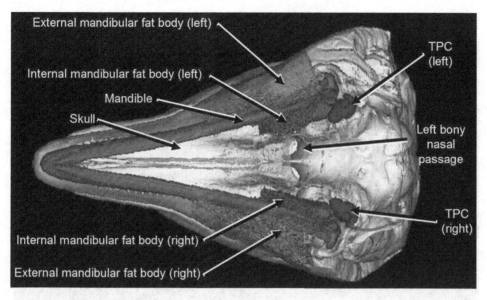

Figure 2.4
Ventral view of the sound reception anatomy in a bottlenose dolphin (*Tursiops truncatus*).

respective TPC at precisely two locations. The ventral branch of the iMFB attaches to a relatively thin portion of the tympanic bulla known as the tympanic plate (Fleischer, 1973). The dorsal, smaller branch is nestled in a bony trough, the medial sulcus of the mallear ridge (fig. 2.5). The existence of these dual fatty attachments to the TPC has now been demonstrated across a broad spectrum of odontocetes, from beaked whales to porpoises (Cranford et al., 2010).

The terminal lipid branches of each iMFB provide vibroacoustic continuity with the associated TPC. Their interface with the dense bones of the TPCs involves an extreme mismatch of acoustic impedance, which thus produces an excellent *reflective* boundary. One could reasonably ask: how can acoustic energy incident on such a drastically mismatched boundary actuate the TPC? The answer appears to be that where the impedance mismatch is the greatest, is also where the greatest force will be exerted when the sound reflects off the interface. That is, the bony floor of the medial sulcus of the mallear ridge, or the interface, which receives the dorsal branch of the iMFB, is bone so thin as to be translucent and is located adjacent to the fusion between the tympanic bone and the malleus of the middle ear. Vibrational analysis shows complex, frequency-dependent movements at this interface and the adjacent ossicular chain if exposed to sounds (Cranford et al., 2010).

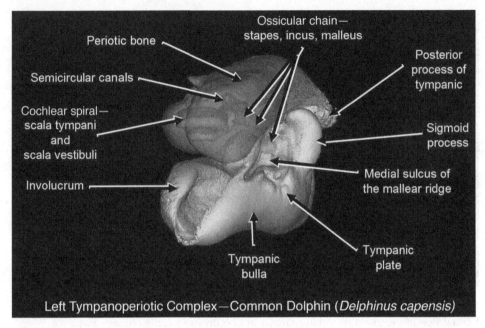

Figure 2.5
Anterolateral view of the left tympanoperiotic complex in the long-beaked common dolphin (*Delphinus capensis*).

The morphology of the TPC across odontocetes is structurally diverse (fig. 2.6), although the basic fluid and nervous transduction mechanisms occurring within the cochlea are broadly similar across mammals. Ketten and colleagues have studied the specialized anatomy of the cochlea and proposed functional implications (Ketten, 1992, 1994, 1997, 2000; Miller, Zosuls, Ketten, & Mountain, 2006; Tubelli, Zosuls, Ketten, & Mountain, 2014). However, since the dynamic range of hearing in delphinoids stretches across more than two orders of magnitude, there are clearly many adaptations within the TPC that allow these animals to perceive sounds with frequencies above 100 kHz.

In all delphinoids, the TPC is essentially acoustically isolated from the skull, primarily by air spaces, but also by some connective tissue boundaries. Each TPC is suspended from the ventral aspect of the skull by numerous ligamentous cords (see fig. 4 in Cranford et al., 2010). Some bony components of the TPC are hypermineralized (Li & Pasteris, 2014), and this appears to be an important factor in determining the essential properties of the multimodal lever arm formed by the tympanic bone. This unique lever arm construction is similar across all cetaceans, and this similarity suggests its

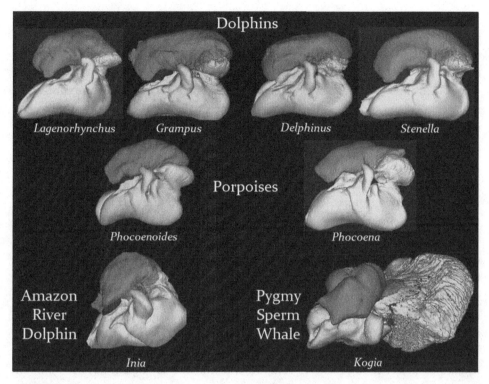

Figure 2.6
Diversity of TPC shapes in selected odontocetes: Five dolphins (*Lagenorhynchus obliquidens, Grampus griseus, Delphinus delphis, Stenella coeruleoalba, Orcinus orca*); two porpoises (*Phocoenoides dalli, Phocoena phocoena*); one river dolphin (*Inia geoffrensis*); one beaked whale (*Mesoplodon carlhubbsi*); one sperm whale (*Kogia breviceps*). Relative sizes of these TPCs are not to scale.

central role in the evolution of underwater hearing in cetaceans. The implication here is that diversity in TPC shape across the Odontoceti is suggestive of differential vibratory patterns and vibroacoustic function.

Lancaster et al. have recently discovered that the tympanic bone of the TPC has grown to adult size soon after birth (precocial development) across a broad range of cetacean taxa, including mysticetes and odontocetes (Lancaster, Ary, Krysl, & Cranford, 2014). These findings support the primacy of this sensory system early in life and lend support to the hypothesis that the increased generation of signature whistles in late-term mothers might imprint the fetus in utero so that it can recognize its mother's whistle immediately after birth (Mello & Amundin, 2005). Preliminary indications are that precocial development of the TPC also existed in archaeocetes.

Measuring Hearing
Auditory Evoked Potentials

In the 1980s, the advent of recording dolphin brain activity within the brain stem and midbrain, referred to as ABR (auditory brain stem response), in response to sound stimuli, sparked a new area of research into dolphin hearing. It is essentially the same technique that is now known as auditory evoked potentials (AEP) or auditory steady-state response (ASSR), and continues to produce results that greatly increase our understanding of biosonar (e.g., Nachtigall, Mooney, Taylor, & Yuen, 2007; Ridgway et al., 1981; Thomas et al., 2003). The brain activity is recorded by electrodes placed on the surface of the head. To discern the low-amplitude brain signals from the noise, researchers must average the recorded signals over thousands of stimulus events, but this takes only seconds for each frequency tested.

Studies that compared audiograms collected using AEP with behaviorally derived data show that the results are comparable, particularly when the differences in stimulus duration and temporal summation are considered (Schlundt, Dear, Green, Houser, & Finneran, 2007; Yuen, Nachtigall, Breese, & Supin, 2005). One of the primary benefits of AEP is that it requires only basic training and cooperation from animals in a laboratory setting, or no requisite cooperation in stranded or temporarily captive animals (Nachtigall et al., 2008; Pacini et al., 2011). Also, the short time (minutes) it takes to collect enough data to compose a reasonable audiogram has been a boon to understanding the parameters that can cause temporary threshold shifts (TTS) (Finneran & Schlundt, 2010; Lucke, Siebert, Lepper, & Blanchet, 2009) and permanent threshold shifts (PTS) in an ever-expanding list of marine mammals (Houser & Moore, 2014), which now includes polar bears, sperm whales, and multiple beaked whale species.

The use of AEP has also proved to be a powerful tool to predict what dolphin brains react to *during* echolocation and has produced the discovery that dolphins can change their hearing sensitivity with changing circumstances (Nachtigall & Supin, 2008; Supin & Nachtigall, 2013; Supin, Nachtigall, & Breese, 2008). In fact, recognizing this capability led to one of the most intriguing new results in dolphin audiology; Nachtigall and Supin (2013) have shown that when a dolphin is presented with a warning signal before being exposed to a loud sound, the animal can reduce its hearing sensitivity, presumably as a buffer against potential deleterious effects of high-intensity sound. The mechanisms by which this occurs are unknown but may involve processes like the "middle ear reflex" mediated by striated muscle.

One major drawback of AEP, as well as conditioned responses to test hearing, is that these techniques are ineffective at showing us *how* the physiological processes work. The complex anatomic structure within the peripheral auditory system has often been treated as a kind of "black box" for which little can be learned about the specific mechanisms involved using these approaches. Fortunately, the recent innovative application of finite element modeling methods has allowed us to begin investigating many of these previously intractable mechanisms (Cranford & Krysl, 2015).

"Jaw Hearing" and Jawphones

The concept of "jaw hearing" prompted Patrick Moore to invent and refine a device he dubbed the "jawphone" to selectively stimulate the assumed binaural sound reception channels (Moore, Pawloski, & Dankiewicz, 1995). The jawphone is a piezoelectric transducer embedded in a silicone rubber suction cup and attached somewhere along the dolphin's jaw. Jawphones have been used for more than two decades, but the mechanisms by which they activate the hearing apparatus in toothed whales remain largely unknown. Laudable work by Brill and others attempted to understand jawphone function (Brill, 1988, 1989, 1999; Brill & Harder, 1991; Brill, Moore, & Dankiewicz, 2001; Brill, Moore, Helweg, & Dankiewicz, 2001). However, despite the lack of understanding about just how jawphones work, they have been replicated and used worldwide by researchers to gather valuable information.

Perhaps the most revealing study involving jawphones was conducted by Møhl and colleagues (Møhl, Au, Pawloski, & Nachtigall, 1999). They determined that the most sensitive location was actually forward of the pan bone, but a place almost as sensitive was located underneath the lower jaw in the gular region. They also demonstrated that small differences in the placement of jawphones could produce widely variable results. A cautionary note about jawphone placement is appropriate here.

David Mann and colleagues (Mann, Hill-Cook, Manire et al., 2010) used a jawphone to stimulate the hearing apparatus in eight different odontocete species (seven delphinids and one ziphiid) while measuring auditory evoked potentials from the brain. They tested 34 animals that had been stranded alive or had been entangled in fishing gear. What they found was alarming: more than half of the bottlenose dolphins and a third of the rough-toothed dolphins (*Steno bredanensis*) demonstrated severe to profound hearing loss. They considered many factors that might have produced these drastic hearing losses in such a large percentage of the animals, including exposure to intense chronic noise, explosions, ototoxic drug exposure, chemical pollution, congenital defects, and age-related hearing loss. Oddly, they did not discuss other potentially confounding factors, such as the inconsistent placement of the jawphone. For example, Møhl et al. (1999) showed that small differences in jawphone position resulted in significantly different hearing sensitivities. Another possible confounding factor that is related to jawphone placement is that the anatomic geometry of the IAP differs across species. The compromised health of stranded animals may also have played a role.

In an attempt to understand jawphone function, we recently used our FEM system to mimic how the jawphones trigger dolphin hearing in a model of a common dolphin. Our preliminary results (summarized in Houser & Moore, 2014) suggest that jawphones actually *create unnatural* or *artificial* jaw hearing, caused by localized "rattling" of the jaws. It appears that these vibrations eventually find their way to the ears by various novel, often anomalous pathways compared to free-field simulations. Perhaps the most significant finding from these preliminary simulations was that a difference of few centimeters in the placement of a jawphone could cause a received amplitude difference of 6 to 9 dB at the ear, depending on the locations compared.

Regardless of the impact on previous studies, we still need to know how and by what mechanism jawphones stimulate the hearing apparatus. We do not suggest that all the literature that has been compiled using jawphones should be discarded; rather, we suggest that (1) our understanding of what jawphones do is incomplete; (2) procedures need to be standardized across at least a few representative species; and (3) a careful reevaluation should be conducted to determine more precisely what, if any, adjustments should be made in our understanding of previous jawphone work. This is particularly important because some decisions about the apparent health of stranded animals are now being made based on jawphone use, integrated with the AEP method (see Mann et al., 2010).

Future Directions

Finite Element Modeling
As we have shown, computational models are valuable tools for understanding underwater bioacoustic pathways and interactions, specifically those between various tissue geometries within the heads and bodies of various toothed whales. We predict that implements from our vibroacoustic tool kit (VATk) will play a major role in understanding bioacoustic mechanisms in aquatic animals in the future. These finite element modeling tools are cost-effective and can provide a test bed for virtual experiments and a means for determining best mitigation strategies.

Head-Related Transfer Functions
Last year we began calculating the head-related transfer function (HRTF) in a dolphin, as we had previously done for a Cuvier's beaked whale. These HRTF maps indicate the directions from which environmental sounds preferentially enter the head and how

various frequencies are transformed by the anatomic geometry before reaching the ears (Krysl & Cranford, 2015).

Since our modeling approach considers the function of the entire head of these animals, this HRTF work is currently the most reliable and economical means of sorting out the directionality of the sound reception system according to frequency. We began by investigating the lower frequencies and plan to move up the scale. The paramount importance of this work is that it reveals the characteristics of this directionality (Mooney et al., 2008). This stands in contrast to some anthropogenic noise impact models that treat the animal as an omnidirectional point receiver (Moretti et al., 2014).

We have made great strides in understanding cetacean bioacoustics over the last half century, but much of the work has been done primarily with stationary animals in controlled laboratory conditions. Continued efforts to understand the full range of cetacean sound generation and reception are warranted, especially considering the continuously increasing levels of ocean noise.

References

Alekseeva, Y. I., Panova, E. M., & Bel'kovich, V. M. (2013). Behavioral and acoustical characteristics of the reproductive gathering of beluga whales (*Delphinapterus leucas*) in the vicinity of Myagostrov, Golyi Sosnovets, and Roganka Islands (Onega Bay, the White Sea). *Biological Bulletin, 40*(3), 307–317.

Amundin, M. (1991). *Sound production in odontocetes with emphasis on the harbour porpoise Phocoena phocoena*. Dissertation, University of Stockholm, Stockholm, Sweden.

Amundin, M., & Andersen, S. H. (1983). Bony nares air pressure and nasal plug muscle activity during click production in the harbour porpoise, *Phocoena phocoena*, and the bottlenosed dolphin, *Tursiops truncatus*. *Journal of Experimental Biology, 105*, 275–282.

Ary, W. J., Cranford, T. W., Berta, A., & Krysl, P. (2015). Functional morphology and symmetry in the odontocete ear complex. In A. N. Popper & A. D. Hawkins (Eds.), *Effects of noise on aquatic life II*. New York: Springer Science+Business Media.

Au, W. W. L. (1993). *The sonar of dolphins*. New York: Springer.

Au, W. W. L., Floyd, R. W., Penner, R. H., & Murchison, A. E. (1974). Measurement of echolocation signals of the Atlantic bottlenose dolphin, *Tursiops truncatus* (Montagu), in open waters. *Journal of the Acoustical Society of America, 56*(4), 1280–1290.

Au, W. W., Houser, D. S., Finneran, J. J., Lee, W. J., Talmadge, L. A., & Moore, P. W. (2010). The acoustic field on the forehead of echolocating Atlantic bottlenose dolphins (*Tursiops truncatus*). *Journal of the Acoustical Society of America, 128*(3), 1426–1434.

Au, W. W. L., Kastelein, R. A., Benoit-Bird, K. J., Cranford, T. W., & McKenna, M. F. (2006). Acoustic radiation from the head of echolocating harbor porpoises (*Phocoena phocoena*). *Journal of Experimental Biology*, *209*(14), 2726–2733.

Au, W. W. L., & Moore, P. W. B. (1984). Receiving beam patterns and directivity indices of the Atlantic bottlenose dolphin *Tursiops truncatus*. *Journal of the Acoustical Society of America*, *75*(1), 255–262.

Au, W. W. L., & Moore, P. W. B. (1990). Critical ratio and critical bandwidth for the Atlantic bottlenose dolphin. *Journal of the Acoustical Society of America*, *88*(3), 1635–1638.

Au, W. W. L., & Snyder, K. J. (1980). Long-range target detection in open waters by an echolocating Atlantic bottlenose dolphin (*Tursiops truncatus*). *Journal of the Acoustical Society of America*, *68*(4), 1077–1083.

Barroso, C., Cranford, T. W., & Berta, A. (2012). Shape analysis of odontocete mandibles: Functional and evolutionary implications. *Journal of Morphology*, *273*(9), 1021–1030.

Benoit-Bird, K. J., & Au, W. W. L. (2003). Prey dynamics affect foraging by a pelagic predator (*Stenella longirostris*) over a range of spatial and temporal scales. *Behavioral Ecology and Sociobiology*, *53*, 364–373.

Benoit-Bird, K. J., & Au, W. W. (2009). Phonation behavior of cooperatively foraging spinner dolphins. *Journal of the Acoustical Society of America*, *125*(1), 539–546.

Branstetter, B. K., Moore, P. W., Finneran, J. J., Tormey, M. N., & Aihara, H. (2012). Directional properties of bottlenose dolphin (*Tursiops truncatus*) clicks, burst-pulse, and whistle sounds. *Journal of the Acoustical Society of America*, *131*(2), 1613–1621.

Brill, R. L. (1988). The jaw-hearing dolphin: Preliminary behavioral and acoustical evidence. In P. E. Nachtigall & P. W. B. Moore (Eds.), *Animal sonar: Processes and performance* (pp. 281–288). New York: Plenum.

Brill, R. L. (1989). *The acoustical function of the lower jaw of the bottlenose dolphin, Tursiops truncatus (Montagu), during echolocation.* Doctoral dissertation, Loyola University of Chicago.

Brill, R. L. (1999). Norris and dolphin echolocation: A paradigm shift for sound production and reception. *Marine Mammal Science*, *15*(4), 936–940.

Brill, R. L., & Harder, P. J. (1991). The effects of attenuating returning echolocation signals at the lower jaw of a dolphin (*Tursiops truncatus*). *Journal of the Acoustical Society of America*, *89*, 2851–2857.

Brill, R. L., Moore, P. W., & Dankiewicz, L. A. (2001). Assessment of dolphin (*Tursiops truncatus*) auditory sensitivity and hearing loss using jawphones. *Journal of the Acoustical Society of America*, *109*(4), 1717–1722.

Brill, R. L., Moore, P. W. B., Helweg, D. A., & Dankiewicz, L. A. (2001). *Investigating the dolphin's peripheral hearing system: Acoustic sensitivity about the head and lower jaw.* San Diego: SPAWAR Systems Center San Diego.

Bullock, T. H., Grinnell, A. D., Ikezono, E., Kameda, K., Katsuki, Y., Nomoto, M., et al. (1968). Electrophysiological studies of central auditory mechanisms in cetaceans. *Zeitschrift für Vergleichende Physiologie*, *59*(2), 117–156.

Bullock, T. H., & Ridgway, S. H. (1972). Evoked potentials in the central auditory system of alert porpoises to their own and artificial sounds. *Journal of Neurobiology*, *3*(1), 79–99.

Capus, C., Pailhas, Y., Brown, K., Lane, D. M., Moore, P. W., & Houser, D. (2007). Bio-inspired wideband sonar signals based on observations of the bottlenose dolphin (*Tursiops truncatus*). *Journal of the Acoustical Society of America*, *121*(1), 594–604.

Chmelnitsky, E. G., & Ferguson, S. H. (2012). Beluga whale, *Delphinapterus leucas*, vocalizations from the Churchill River, Manitoba, Canada. *Journal of the Acoustical Society of America*, *131*(6), 4821–4835.

Cranford, T. W. (1992). *Functional morphology of the odontocete forehead: Implications for sound generation*. Doctoral dissertation, University of California, Santa Cruz.

Cranford, T. W., & Amundin, M. E. (2004). Biosonar pulse production in odontocetes: The state of our knowledge. In J. A. Thomas, C. F. Moss, & M. Vater (Eds.), *Echolocation in bats and dolphins* (pp. 27–35). Chicago: University of Chicago Press.

Cranford, T. W., Amundin, M., & Norris, K. S. (1996). Functional morphology and homology in the odontocete nasal complex: Implications for sound generation. *Journal of Morphology*, *228*(3), 223–285.

Cranford, T. W., Elsberry, W. R., Van Bonn, W. G., Jeffress, J. A., Chaplin, M. S., Blackwood, D. J., et al. (2011). Observation and analysis of sonar signal generation in the bottlenose dolphin (*Tursiops truncatus*): Evidence for two sonar sources. *Journal of Experimental Marine Biology and Ecology*, *407*(1), 81–96.

Cranford, T. W., & Krysl, P. (2012). Acoustic function in the peripheral auditory system of Cuvier's beaked whale (*Ziphius cavirostris*). In A. N. Popper & A. D. Hawkins (Eds.), *Effects of noise on aquatic life* (Vol. 730, pp. 69–72). New York: Springer Science+Business Media.

Cranford, T. W., Krysl, P., & Amundin, M. (2010). A new acoustic portal into the odontocete ear and vibrational analysis of the tympanoperiotic complex. *PLoS ONE*, *5*(8), e11927.

Cranford, T. W., Krysl, P., & Hildebrand, J. A. (2008). Acoustic pathways revealed: Simulated sound transmission and reception in Cuvier's beaked whale (*Ziphius cavirostris*). *Bioinspiration and Biomimetics*, *3*(1), 1–10.

Cranford, T. W., Trijoulet, V., Smith, C. R., & Krysl, P. (2014). Validation of a vibroacoustic finite element model using bottlenose dolphin simulations: The dolphin biosonar beam is focused in stages. *Bioacoustics*, *23*(2), 161–194.

Cranford, T. W., & Krysl, P. (2015). Fin whale sound reception mechanisms: Skull vibration enables low-frequency hearing. *PLoS ONE* 10(1): e0116222. doi:10.1371/journal.pone.0116222.

Dormer, K. J. (1979). Mechanism of sound production and air recycling in delphinids: Cineradiographic evidence. *Journal of the Acoustical Society of America, 65*(1), 229–239.

Dudzinski, K. M. (1996). *Communication and behavior in the Atlantic spotted dolphin (Stenella frontalis): Relationships between vocal and behavioral activities*. Doctoral dissertation, Texas A&M University.

Evans, W. E. (1973). Echolocation by marine delphinids and one species of fresh-water dolphin. *Journal of the Acoustical Society of America, 54*(1), 191–199.

Evans, W. E., & Prescott, J. H. (1962). Observations of the sound production capabilities of the bottlenose porpoise: A study of whistles and clicks. *Zoologica, 47*(11), 121–128.

Finneran, J. J., & Schlundt, C. E. (2010). Frequency-dependent and longitudinal changes in noise-induced hearing loss in a bottlenose dolphin (*Tursiops truncatus*). *Journal of the Acoustical Society of America, 128*(2), 567–570.

Fleischer, G. (1973). Structural analysis of the tympanicum complex in the bottle-nosed dolphin (*Tursiops truncatus*). *Journal of Auditory Research, 13*, 178–190.

Herman, L. M., & Arbeit, W. R. (1972). Frequency difference limens in the bottlenose dolphin: 1–70 Kc/s. *Journal of Auditory Research, 12*(2), 109–120.

Herzing, D. L. (1996). Vocalizations and associated underwater behavior of free-ranging Atlantic spotted dolphins, *Stenella frontalis*, and bottlenose dolphins, *Tursiops truncatus. Aquatic Mammals, 22*(2), 61–79.

Herzing, D. L., & Santos, M. E. (2004). Functional aspects of echolocation in dolphins. In J. A. Thomas, C. F. Moss, & M. Vater (Eds.), *Echolocation in bats and dolphins* (pp. 404–410). Chicago: University of Chicago Press.

Houser, D. S., & Finneran, J. J. (2006). A comparison of underwater hearing sensitivity in bottlenose dolphins (*Tursiops truncatus*) determined by electrophysiological and behavioral methods. *Journal of the Acoustical Society of America, 120*(3), 1713–1722.

Houser, D. S., & Moore, P. W. B. (2014). Report on the current status and future of underwater hearing research. San Diego: National Marine Mammal Foundation.

Janik, V., Sayigh, L., & Wells, R. (2006). Signature whistle shape conveys identity information to bottlenose dolphins. *Proceedings of the National Academy of Sciences of the United States of America, 103*(21), 5.

Jensen, F. H., Perez, J. M., Johnson, M., Soto, N. A., & Madsen, P. T. (2011). Calling under pressure: Short-finned pilot whales make social calls during deep foraging dives. *Proceedings of the Royal Society B, 278*(1721), 3017–3025.

Johnson, C. S. (1967). Sound detection thresholds in marine mammals. In W. N. Tavolga (Ed.), *Marine bio-acoustics* (Vol. 2, pp. 247–260). Oxford: Pergamon.

Kastelein, R. A., Au, W. W. L., Rippe, H. T., & Schooneman, N. M. (1999). Target detection by an echolocating harbor porpoise (*Phocoena phocoena*). *Journal of the Acoustical Society of America*, *105*(4), 6.

Kastelein, R. A., Janssen, M., Verboom, W. C., & de Haan, D. (2005). Receiving beam patterns in the horizontal plane of a harbor porpoise (*Phocoena phocoena*). *Journal of the Acoustical Society of America*, *118*, 1172–1179.

Kellogg, W. N., & Kohler, R. (1952). Reactions of the porpoise to ultrasonic frequencies. *Science*, *116*, 250–252.

Kellogg, W. N., Kohler, R., & Morris, H. N. (1953). Porpoise sounds as sonar signals. *Science*, *117*, 239–243.

Ketten, D. R. (1992). The cetacean ear: Form, frequency, and evolution. In R. A. Kastelein, A. Y. Supin, & J. A. Thomas (Eds.), *Marine mammal sensory systems* (pp. 53–75). New York: Plenum.

Ketten, D. R. (1994). Functional analysis of whale ears: Adaptations for underwater hearing. *IEEE Proceedings of Underwater Acoustics*, *1*, 264–270.

Ketten, D. R. (1997). Structure and function in whale ears. *Bioacoustics*, *8*(1), 103–135.

Ketten, D. R. (2000). Cetacean ears. In W. W. L. Au, A. N. Popper, & R. R. Fay (Eds.), *Hearing by whales and dolphins* (pp. 43–108). New York: Springer.

Kloepper, L. N., Nachtigall, P. E., Donahue, M. J., & Breese, M. (2012). Active echolocation beam focusing in the false killer whale, *Pseudorca crassidens*. *Journal of Experimental Biology*, *215*(8), 1306–1312.

Kloepper, L. N., Nachtigall, P. E., Quintos, C., & Vlachos, S. A. (2012). Single-lobed frequency-dependent beam shape in an echolocating false killer whale (*Pseudorca crassidens*). *Journal of the Acoustical Society of America*, *131*(1), 577–581.

Koopman, H. N., Budge, S. M., Ketten, D. R., & Iverson, S. J. (2006). Topographical distribution of lipids inside the mandibular fat bodies of odontocetes: Remarkable complexity and consistency. *IEEE Journal of Ocean Engineering*, *31*(1).

Krysl, P., & Cranford, T. W. (2015). Directional hearing and head-related transfer function in the common dolphin (*Delphinus capensis*). In A. N. Popper & A. D. Hawkins (Eds.), *Effects of noise on aquatic life II*. New York: Springer Science+Business Media.

Kyhn, L. A., Tougaard, J., Beedholm, K., Jensen, F. H., Ashe, E., Williams, R., et al. (2013). Clicking in a killer whale habitat: Narrow-band, high-frequency biosonar clicks of harbour porpoise (*Phocoena phocoena*) and Dall's porpoise (*Phocoenoides dalli*). *PLoS ONE*, *8*(5), e63763.

Lammers, M. O., & Au, W. W. L. (2003). Directionality in the whistles of Hawaiian spinner dolphins (*Stenella longirostris*): A signal feature to cue direction of movement? *Marine Mammal Science*, *19*, 249–264.

Lammers, M. O., Au, W. W. L., & Herzing, D. L. (2003). The broadband social acoustic signaling behavior of spinner and spotted dolphins. *Journal of the Acoustical Society of America*, *114*(3), 1629–1639.

Lammers, M. O., & Castellote, M. (2009). The beluga whale produces two pulses to form its sonar signal. *Biology Letters*, *5*, 297–301.

Lancaster, W. C., Ary, W. J., Krysl, P., & Cranford, T. W. (2014). Precocial development within the tympanoperiotic complex in cetaceans. *Marine Mammal Science*. doi:10.1111/mms.12145.

Li, Z., & Pasteris, J. D. (2014). Tracing the pathway of compositional changes in bone mineral with age: Preliminary study of bioapatite aging in hypermineralized dolphin's bulla. *Biochimica et Biophysica Acta*, *1840*, 2331–2339.

Lilly, J. C. (1962). Vocal behavior of the bottlenose dolphin. *Proceedings of the American Philosophical Society*, *106*(6), 520–529.

Lilly, J. C. (1978). *Communication between man and dolphin: The possibility of talking with other species*. New York: Crown.

Lucke, K., Siebert, U., Lepper, P. A., & Blanchet, M. A. (2009). Temporary shift in masked hearing thresholds in a harbor porpoise (*Phocoena phocoena*) after exposure to seismic airgun stimuli. *Journal of the Acoustical Society of America*, *125*(6), 4060–4070.

Mackay, R. S., & Liaw, H. M. (1981). Dolphin vocalization mechanisms. *Science*, *212*(4495), 676–678.

Madsen, P. T., Jensen, F. H., Carder, D., & Ridgway, S. H. (2012). Dolphin whistles: A functional misnomer revealed by heliox breathing. *Biology Letters*, *8*, 211–213.

Madsen, P. T., Lammers, M., Wisniewska, D., & Beedholm, K. (2013). Nasal sound production in echolocating delphinids (*Tursiops truncatus* and *Pseudorca crassidens*) is dynamic, but unilateral: Clicking on the right side and whistling on the left side. *Journal of Experimental Biology*, *216*(21), 4091–4102.

Madsen, P. T., Wisniewska, D., & Beedholm, K. (2010). Single source sound production and dynamic beam formation in echolocating harbour porpoises (*Phocoena phocoena*). *Journal of Experimental Biology*, *213*, 3105–3110.

Mann, D., Hill-Cook, M., Manire, C., Greenhow, D., Montie, E., Powell, J., et al. (2010). Hearing loss in stranded odontocete dolphins and whales. *PLoS ONE*, *5*(11), e13824.

McBride, A. F. (1956). Evidence for echolocation by cetaceans. *Deep-Sea Research*, *3*, 153–154.

McCormick, J. G., Wever, E. G., Palin, J., & Ridgway, S. H. (1970). Sound conduction in the dolphin ear. *Journal of the Acoustical Society of America*, *48*(6), 1418–1428.

McCormick, J. G., Wever, E. G., Ridgway, S. H., & Palin, J. (1980). Sound reception in the porpoise as it relates to echolocation. In R. G. Busnel & J. F. Fish (Eds.), *Animal sonar systems* (pp. 449–467). New York: Plenum.

Mello, I., & Amundin, M. (2005). Whistle production pre- and post-partum in bottlenose dolphins (*Tursiops truncatus*) in human care. *Aquatic Mammals, 31*(2), 169–175.

Miller, B. S., Zosuls, A. L., Ketten, D. R., & Mountain, D. C. (2006). Middle-ear stiffness of the bottlenose dolphin, *Tursiops truncatus*. *IEEE Journal of Oceanic Engineering, 31*(1), 87–94.

Miller, P. J. O., Smarra, F. I. P., & Perthuison, A. D. (2007). Caller sex and orientation influence spectral characteristics of "two-voice" stereotyped calls produced by free-ranging killer whales. *Journal of the Acoustical Society of America, 121*(6), 3932–3937.

Møhl, B., Au, W. W. L., Pawloski, J., & Nachtigall, P. E. (1999). Dolphin hearing: Relative sensitivity as a function of point of application of a contact sound source in the jaw and head region. *Journal of the Acoustical Society of America, 105*(6), 3421–3424.

Mooney, T. A., Li, S., Ketten, D. R., Wang, K., & Wang, D. (2011). Auditory temporal resolution and evoked responses to pulsed sounds for the Yangtze finless porpoises (*Neophocaena phocaenoides asiaeorientalis*). *Journal of Comparative Physiology A: Neuroethology, Sensory, Neural, and Behavioral Physiology, 197*, 1149–1158.

Mooney, T. A., Nachtigall, P. E., Castellote, M., Taylor, K. A., Pacini, A. F., & Esteban, J. A. (2008). Hearing pathways and directional sensitivity of the beluga whale, *Delphinapterus leucas*. *Journal of Experimental Marine Biology and Ecology, 362*, 108–116.

Mooney, T. A., Nachtigall, P. E., Taylor, K. A., Rasmussen, M. H., & Miller, L. A. (2009). Auditory temporal resolution of a wild white-beaked dolphin (*Lagenorhynchus albirostris*). *Journal of Comparative Physiology A: Neuroethology, Sensory, Neural, and Behavioral Physiology, 195*, 375–384.

Mooney, T. A., Nachtigall, P. E., & Yuen, M. M. L. (2006). Temporal resolution of the Risso's dolphin, *Grampus griseus*, auditory system. *Journal of Comparative Physiology A: Neuroethology, Sensory, Neural, and Behavioral Physiology, 192*(4), 373–380.

Moore, P. W., Dankiewicz, L. A., & Houser, D. S. (2008). Beamwidth control and angular target detection in an echolocating bottlenose dolphin (*Tursiops truncatus*). *Journal of the Acoustical Society of America, 124*(5), 3324–3332.

Moore, P. W., Pawloski, D. A., & Dankiewicz, L. (1995). Interaural time and intensity difference thresholds in the bottlenose dolphin (*Tursiops truncatus*). In R. A. Kastelein, J. A. Thomas, & P. E. Nachtigall (Eds.), *Sensory systems of aquatic mammals* (pp. 11–23). Woerden, the Netherlands: De Spil.

Moretti, D., Thomas, L., Marques, T., Harwood, J., Dilley, A., Neales, B., et al. (2014). A risk function for behavioral disruption of Blainville's beaked whales (*Mesoplodon densirostris*) from mid-frequency active sonar. *PLoS ONE, 9*(1), e85064.

Murray, S. O., Mercado, E., & Roitblat, H. L. (1998). Characterizing the graded structure of false killer whale (*Pseudorca crassidens*) vocalizations. *Journal of the Acoustical Society of America, 104*(3), 1679–1688.

Nachtigall, P. E., Mooney, T. A., Taylor, K. A., Miller, L. A., Rasmussen, M. H., Akamatsu, T., et al. (2008). Shipboard measurements of the hearing of the white-beaked dolphin *Lagenorhynchus albirostris*. *Journal of Experimental Biology, 211*, 642–647.

Nachtigall, P. E., Mooney, T. A., Taylor, K. A., & Yuen, M. M. L. (2007). Hearing and auditory evoked potential methods as applied to odontocete cetaceans. *Aquatic Mammals, 33*(1), 93–99.

Nachtigall, P. E., & Supin, A. Y. (2008). A false killer whale adjusts its hearing when it echolocates. *Journal of Experimental Biology, 211*, 1714–1718.

Nachtigall, P. E., & Supin, A. Y. (2013). A false killer whale reduces its hearing sensitivity when a loud sound is preceded by a warning. *Journal of Experimental Biology, 216*(16), 3062–3070.

Norris, K. S. (1964). Some problems of echolocation in cetaceans. In W. N. Tavolga (Ed.), *Marine bio-acoustics* (pp. 317–336). New York: Pergamon Press.

Norris, K. S. (1968). The evolution of acoustic mechanisms in odontocete cetaceans. In E. T. Drake (Ed.), *Evolution and environment* (pp. 297–324). New Haven: Yale University Press.

Norris, K. S. (1975). Cetacean biosonar: Part 1—anatomical and behavioral studies. In D. C. Malins & J. R. Sargent (Eds.), *Biochemical and biophysical perspectives in marine biology* (pp. 215–234). New York: Academic Press.

Norris, K. S. (1980). Peripheral sound processing in odontocetes. In R. G. Busnel & J. F. Fish (Eds.), *Animal sonar systems* (pp. 495–509). New York: Plenum.

Norris, K. S., Harvey, G. W., Burzell, L. A., & Kartha, T. D. K. (1972). Sound production in the freshwater porpoises *Sotalia* cf. *fluviatilis* (Gervais and Deville) and *Inia geoffrensis* (Blainville), in the Rio Negro, Brazil. In G. Pilleri (Ed.), *Investigations on cetacea* (Vol. 4, pp. 251–260). Bern: Hirnanatomisches Institut der Universität Bern.

Pacini, A. F., Nachtigall, P. E., Quintos, C. T., Schofield, T. D., Look, D. A., Levine, G. A., et al. (2011). Audiogram of a stranded Blainville's beaked whale (*Mesoplodon densirostris*) measured using auditory evoked potentials. *Journal of Experimental Biology, 214*, 2409–2415.

Penner, R. H. (1988). Attention and detection in dolphin echolocation. In P. E. Nachtigall & P. W. B. Moore (Eds.), *Animal sonar: Processes and performance* (Vol. 156, pp. 707–713). New York: Plenum.

Popov, V. V., Supin, A. Y., Klishin, V. O., Tarakanov, M. B., & Pletenko, M. G. (2008). Evidence for double acoustic windows in the dolphin, *Tursiops truncatus*. *Journal of the Acoustical Society of America, 123*(1), 552–560.

Rasmussen, M., & Miller, L. (2002). Whistles and echolocation signals from white-beaked dolphins, *Lagenorhynchus albirostris*, recorded in Faxaflói Bay, Iceland. *Aquatic Mammals, 28*(1), 78–89.

Rasmussen, M. H., Wahlberg, M., & Miller, L. A. (2004). Estimated transmission beam pattern of clicks recorded from free-ranging white-beaked dolphins (*Lagenorhynchus albirostris*). *Journal of the Acoustical Society of America, 116*(3), 1826–1831.

Ridgway, S. H. (1980). Electrophysiological experiments on hearing in odontocetes. In R. G. Busnel & J. F. Fish (Eds.), *Animal sonar systems* (pp. 483–493). New York: Plenum.

Ridgway, S. H. (1983). Dolphin hearing and sound production in health and illness. In R. R. F. G. Gourevitch (Ed.), *Hearing and other senses: Presentations in honor of E. G. Wever* (pp. 247–296). Groton, CT: Amphora Press.

Ridgway, S. H. (1985). The cetacean central nervous system. In G. Adelman (Ed.), *Encyclopedia of neuroscience* (p. 22). Boston: Birkhauser.

Ridgway, S. H. (2011). Neural time and movement time in choice of whistle or pulse burst responses to different auditory stimuli by dolphins. *Journal of the Acoustical Society of America*, *129*, 1073–1080.

Ridgway, S. H., Bullock, T. H., Carder, D. A., Seeley, R. L., Woods, D., & Galambos, R. (1981). Auditory brainstem response in dolphins. *Proceedings of the National Academy of Sciences of the United States of America*, *78*(3), 1943–1947.

Ridgway, S. H., & Carder, D. A. (1988). Nasal pressure and sound production in an echolocating white whale, *Delphinapterus leucas*. In P. E. Nachtigall & P. W. B. Moore (Eds.), *Animal sonar systems: Processes and performance* (pp. 53–60). New York: Plenum.

Ridgway, S. H., Carder, D. A., Green, R. F., Gaunt, A. S., Gaunt, S. L. L., & Evans, W. E. (1980). Electromyographic and pressure events in the nasolaryngeal system of dolphins during sound production. In R. G. Busnel & J. F. Fish (Eds.), *Animal sonar systems* (pp. 239–250). New York: Plenum.

Ridgway, S. H., Carder, D. A., Kamolnick, T., Smith, R. R., Schlundt, C. E., & Elsberry, W. R. (2001). Hearing and whistling in the deep sea: Depth influences whistle spectra but does not attenuate hearing by white whales (*Delphinapterus leucas*) (*Odontoceti, Cetacea*). *Journal of Experimental Biology*, *204*, 3829–3841.

Ridgway, S. H., McCormick, J. G., & Wever, E. G. (1974). Surgical approach to the dolphin's ear. *Journal of Experimental Biology*, *188*, 266–276.

Ridgway, S. H., Scronce, B. L., & Kanwisher, J. (1969). Respiration and deep diving in the bottlenose porpoise. *Science*, *166*, 1651–1654.

Rossbach, K. A., & Herzing, D. L. (1997). Underwater observations of benthic-feeding bottlenose dolphins near Grand Bahama Island. *Marine Mammal Science*, *13*(3), 7.

Schevill, W. E., & Lawrence, B. (1953). Auditory response of a bottlenosed porpoise, *Tursiops truncatus*, to frequencies above 100KC. *Journal of Experimental Zoology*, *124*(1), 147–165.

Schlundt, C. E., Dear, R. L., Green, L., Houser, D. S., & Finneran, J. J. (2007). Simultaneously measured behavioral and electrophysiological hearing thresholds in a bottlenose dolphin (*Tursiops truncatus*). *Journal of the Acoustical Society of America*, *122*(1), 615–622.

Scott, M. D., Wells, R. S., & Irvine, A. B. (1990). A long-term study of bottlenose dolphins on the west coast of Florida. In S. Leatherwood & R. R. Reeves (Eds.), *The bottlenose dolphin* (pp. 235–244). New York: Academic Press.

Starkhammar, J., Moore, P. W., Talmadge, L., & Houser, D. S. (2011). Frequency-dependent variation in the 2-dimensional beam pattern of an echolocating dolphin. *Biology Letters, 7*, 836–839.

Supin, A. Y., & Nachtigall, P. E. (2013). Gain control in the sonar of odontocetes. *Journal of Comparative Physiology A: Neuroethology, Sensory, Neural, and Behavioral Physiology, 199*(6), 471–478.

Supin, A. Y., Nachtigall, P. E., & Breese, M. (2008). Hearing sensitivity during target presence and
absence while a whale echolocates. *Journal of the Acoustical Society of America, 123*(1), 534–541.

Thomas, J. A., Moss, C. F., & Vater, M. (2003). *Echolocation in bats and dolphins.* Chicago: University of Chicago Press.

Tubelli, A. A., Zosuls, A., Ketten, D. R., & Mountain, D. C. (2014). Elastic modulus of cetacean
auditory ossicles. *Anatomical Record, 297*(5), 892–900.

Verfuß, U. K., Miller, L. A., Pilz, P. K., & Schnitzler, H. U. (2009). Echolocation by two foraging
harbor porpoises (*Phocoena phocoena*). *Journal of Experimental Biology, 212*, 823–834.

Verfuß, U. K., Miller, L. A., & Schnitzler, H. U. (2005). Spatial orientation in echolocating harbour
porpoises (*Phocoena phocoena*). *Journal of Experimental Biology, 208*, 3385–3394.

Villadsgaard, A., Wahlberg, M., & Tougaard, J. (2007). Echolocation signals of wild harbour porpoises, *Phocoena phocoena. Journal of Experimental Biology, 210*, 56–64.

Wahlberg, M., Jensen, F. H., Aguilar Soto, N., Beedholm, K., Bejder, L., Oliveira, C., et al. (2011).
Source parameters of echolocation clicks from wild bottlenose dolphins (*Tursiops aduncus* and
Tursiops truncatus). *Journal of the Acoustical Society of America, 130*(11), 2263–2274.

Wever, E. G., McCormick, J. G., Palin, J., & Ridgway, S. H. (1971a). The cochlea of the dolphin,
Tursiops truncatus: Hair cells and ganglion cells. *Proceedings of the National Academy of Sciences of
the United States of America, 68*(12), 2908–2912.

Wever, E. G., McCormick, J. G., Palin, J., & Ridgway, S. H. (1971b). The cochlea of the dolphin,
Tursiops truncatus: The basilar membrane. *Proceedings of the National Academy of Sciences of the
United States of America, 68*(11), 2708–2711.

Wever, E. G., McCormick, J. G., Palin, J., & Ridgway, S. H. (1972). Cochlear structure in the dolphin, *Lagenorhynchus obliquidens. Proceedings of the National Academy of Sciences of the United States
of America, 69*, 657–661.

Wood, F. G., & Evans, W. E. (1980). Adaptiveness and ecology of echolocation in toothed whales.
In R. G. Busnel & J. F. Fish (Eds.), *Animal sonar systems* (pp. 381–426). New York: Plenum.

Yovel, Y., Falk, B., Moss, C. F., & Ulanovsky, N. (2010). Optimal localization by pointing off axis.
Science, 327(5966), 701–704.

Yuen, M. M. L., Nachtigall, P. E., Breese, M., & Supin, A. Y. (2005). Behavioral and auditory
evoked potential audiograms of a false killer whale (*Pseudorca crassidens*). *Journal of the Acoustical
Society of America, 118*(4), 2688–2695.

3   Ecology and Evolution of Dolphin Sensory Systems

Wolf Hanke and Nicola Erdsack

Introduction

Dolphins have a long history of evolutionary adaptation to the aquatic environment. This includes adaptations of their sensory organs and sensorimotor systems, which in turn may have influenced their cognitive skills, and certainly must be taken into account when designing experiments to investigate cognition. The best-studied dolphin, regarding both sensorimotor systems and cognitive skills, is the bottlenose dolphin, *Tursiops truncatus*, a member of the family Delphinidae. This data situation results from the fact that the bottlenose dolphin is the species most frequently held in captivity. The bottlenose dolphin has for practical reasons become a model organism for dolphins to some degree. It should, however, be kept in mind that dolphins, even within the family Delphinidae, are morphologically highly diverse, and fundamental aspects of the behavior, ecology, and physiology of most species are essentially unknown (LeDuc, 2009). There is a broad consensus that comparative work in more species of dolphins and whales is needed to assess the full range of sensory abilities of this group.

Dolphins in a narrower sense are the members of the family Delphinidae, or delphinids. A broader, also commonly used, definition includes the families Platanistidae, Iniidae, and Pontoporiidae, which are collectively termed "river dolphins." The more than thirty species of delphinids are, with two exceptions, marine. The four species of river dolphins have evolved in freshwater and, with one exception, live in freshwater. River dolphins are a paraphyletic group (de Muizon, 1994; Messenger & McGuire, 1998). Likewise, the dolphins in the broader sense are a paraphyletic group, because the narwhale, the beluga, and the porpoises, which together with the delphinids make up the suprafamily Delphinoidea, are somewhat arbitrarily left out.

Earlier reviews of dolphin sensory systems are found in books by Supin, Popov, and Mass (2001), and books edited by Herman (1980), by Schusterman, Thomas, and Wood

(1986), and by Thomas and Kastelein (1990). Dehnhardt (2002) provides an overview of sensory systems of marine mammals, and the book edited by Thewissen and Nummela (2008) provides an overview of sensory systems of secondarily aquatic vertebrates.

This chapter discusses the sensory systems of dolphins, including river dolphins. It is structured by the sensory systems of vision; hearing and echolocation; chemoreception; magnetoreception; mechanoreception, including sense of touch and hydrodynamic reception; and electroreception. We discuss how the aquatic environment and the transformation of the limbs to fins may have influenced the evolution of sensory systems, and how sensory systems have been shaped by the dolphins' ecology.

Vision

The eyes of dolphins and other whales are, with one exception (the "blind" river dolphin, *Platanista gangetica*), well developed. They possess oculomotor muscles that make them quite mobile (Supin et al., 2001). The basic design of the eye corresponds to that of terrestrial mammals; however, several modifications have evolved in adaptation to the aquatic environment.

The eyeball of dolphins and other cetaceans is not round; its anterior part is flattened, as the anterior chamber is small (Mass & Supin, 2007, 2009; Mass, Supin, Abramov, Mukhametov, & Rozanova, 2013). Notable exceptions are the Amazon river dolphin, *Inia geoffrensis*, whose eyeball is round (fig. 1C in Mass & Supin, 2009), and the blind river dolphin, *Platanista*, whose eyes are largely reduced (Herald et al., 1969).

In addition to the four oblique and two straight oculomotor muscles found in all mammals, the cetacean eye has a strong retractor muscle (musculus retractor bulbi, present in most mammals, but not in humans) that allows axial displacement (in and out movements) of the eyeball (Supin et al., 2001). Bottlenose dolphins can move their eyes independently in all directions (Dawson, 1980). In an experiment where moving stripe patterns were used as a visual stimulus, eye movements were smaller and slower than in humans and very weakly correlated (Dawson, Carder, Ridgway, & Schmeisser, 1981), although this finding does not necessarily hold for all other stimulus conditions. Axial movements can be largely independent, as well; the range of axial movements is at least several centimeters (Dawson et al., 1981).

The visual field of whales and dolphins has been estimated from eye morphology. Yablokov, Belkovich, and Borisow (1972) conclude from the large ratio of the corneal diameter to the physiological axis of the eye that cetaceans have a "large field of vision." The same authors note that the lower eyelid in cetaceans, in contrast to the upper eyelid, is narrow, shifting the visual field somewhat toward the ventral side.

Dawson, Schroeder, and J. F. Dawson (1987) use photographs of the eye fundus of *Tursiops truncatus* and *Grampus griseus* to estimate visual fields. Assuming similarities to the terrestrial mammal fundus, they conclude that the nasal, superior, temporal, and inferior field limits are 48, 40, 80, and 64 degrees of arc in *Tursiops*, and 52, 36, 50, and 80 degrees of arc in *Grampus*. These findings would be in line with the aforementioned studies. Psychophysical studies on the visual field are not yet available.

Another indicator for the extent of the visual field is the distribution of ganglion cells in the retina. Dral (1977), without stating angles, presents a sketch of the eye of the bottlenose dolphin, including the distribution of ganglion cells that strongly indicates that the visual field approximates 130 degrees of arc or more in a close-to-horizontal plane. Supin et al. (2001) present counts of ganglion cells (densities of ganglion cells per spatial unit) on hemispherical projections of the retina centered on the center of the lens. These are strongly indicative of the angles in space that the (motionless) eye can see, as far as the refractive power of the cornea is neglected, a good approximation at least underwater. Visual field sizes obtained from these studies are, again, large: at least 130 degrees in the horizontal and 100 degrees in the vertical plane for the bottlenose dolphin.

A striking peculiarity of all cetaceans studied so far, with the exception of the Amazon river dolphin, *Inia geoffrensis*, and the blind river dolphin, *Platanista gangetica*, is that they have two areas of best vision per eye. The density of ganglion cells across the retina, as in almost all vertebrates, is not uniform. Areas with many ganglion cells per square millimeter indicate a high visual acuity and are identified with an area of best vision. While many mammals, such as primates and carnivores, have one relatively small area of best vision (the fovea), others, such as ungulates, have a broad horizontal stripe of best vision (the visual streak) that covers a broad horizontal angle. In cetaceans, one temporal area and one nasal area of best vision are present. This corresponds to the fact that the cetacean's pupil, which is rather round in dim light, in bright light constricts to two small openings. The two small openings, effectively one temporal and one nasal pupil, are separated by the operculum, a specialized part of the iris. Each one forms an image on one of the areas of best vision. The advantage of a small pupil over a large one, apart from its function in adapting the light level, is that it adds a pinhole effect (camera obscura effect) to the dioptrical apparatus of the eye. This pinhole effect can largely overcome refractive errors caused by a change from water to air or by limited accommodation capabilities.

The two areas of best vision are often, but not exclusively, used differently in air and underwater. In air, bottlenose dolphins tend to fixate objects with both eyes simultaneously, positioning so that the object is in front of them, thus using their temporal area

Figure 3.1
In bottlenose dolphins, the visual fields of both eyes overlap to a small extent. This bottlenose dolphin appears to be looking at the camera with both eyes simultaneously. Courtesy of Zoological Garden Nuremberg (Tiergarten Nürnberg), Germany.

of best vision. Underwater, they often position laterally to an object, using their nasal area of best vision. However, also underwater, dolphins have been observed appearing to fixate objects in front of them using both eyes (fig. 3.1). This vision with both eyes simultaneously, or binocular vision, would in principle enable stereoscopic vision.

Binocular vision occurs in a relatively narrow sector in front of the dolphin (published estimates refer specifically to the bottlenose dolphin; the large variety of head shapes in dolphins [Shirihai & Jarret, 2006] would suggest a considerable interspecific variability of binocular visual fields). Dral (1972) reported measurements with a Heine-May ophthalmoscope that are in line with his impression that bottlenose dolphins fixate objects in front with both eyes simultaneously; however, he did not estimate the binocular field of vision (the overlap of the visual fields of both eyes) but stated merely that emmetropic vision (i.e., normally sighted as opposed to near- or farsighted vision) in air should be possible within approximately 40 degrees from the body axis horizontally, and up to 20 degrees ventrally. Supin et al. (2001) estimate that the visual fields of both eyes can overlap by 20 to 30 degrees in front of the animal by using morphological data plus the information that bottlenose dolphins can move their eyes forward by some 10 to 15 mm (Dral, 1977). This is a quite narrow binocular field of view compared with other predatory animals such as cats (100 degrees; Hughes, 1976), dogs (60–80 degrees; Hughes, 1976), or harbor seals (67 degrees; Hanke, Römer, & Dehnhardt, 2006). The narrowness of the binocular visual field is even more effective because bottlenose dolphins can move their heads only very little.

A narrow binocular field of vision in dolphins is in line with the concept of the praxic space of an animal that has been put forward by Trevarthen (1968). The praxic space is the space in which an animal manipulates objects (Hughes, 1977; Trevarthen, 1968). For example, the larger binocular field of the cat as compared to that of the dog would be due to the fact that the cat catches prey with its well-abductable forelimbs, whereas the dog catches prey mainly with its jaws with only occasional help of the forelimbs. Binocular vision is most relevant within this praxic space. The dolphin catches prey exclusively with its jaws and thus has a narrow praxic space, resulting in a narrow binocular field of view.

This binocular visual field provides the prerequisite for stereoscopic vision, although true stereoscopic vision remains to be shown behaviorally in dolphins. It is remarkable in this context that dolphins are an extreme example of the so-called law of Newton-Müller-Gudden (e.g., Walls, 1963). This law, or better rule, signifies that in mammals—and only in mammals—the degree of decussation of optical fibers in the chiasma opticum is inversely proportional to the degree of frontality of the mammal's eyes. For example, in horses, about one-eighth to one-sixth of the optical nerve fibers remain uncrossed in the chiasma opticum and project to the ipsilateral side of the brain. The portions of uncrossed fibers are about one-fourth in dogs, one-third in cats, and up to one-half in primates (Walls, 1963). In dolphins however, all the optical fibers cross to the contralateral side, and no uncrossed fibers have been found (Tarpley, Gelderd, Bauserman, & Ridgway, 1994). Considering the rule of Newton-Müller-Gudden, researchers have discussed whether uncrossed fibers may be essential for stereoscopic vision (Walls, 1963). However, behavioral experiments in barn owls (*Tyto alba*) indicate that birds, which like dolphins have no uncrossed fibers, possess stereoscopic vision based on retinal disparity comparable to humans (van der Willigen, Frost, & Wagner, 1998). Behavioral experiments on stereoscopic vision in dolphins would be desirable.

Visual acuity, the measure for the ability to visually resolve fine patterns, is generally relatively high in dolphins, and somewhat higher underwater than in air. Herman, Peacock, Yunker, and Madsen (1975) psychophysically measured the visual acuity of bottlenose dolphins both in air and underwater. Striped patterns (grids) were presented to the dolphin at distances from 1 to 2.5 m. In a behavioral task, the dolphin had to discriminate patterns of various stripe widths from patterns that were so fine that they were assumed to give the impression of gray surfaces. Acuity was defined as the visual angle subtended by one stripe of the grid at the behavioral threshold of the dolphin. Acuity ranged from 12' (12 minutes of arc) to 19' in air, and from 8' to 14' in water, depending on viewing distance. This compares rather favorably to other predatory mammals, such as pinnipeds, which have ranges of 5.5' to 12.3' at 2.5 m distance underwater (Weiffen, Möller, Mauck, & Dehnhardt, 2006) and 5.3' to 6.3' at 0.6 m distance in air, decreasing to 10' and more as pupil size increases (Hanke & Dehnhardt, 2009), or cats, near 5' (Blake, Cool, & Crawford, 1974). In dolphins, the double-slit pupil that forms in bright light plays a major role in achieving this good acuity (Herman et al., 1975), specifically in air, where the optics of the eye would otherwise lead to severe myopia (Dral, 1972). The double-slit pupil possibly also explains why acuity was better at larger viewing distances in air, but better at shorter viewing distances underwater (Herman et al., 1975). Another possible mechanism to adapt the eye to aerial vision has been proposed and supported by preliminary measurements by Kröger and Kirschfeld (1989): the cornea appears to flatten locally when a relaxation of the musculus retractor bulbi causes the intraocular pressure to drop.

An estimate for the upper limit for visual acuity (i.e., the acuity that may be reached provided that the dioptrical apparatus adds no errors) can be derived from the spacing of retinal ganglion cells (Supin et al., 2001). Estimates of visual acuity using this approach are in line with the few psychophysical studies that are available. These findings support the view that estimates based on ganglion cell density are good approximations for the visual acuity of several species that have not yet been investigated psychophysically (Mass & Supin, 1989; Mass, Supin, & Popov, 1988; Murayama & Somiya, 1998; Murayama, Somiya, Aoki, & Ishii, 1995; Supin et al., 2001), or psychophysical data is extremely limited (Mass et al., 2013; White, Cameron, Spong, & Bradford, 1971). Visual acuity of marine dolphins by these accounts lies between 8' and 12' underwater and 12' to 15' in air (the difference between air and water resulting from an enlargement factor due to corneal refraction that depends on the medium). For bottlenose dolphins, 9' (water) and 12' (air) of visual acuity were derived from ganglion cell spacing (Mass & Supin, 1995), consistent with the psychophysical results of Herman et al. (1975). For river dolphins, visual acuity is substantially lower. Mass and Supin found a visual acuity of 25' in water for the tucuxi, *Sotalia fluviatilis* (Mass & Supin, 1999), and 40' in water for the Amazon river dolphin, *Inia geoffrensis* (Mass & Supin, 1989). While indicating much poorer vision, these measurements still attest that the visual system is put to use in these animals.

Accommodation—the active alteration of focus by the lens—appears to play at most a minor, if any, role in the bottlenose dolphin. This is indicated by photoretinoscopic measurements in water (Litwiler & Cronin, 2001), as well as by the fact that ciliary muscles are rudimentary (West, Sivak, Murphy, & Kovacs, 1991), excluding the usual mammalian accommodation mechanism by lens deformation. It has been suggested that bottlenose or other dolphins may be able to accommodate indirectly by using the musculus retractor bulbi, by changing the intraocular pressure and, in that way, shifting the lens relative to the retina (Mass & Supin, 2009).

In summary, (marine) dolphins have good visual abilities. This finding is obviously important for all kinds of experiments on learning and cognition that involve the visual system. For example, researchers have recently doubted that the visual acuity of bottlenose dolphins is sufficient to perform a meaningful mirror recognition experiment. Manger (2013), in his highly interesting review of cetacean behavior in comparison with other vertebrate and invertebrate animals, concludes that a triangle of 6 cm diameter painted on the skin of a bottlenose dolphin would be hardly recognizable by the dolphin in front of a mirror. However, the author's estimate is based on the size of the printed images and fails to take into account the angle that the painted triangle subtends in the real world. With the dolphin swimming 0.5 m in front of a mirror (estimated from supplementary video material published with Reiss & Marino, 2001), the triangle would be virtually 1 m away, and would subtend 206' of arc. A visual acuity of 9', as shown in bottlenose dolphins, would be sufficient.

Hearing and Echolocation

A large body of literature, including several reviews (Au, 1993; Au, Popper, & Fay, 2000; Supin & Nachtigall, 2013; Supin et al., 2001), covers the senses of hearing and echolocation in whales and dolphins. Hearing is generally a highly effective sensory modality in the aquatic environment, as sound propagates through the water quickly and with little attenuation (Au & Hastings, 2008), whereas vision is always limited in range and often precluded in dark or turbid waters.

Dolphins have an acute sense of hearing underwater and share some common characteristics. Generally they hear in a frequency range from a few 100 Hz or less (there is a lack of low-frequency hearing data on many species) to far into the ultrasound range, beyond 100 kHz. Toothed whales, including the dolphins, have the broadest hearing range of all mammals, exceeding nine octaves (Au, 2000; Nachtigall, Lemonds, & Roitblat, 2000).

Best hearing occurs mostly in the ultrasound range, roughly between 20 and 70 kHz, with hearing thresholds between 35 and 70 dB *re* 1 µPa. The transition from land to water required extensive modifications of the ear. The inner ear of dolphins is still similar to the inner ear of terrestrial mammals, with adaptations to high-frequency hearing.

The outer and middle ear, however, have been modified to match the acoustic impedance of water (Hemilä, Nummela, & Reuter, 2010; Ketten, 2000; Nummela, Thewissen, Bajpai, Hussain, & Kumar, 2007). For a review of dolphin hearing anatomy and physics, see Hemilä et al. (2010).

Echolocation, or biosonar, is an active sensing process where an animal emits sound and perceives the echoes that return from the outer world. Echolocation has developed convergently and shows generally similar characteristics in bats and toothed whales, including dolphins (for an overview, see Thomas, Moss, & Vater, 2004). Absolute sensitivity of dolphin echolocation is impressive: in a yes/no experimental procedure, a bottlenose dolphin could detect a 7.62 cm large sphere at distances exceeding 100 m (the 50 percent correct detection threshold in target present trials was at 113 m) in open water (Au & Snyder, 1980). The range at which prey fish can be detected has been calculated using measured echoes from prey fish; it ranged from 70 to 173 m for an echolocating bottlenose dolphin, depending on background noise and fish orientation (Au, Benoit-Bird, & Kastelein, 2007).

Dolphins can also discriminate between objects using echolocation based on object shape (Herman, Pack, & Hoffmann-Kuhnt, 1998; Nachtigall, Murchinson, & Au, 1980), surface texture (DeLong, Au, Lernonds, Harley, & Roitblat, 2006), and material (DeLong et al., 2006; Nachtigall, 1980). While the term *echolocation* was initially formed because animals use it to locate objects (Griffin, 1944), it has been expanded to include the discrimination of objects (Griffin, Friend, & Webster, 1965). Most experiments have involved artificial objects; an extension toward more natural objects, specifically prey fish, has been proposed (Au, Branstetter, Benoit-Bird, & Kastelein, 2009).

Chemical Senses

Olfaction

In cetaceans the sense of smell has been totally lost or largely reduced. However, examinations of fossils suggest that olfaction still persisted in odontocetes from the Late Oligocene (Hoch, 2000) and Middle Miocene (Godfrey, 2013). Anosmia, the lack of a sense of smell, in toothed whales is supported by several morphological, neuroanatomical, and biomolecular investigations. In baleen whales, the peripheral olfactory apparatuses are reduced but still existent during the entire lifetime (Godfrey, Geisler, & Fitzgerald, 2013; Thewissen, George, Rosa, & Kishida, 2011). By contrast, in recent toothed whales, the peripheral olfactory structures (bulbs, tracts, and nerves) have been lost or extremely reduced during evolutionary remodeling and shifts of the nasal apparatus (Breathnach, 1960; Kellogg, 1928; Kruger, 1959; Morgane & Jacobs, 1972). Although the anlagen (the rudimentary embryonic bases) of a peripheral olfactory apparatus (olfactory bulbs and tracts) are still apparent during early fetal development, as found in striped dolphins (*Stenella coeruleoalba*) and harbor porpoises (*Phocoena phocoena*), they degenerate until birth (Kamiya & Pirlot, 1974; Oelschläger & Buhl, 1985a, 1985b; Sinclair, 1966). Nevertheless neuroanatomical studies in several cetacean species showed that olfactory brain structures are largely present (Breathnach, 1960; Jacobs, Morgane, & McFarland, 1971), a fact that has led to speculations about potential nonolfactory functions of these brain structures in cetaceans (Buhl & Oelschläger, 1986; Jacobs et al., 1971; Turner, 1890).

Biomolecular examinations in whales have revealed a significantly reduced number of olfactory receptor (OR) genes and other specific genes for chemoreception, or a significantly higher rate of OR pseudogenes, respectively, compared to semiaquatic and terrestrial animals (Go, Satta, Takenaka, & Takahata, 2004; Kishida, Kubota, Shirayama, & Fukami, 2007; Yu et al., 2010). Furthermore, OR genes were found to be reduced in odontocetes compared to mysticetes (McGowen, Clark, & Gatesy, 2008). By contrast, the number of OR genes in sea lions was comparable to that of terrestrial mammals (Kishida et al., 2007), which is consistent with the high olfactory sensitivity in pinnipeds shown in behavioral studies (Kim, Amundin, & Laska, 2013; Kowalewsky, Dambach, Mauck, & Dehnhardt, 2006; Laska, Lord, Selin, & Amundin, 2010; Laska, Svelander, & Amundin, 2008). This underlines the fact that anosmia is not a general characteristic of aquatic mammals.

While biomolecular and morphological findings speak strongly against an olfactory sense in toothed whales, behavioral evidence for its absence is still lacking, just as for the presumed presence of olfaction in baleen whales. The only behavioral results indicating chemoreception related to olfaction were obtained by Kuznetsov (1990, 1992), testing taste reception in dolphins. He found a higher taste sensitivity in dolphins for odorous substances than in other mammals (Kuznetsov, 1990, 1992). He therefore described the chemical sense in dolphins as "quasi-olfaction," meaning chemoreception other than olfaction and gustation.

Two hypotheses have been considered for the absence of an olfactory sense in odontocetes compared to the presence of a sense of smell in mysticetes. The "echolocation-priority" hypothesis (Hoch, 2000) proposes that the importance of olfaction had decreased with the development of the sonar system, which led to degeneration and to total loss of the sense of smell in cetaceans that use echolocation. The "filter-feeder" hypothesis (Cave, 1988; Godfrey et al., 2013; Thewissen et al., 2011) suggests that in mysticetes, which do not rely on echolocation, the olfactory sense has been maintained due to the importance of locating planktonic prey by its characteristic odor.

Kishida and Thewissen (2012) found that evolutionary changes of the selective pressure on olfactory marker protein genes of six cetacean species fit better to the filter-feeder hypothesis than to the echolocation-priority hypothesis. Nevertheless both hypotheses are plausible and do not exclude each other.

Taste Reception

Taste reception is an important factor for distinguishing between tolerable and indigestible or toxic food, particularly in totally aquatic animals, like dolphins, foraging underwater. Anatomical structures for taste reception—that is, innervated taste buds— were found in the roots of the tongues of common dolphins (*Delphinus delphis*) and bottlenose dolphins (Donaldson, 1977; Suchowskaja, 1972). The tongues of striped dolphins (*Stenella coeruleoalba*), harbor porpoises, and Amazon river dolphins (*Inia geoffrensis*) also possess taste buds, but only in fetuses and juveniles, since the taste buds degenerate in adults (Komatsu & Yamasaki, 1980; Kuznetsov, 1990, 1992; Yamasaki, Komatsu, & Kamiya, 1978a, 1978b; Yamasaki, Satomi, & Kamiya, 1976). Furthermore, Kruger (1959) suggested a well-developed gustatory sense owing to the neural structure and development of the thalamus in bottlenose dolphins.

Behavioral experiments in dolphins also provide evidence for the ability to discriminate certain tastes (Kuznetsov, 1990, 1992; Nachtigall & Hall, 1984). Nachtigall and Hall (1984) found the sensitivity for citric acid (sour) and quinine sulfate (bitter) in bottlenose dolphins (detection thresholds between 0.026 and 0.016 M for citric acid and between $2.86*10^{-5}$ M and $0.81*10^{-5}$ M for quinine sulfate) to be nearly as good as in humans. Similar results were obtained by Kuznetsov (1990, 1992), who investigated chemoreception using conditioning experiments in bottlenose dolphins, harbor porpoises (*Phocoena phocoena*), and common dolphins (*Delphinus delphis*). He found that all three species were able to distinguish several chemical-freshwater solutions poured into the animal's oral cavity. But in contrast to Nachtigall (1986) and Friedl et al. (1990), who found, besides the ability to distinguish different salinities, also a sensitivity for sucrose in bottlenose dolphins, Kuznetsov (1990, 1992) reported an insensitivity to sucrose and glucose in bottlenose dolphins. Further indication for taste reception was given by behavioral observations of captive dolphins that rejected certain fish species even when cut in pieces, or refused fish with inserted tablets (Donaldson, 1977; Herman & Tavolga, 1980).

A further function of taste reception in aquatic animals might be communication. Herman and Tavolga (1980) speculated about the role of taste reception in cetacean communication. Excretions of perianal and preputial glands in male dolphins may serve as chemical signals detected by conspecifics. Indication for this is given by Ceruti,

Fennessey, and Tjoa (1985), who analyzed dolphin blood and glandular secretions and found high abundances of chemoreceptive compounds, detectable by humans. Several anecdotes report cetacean behavior that implicates a sensitivity to detecting blood or glandular secretions: for example, beluga whales (*Delphinapterus leucas*) panicking when entering areas where conspecifics had been killed or where others had panicked as well (Yablokov, 1961; cited by Kuznetsov, 1990, and Nachtigall, 1986). Further indication was given by Herzing (2000), who reported "open-mouth behavior" in dolphins. Since this behavior was observed not only in interacting dolphins, most likely serving as a visual signal, but in solitary individuals as well, it might serve the perception of chemical signals (Herzing, 2000).

All these findings suggest that whatever evolutionary pressures might have been responsible for the reduction of the sense of smell did not also act to reduce the dolphins' sense of taste.

Magnetoreception

Cetaceans migrate over long distances, including traveling across the open ocean, where no landmarks are available for navigation. This raises the question of how they orientate in open waters. Since other migrating vertebrates and invertebrates, such as pigeons and honeybees, use, among other cues, the earth's geomagnetic field to find their way, researchers have proposed that cetaceans as well may use magnetism for orientation. Magnetic material carrying natural remanent magnetization (NRM) was found in the heads of common Pacific dolphins (*Delphinus delphis*). NRM was particularly strong in some parts of the falx cerebri or the tentorium cerebelli, which are folds of the dura mater between the brain hemispheres or between the cerebrum and cerebellum, respectively (Zoeger, Dunn, & Fuller, 1981). Areas of highest NRM were not consistent across animals. This material was at least in part magnetite (Fe_3O_4). In one dolphin, a magnetite particle visible to the naked eye was found that was associated with fibers that "appear to be nerve fibers" (Zoeger et al., 1981). The particle was magnetically soft and was substantially demagnetized in fields of a few hundred A/m. Bauer, Fuller, Perry, Dunn, and Zoeger (1985) investigated samples from the falx cerebri and tentorium cerebelli in bottlenose dolphins, Dall's porpoise (*Phocoenoides dalli*), and Cuvier's beaked whale, as well as one mysticete, a humpback whale (*Megaptera novae-angliae*). They improved the methods used by Zoeger et al. (1981) to further reduce any risk of contamination or unintended magnetization of the samples during preparation. NRM was considerably lower in the Dall's porpoise and in the humpback whale than in the common dolphins investigated by Zoeger et al. (1981). In the bottlenose dolphin and Cuvier's beaked whale, saturation isothermal remanence (sIRM; a measure of how well artificially inflicted magnetization is conserved) was measured and was found to be considerable, demonstrating the presence of magnetic material.

The nature of this magnetic material is incompletely determined, but observations are consistent with the theory that it is biogenic magnetite. Magnetite has been proposed to play a role in magnetoreception in animals in general. It has been found in magnetoreceptive species such as honeybees (Gould, Kirschvink, & Deffeyes, 1978; Walcott, Gould, & Kirschvink, 1979), rainbow trout (Walker, 1997), and several more fish species, as well as homing pigeons (Diebel, Proksch, Green, Neilson, & Walker, 2000; Gould, 1980; Kirschvink, 1989; Kirschvink & Gould, 1981). However, the presence of magnetic material in an animal does not necessarily imply that it uses magnetoreception (Kirschvink, 1989); nor do other iron-rich structures, as iron is also accumulated, for example, in the course of iron homeostasis (compare the discussion in Treiber et al., 2012). If the magnetic material could be associated clearly with specialized nerve endings, it would strengthen the hypothesis that the magnetic material is involved in a sensory system, a question that has not yet been answered in cetaceans.

Indications of orientation by a magnetic sense in cetaceans arise from the recurrence of live strandings in various cetacean species. In particular, researchers have found that most cetacean live strandings occur in areas with low geomagnetic intensity (Kirschvink, 1990; Kirschvink, Dizon, & Westphal, 1986; Klinowska, 1985, 1986). Walker, Kirschvink, Ahmed, and Dizon (1992) found correlations between sighting positions of fin whales in the open ocean and geomagnetic lows as well. However, it is generally difficult to separate magnetic field effects on whale sensory systems from other parameters, such as bottom topography, including the possibility that whales might have been tracking magnetosensitive fish.

No such indication for magnetic sensing was found by Hui (1994), who analyzed the distribution of free-ranging common dolphins (*Delphinus delphis*) in relation to magnetic intensity patterns. Distribution and swimming directions of common dolphins in the Southern California continental borderland were not significantly associated with magnetic field gradients but with bottom topography. Nevertheless, this negative result does not disprove a magnetic sense in dolphins.

A physiological laboratory experiment was conducted by Kuznetsov (1999), who measured galvanic skin responses, electrocardiograms, and respiration in a captive bottlenose dolphin in a seawater tank exposed to changing magnetic fields. The dolphin responded with changes of these vegetative parameters (Kuznetsov, 1999); apparently these changes occurred in all the measured parameters and were associated with movements of the dolphin, which would indicate quite strong responses. However, it is remarkable that the dolphin responded with latencies of several seconds up to about a minute and response latencies were grouped in peaks. As this is the only available psychophysical study, a replication of the experiment would be highly desirable. One way to improve control experiments is to use double-wrapped electromagnetic coils in which current flows in either parallel or antiparallel directions, generating a magnetic field or not, with all other parameters unchanged (Kirschvink, Winklhofer, & Walker, 2010).

The only behavioral experiments so far on magnetoreception in cetaceans were conducted by Bauer et al. (1985) in bottlenose dolphins (*Tursiops truncatus*). In these experiments, no indication of magnetoreception was found. Again, this negative result does not disprove a magnetic sense in dolphins. In summary, clear and reproducible experimental evidence for magnetic sensitivity in cetaceans is still lacking.

Mechanoreception

Sense of Touch

Although odontocetes rely mainly on their sonar and visual systems for orientation and foraging, they have a keen sense of touch as well. Dolphin skin is highly sensitive to mechanical stimuli, in particular in the areas around the eyes, snout, melon, and blowhole (Kolchin & Belkovich, 1973; Ridgway & Carder, 1990), due to the skin's rich endowment with mechanoreceptors. In bottlenose dolphins (Ridgway & Carder, 1990) and common dolphins (Kolchin & Belkovich, 1973), touch sensitivity in these areas is comparable to the sensitivity of human lips and fingertips. This plays an important role in social contact and communication with conspecifics (Herman & Tavolga, 1980; Ling, 1974; Paulos, Dudzinski, & Kuczaj, 2008). The skin of bottlenose dolphins contains huge numbers of nerve endings and Pacinian corpuscle-like formations, especially at the snout and the genital regions (Palmer & Weddell, 1964). Belkovich and Dubrovskiy (1976) report Krause bulbs, Pacini corpuscles, and Schwann cells in these sensitive areas but do not describe the methods or name the sources of these findings. The snout as well as the genital regions are prime areas of body contact between individuals (Herman & Tavolga, 1980).

Physical contact is a common expression of affection between dolphins, but it can also express aggression (Pryor, 1990). In bottlenose dolphins and Atlantic spotted dolphins (*Stenella frontalis*), the genital region might serve as a tactile receptor for low-frequency echolocation trains, so-called "buzzes," in social contexts such as mating and mother-calf interactions (Herzing, 2000, 2004). Sound pressure levels of these buzzes are high, such as 200 to 210 dB *re* 1 µPa in Atlantic spotted dolphins (Au & Herzing,

2003), and above the threshold of tactile sensitivity, measured in common dolphins (*Delphinus delphis*) (Kolchin & Belkovich, 1973, cited by Herzing, 2004).

Most mammals possess sinus hairs, also known as vibrissae, as important tactile organs. Vibrissae on the head and face are commonly called "whiskers"; in the literature on cetaceans, the term "bristle" is sometimes used. The semiaquatic pinnipeds possess the most-developed vibrissae known so far (largest blood supply, innervation, and endowment with mechanoreceptors; Dehnhardt, 2002), which they use for active touch, but also as hydrodynamic sense. All members of the order of the totally aquatic Sirenia have highly sensitive vibrissae also on the entire body surface (Ling, 1974, 1977; Reep et al., 2011; Reep, Marshall, & Stoll, 2002). All cetaceans possess sensory hairs as well, at least during some period in their development. As in the pinnipeds, these occur exclusively at the head but are less well developed. Whereas baleen whales have short whiskers at the upper and lower jaws (Ling, 1977), in toothed whales, the vibrissae are mostly limited to the upper jaws and are lost during fetal development or shortly after birth (Ling, 1977).

At the location of the vibrissae, empty pits remain. In bottlenose dolphins, the remaining empty pits were found to be richly innervated and were hypothesized to serve as sensory organs for perceiving changes in water pressure (Palmer & Weddell, 1964). However, Ridgway and Carder (1990) did not find greater responses to vibratory stimuli in the area of the vibrissal pits than in adjacent skin areas of the head. The only odontocete maintaining bristles as adult known so far is the Amazon river dolphin (*Inia geoffrensis*), which presumably uses them for detecting prey while digging in the mud (Layne & Caldwell, 1964). In the Guiana dolphin (*Sotalia guianensis*), the vibrissal pits transformed to electrosensitive organs (Czech-Damal et al., 2012).

At first glance, a tactile sense does not seem to be necessary for foraging in dolphins, owing to their well-developed visual and sonar systems. However, benthic feeding presents a different situation. In fact, in some populations of bottlenose dolphins, researchers have observed special benthic foraging techniques with a possible involvement of the tactile sense such as "crater-feeding" (Rossbach & Herzing, 1997) and "sponging" (Mann et al., 2008; Smolker, Richards, Connor, Mann, & Berggren, 1997).

Hydrodynamic Perception
Hydrodynamic perception is the reception and analysis of water movements. It has evolved several times independently in aquatic and semiaquatic species (Bleckmann, 1994). Water movements are caused by many sources such as abiotic sources, predators, conspecifics, and prey (Hanke, 2014; Hanke & Bleckmann, 2004; Hanke, Brücker, & Bleckmann, 2000; Niesterok & Hanke, 2013). Phocid seals (Phocidae) and eared seals

(Otatriidae) have accomplished skills in detecting and using water movements (Hanke, Wieskotten, Marshall, & Dehnhardt, 2013). They do so using their vibrissae (Dehnhardt, Mauck, Hanke, & Bleckmann, 2001; Gläser, Wieskotten, Otter, Dehnhardt, & Hanke, 2011; Wieskotten, Mauck, Miersch, Dehnhardt, & Hanke, 2011).

The vibrissae of manatees are highly sensitive to water movements as well (Gaspard et al., 2013). Since mysticetes also have vibrissae (Ling, 1977), the detection of water movements by mysticetes appears likely, although it has not yet been demonstrated. Odontocetes develop vibrissae in utero but shed them before or shortly after birth, with the exception of the Amazon river dolphin, *Inia geoffrensis* (Layne & Caldwell, 1964). Detection of water movements by these vibrissae is likely but has not been investigated. Even as adults, most dolphins maintain reduced forms of vibrissal follicle-sinus complexes (F-SCs), the sensory structures at the bases of vibrissae. These reduced forms of F-SCs are called hair pits or vibrissal crypts (Czech-Damal et al., 2012). Vibrissal crypts, hypothetically, may be used for different purposes in different odontocete species. We do not know if some of them are organs of direct touch reception. Some are electrosensory organs (see the following section). There is no indication so far that vibrissal crypts are used to receive water movements. Any reception of water movements by dolphins would most likely occur via the skin (see mechanoreception section); this sensory modality requires further investigation.

Electrical Senses

Czech-Damal et al. (2012) found that the vibrissal crypts of the Guiana dolphin (*Sotalia guianensis*) have evolved into electroreceptors. The vibrissal crypts are remnants of the vibrissal follicle-sinus complexes (F-SCs), the specialized sensory organs at the bases of the vibrissae of mammals, that have been largely reduced in toothed whales. The Guiana dolphin's vibrissal crypts are located on the snout, one row of approximately four to seven crypts on each side. The vibrissal crypts were found to be ampulla-like structures that open to the environment. They contain a gel-like substance and a diffuse network of corneocytes and keratinous fibers. The vibrissa is completely absent, and the vibrissal crypts are highly innervated. Fine nerve endings reach into the epidermis (in contrast with the situation in the vibrissal follicle-sinus complex of all other mammals investigated) and come close to the lumen of the ampulla. The histological findings are consistent with the hypothesis that an electric field gradient between the lumen of the ampulla and the tissue of the rostrum is measured via minute currents at the location of the nerve endings, while a higher resistance between the lumen and the tissue is maintained at all other locations. The same basic principle appears to be responsible for the detection of electric fields in the ampullary organs of fish (Andres & von Düring, 1988; Manger & Pettigrew, 1996), and the gland receptors of the duck-billed platypus (*Ornithorhynchus anatinus*) (Andres & von Düring, 1984, 1988).

Psychophysical assessment of electrosensory reception in a Guiana dolphin resulted in a detection threshold of 4.6 μV/cm electric field strength (Czech-Damal et al., 2012). In a go/no-go response paradigm, a Guiana dolphin was trained to respond to the electric field generated by a pair of electrodes in the water. The resulting behavioral threshold lies clearly in the range of the electric field strengths generated by fishes (Czech-Damal, Dehnhardt, Manger, & Hanke, 2013; Peters, Eeuwes, & Bretschneider, 2007).

Conclusions

Cetaceans, and with them the paraphyletic group known as dolphins, have adapted to the aquatic environment in the most accomplished way of all mammals. Vision, hearing, and echolocation in dolphins have been studied more extensively than chemical senses, magnetic senses, sense of touch, and electrical senses. Nonetheless all these areas still present intriguing research questions. For example, in vision, what is the role of global movement (optic flow)? Does it aid in path integration during navigation? Does binocular vision in dolphins enable true stereoscopic vision? In hearing and echolocation, what are the physiological correlates of auditory gain control, and what are the limits of object recognition and object discrimination? Chemical senses need to be assessed in psychophysical studies. Magnetic senses, while several indications for their existence have been collected, have not yet been unequivocally and reproducibly demonstrated in dolphins. If they are found, what is the underlying mechanism? Have electrical senses developed in species other than the Guiana dolphin?

Dolphins are a diverse group of specialized mammals that not only are interesting in themselves but can also serve as models for evolutionary processes with regard to morphology, physiology, and psychology. Understanding the sensory abilities of dolphins constitutes an essential basis for understanding their cognitive abilities.

References

Andres, K. H., & von Düring, M. (1984). The platypus bill: A structural and functional model of a pattern-like arrangement of different cutaneous sensory receptors. In W. Hamann & A. Iggo (Eds.), *Sensory receptor mechanisms* (pp. 81–89). Singapore: World Scientific Publishing.

Andres, K. H., & von Düring, M. (1988). Comparative anatomy of vertebrate electroreceptors. *Progress in Brain Research, 74*, 113–131.

Au, W. W. L. (1993). *The sonar of dolphins*. New York: Springer.

Au, W. W. L. (2000). Hearing in whales and dolphins: An overview. In W. W. L. Au, A. N. Popper, & R. R. Fay (Eds.), *Hearing by whales and dolphins*. New York: Springer.

Au, W. W. L., Benoit-Bird, K. J., & Kastelein, R. A. (2007). Modeling the detection range of fish by echolocating bottlenose dolphins and harbor porpoises. *Journal of the Acoustical Society of America, 121*(6), 3954–3962.

Au, W. W. L., Branstetter, B. K., Benoit-Bird, K. J., & Kastelein, R. A. (2009). Acoustic basis for fish prey discrimination by echolocating dolphins and porpoises. *Journal of the Acoustical Society of America, 126*(1), 460–467.

Au, W. W. L., & Hastings, M. C. (2008). *Principles of marine bioacoustics*. New York: Springer.

Au, W. W. L., & Herzing, D. L. (2003). Echolocation signals of wild Atlantic spotted dolphin (*Stenella frontalis*). *Journal of the Acoustical Society of America, 113*(1), 598–604.

Au, W. W. L., Popper, A. N., & Fay, R. R. (Eds.). (2000). *Hearing by whales and dolphins*. New York: Springer.

Au, W. W. L., & Snyder, K. J. (1980). Long-range target detection in open waters by an echolocating Atlantic bottlenose dolphin (*Tursiops truncatus*). *Journal of the Acoustical Society of America, 68*(4), 1077–1084.

Bauer, G. B., Fuller, M., Perry, A., Dunn, J. R., & Zoeger, J. (1985). Magnetoreception and biomineralization of magnetite in cetaceans. In J. L. Kirschvink, D. L. Jones, & B. J. MacFadden (Eds.), *Magnetite biomineralization and magnetoreception in organisms: A new biomagnetism* (pp. 489–507). New York: Plenum.

Belkovich, V. M., & Dubrovskiy, N. A. (1976). *Sensory bases of cetacean orientation*. Leningrad: Nauka.

Blake, R., Cool, S. J., & Crawford, M. L. (1974). Visual resolution in the cat. *Vision Research, 14*(11), 1211–1217.

Bleckmann, H. (1994). *Reception of hydrodynamic stimuli in aquatic and semiaquatic animals*. Stuttgart: Gustav Fischer.

Breathnach, A. S. (1960). The cetacean central nervous system. *Biological Reviews of the Cambridge Philosophical Society, 35*(2), 187–230.

Buhl, E. H., & Oelschläger, H. A. (1986). Ontogenetic development of the nervus terminalis in toothed whales: Evidence for its non-olfactory nature. *Anatomy and Embryology, 173*(3), 285–294.

Cave, A. J. E. (1988). Note on olfactory activity in mysticetes. *Journal of Zoology, 214*, 307–311.

Ceruti, M. G., Fennessey, P. V., & Tjoa, S. S. (1985). Chemoreceptively active compounds in secretions, excretions, and tissue extracts of marine mammals. *Comparative Biochemistry and Physiology A: Comparative Physiology, 82*(3), 505–514.

Czech-Damal, N. U., Dehnhardt, G., Manger, P., & Hanke, W. (2013). Passive electroreception in aquatic mammals. *Journal of Comparative Physiology A: Neuroethology, Sensory, Neural, and Behavioral Physiology, 199*(6), 555–563.

Czech-Damal, N. U., Liebschner, A., Miersch, L., Klauer, G., Hanke, F. D., Marshall, C. D., … Hanke, W. (2012). Electroreception in the Guiana dolphin (*Sotalia guianensis*). *Proceedings of the Royal Society B, 279*(1729), 663–668.

Dawson, W. W. (1980). The cetacean eye. In L. M. Herman (Ed.), *Cetacean behavior: Mechanisms and functions* (pp. 53–100). New York: Wiley.

Dawson, W. W., Carder, D. A., Ridgway, S. H., & Schmeisser, E. T. (1981). Synchrony of dolphin eye-movements and their power-density spectra. *Comparative Biochemistry and Physiology A: Comparative Physiology, 68*(3), 443–449.

Dawson, W. W., Schroeder, J. P., & Dawson, J. F. (1987). The ocular fundus of 2 cetaceans. *Marine Mammal Science, 3*(1), 1–13.

de Muizon, C. (1994). Are the squalodonts related to the platanistoids? *Proceedings of the San Diego Society of Natural History, 29*, 135–146.

Dehnhardt, G. (2002). Sensory systems. In A. R. Hoelzel (Ed.), *Marine mammal biology* (pp. 116–141). Oxford: Blackwell.

Dehnhardt, G., Mauck, B., Hanke, W., & Bleckmann, H. (2001). Hydrodynamic trail following in harbor seals (*Phoca vitulina*). *Science, 293*, 102–104.

DeLong, C. M., Au, W. W. L., Lernonds, D. W., Harley, H. E., & Roitblat, H. L. (2006). Acoustic features of objects matched by an echolocating bottlenose dolphin. *Journal of the Acoustical Society of America, 119*(3), 1867–1879.

Diebel, C. E., Proksch, R., Green, C. R., Neilson, P., & Walker, M. M. (2000). Magnetite defines a vertebrate magnetoreceptor. *Nature, 406*(6793), 299–302.

Donaldson, B. J. (1977). The tongue of the bottlenosed dolphin (*Tursiops truncatus*). In R. J. Harrison (Ed.), *Functional anatomy of marine mammals* (Vol. 3, pp. 175–197). London: Academic Press.

Dral, A. D. G. (1972). Aquatic and aerial vision in the bottle-nosed dolphin. *Netherlands Journal of Sea Research, 5*, 510–513.

Dral, A. D. G. (1977). On the retinal anatomy of cetacea (mainly *Tursiops truncatus*). In R. J. Harrison (Ed.), *Functional anatomy of marine mammals* (Vol. 3, pp. 81–134). London: Academic Press.

Friedl, W. A., Nachtigall, P. E., Moore, P. W. B., Chun, N. K. W., Haun, J. E., Hall, R. W., et al. (1990). Taste reception in the Pacific bottlenose dolphin (*Tursiops truncatus gilli*) and the California sea lion (*Zalophus californianus*). In J. A. Thomas & R. A. Kastelein (Eds.), *Sensory abilities of cetaceans: Laboratory and field evidence* (Vol. 196, pp. 447–454). New York: Plenum.

Gaspard, J. C., III, Bauer, G. B., Reep, R. L., Dziuk, K., Read, L., & Mann, D. A. (2013). Detection of hydrodynamic stimuli by the Florida manatee (*Trichechus manatus latirostris*). *Journal of Comparative Physiology A: Neuroethology, Sensory, Neural, and Behavioral Physiology, 199*(6), 441–450.

Gläser, N., Wieskotten, S., Otter, C., Dehnhardt, G., & Hanke, W. (2011). Hydrodynamic trail following in a California sea lion (*Zalophus californianus*). *Journal of Comparative Physiology A: Neuroethology, Sensory, Neural, and Behavioral Physiology, 197*(2), 141–151.

Go, Y., Satta, Y., Takenaka, O., & Takahata, N. (2004). Pseudogenization of olfactory receptor genes and deterioration of olfaction in cetacea. *Genes and Genetic Systems, 79*(6), 418.

Godfrey, S. J. (2013). On the olfactory apparatus in the Miocene odontocete *Squalodon* sp. (Squalodontidae). *Comptes Rendus Palevol, 12*, 519–530.

Godfrey, S. J., Geisler, J., & Fitzgerald, E. M. G. (2013). On the olfactory anatomy in an archaic whale (Protocetidae, Cetacea) and the minke whale *Balaenoptera acutorostrata* (Balaenopteridae, Cetacea). *Anatomical Record, 296*(2), 257–272.

Gould, J. L. (1980). The case for magnetic sensitivity in birds and bees (such as it is). *American Scientist, 68*(3), 256–267.

Gould, J. L., Kirschvink, J. L., & Deffeyes, K. S. (1978). Bees have magnetic remanence. *Science, 201*(4360), 1026–1028.

Griffin, D. R. (1944). Echolocation by blind man, bats and radar. *Science, 100*(2609), 589–590.

Griffin, D. R., Friend, J. H., & Webster, F. A. (1965). Target discrimination by echolocation of bats. *Journal of Experimental Zoology, 158*(2), 155–168.

Hanke, F. D., & Dehnhardt, G. (2009). Aerial visual acuity in harbor seals (*Phoca vitulina*) as a function of luminance. *Journal of Comparative Physiology A: Neuroethology, Sensory, Neural, and Behavioral Physiology, 195*(7), 643–650.

Hanke, W. (2014). Natural hydrodynamic stimuli. In H. Bleckmann, J. Mogdans, & S. Coombs (Eds.), *Flow sensing in air and water: Behavioural, neural, and engineering principles of operation* (pp. 3–29). Berlin: Springer.

Hanke, W., & Bleckmann, H. (2004). The hydrodynamic trails of *Lepomis gibbosus* (Centrarchidae), *Colomesus psittacus* (Tetraodontidae) and *Thysochromis ansorgii* (Cichlidae) measured with scanning particle image velocimetry. *Journal of Experimental Biology, 207*, 1585–1596.

Hanke, W., Brücker, C., & Bleckmann, H. (2000). The ageing of the low-frequency water disturbances caused by swimming goldfish and its possible relevance to prey detection. *Journal of Experimental Biology, 203*, 1193–1200.

Hanke, W., Römer, R., & Dehnhardt, G. (2006). Visual fields and eye movements in a harbor seal (*Phoca vitulina*). *Vision Research, 46*(17), 2804–2814.

Hanke, W., Wieskotten, S., Marshall, C. D., & Dehnhardt, G. (2013). Hydrodynamic perception in true seals (Phocidae) and eared seals (Otariidae). *Journal of Comparative Physiology A: Neuroethology, Sensory, Neural, and Behavioral Physiology, 199*, 421–440.

Hemilä, S., Nummela, S., & Reuter, T. (2010). Anatomy and physics of the exceptional sensitivity of dolphin hearing (Odontoceti: Cetacea). *Journal of Comparative Physiology A: Neuroethology, Sensory, Neural, and Behavioral Physiology, 196*(3), 165–179.

Herald, E. S., Brownell, R. L., Frye, F. L., Morris, E. J., Evans, W. E., & Scott, A. B. (1969). Blind river dolphin: First side-swimming cetacean. *Science, 166*(3911), 1408–1410.

Herman, L. M. (Ed.). (1980). *Cetacean behavior: Mechanisms and functions.* New York: Wiley.

Herman, L. M., Pack, A. A., & Hoffmann-Kuhnt, M. (1998). Seeing through sound: Dolphins (*Tursiops truncatus*) perceive the spatial structure of objects through echolocation. *Journal of Comparative Psychology, 112*(3), 292–305.

Herman, L. M., Peacock, M. F., Yunker, M. P., & Madsen, C. J. (1975). Bottlenosed dolphin: Double-slit pupil yields equivalent aerial and underwater diurnal acuity. *Science, 189*(4203), 650–652.

Herman, L. M., & Tavolga, W. N. (1980). The communication systems of cetaceans. In L. M. Herman (Ed.), *Cetacean behavior: Mechanisms and functions* (pp. 149–209). New York: Wiley.

Herzing, D. L. (2000). Acoustics and social behavior of wild dolphins: Implications for a sound society. In W. W. L. Au, A. N. Popper, & R. R. Fay (Eds.), *Hearing by whales and dolphins* (pp. 225–272). New York: Springer.

Herzing, D. L. (2004). Social and nonsocial uses of echolocation in free-ranging *Stenella frontalis* and *Tursiops truncatus*. In J. A. Thomas, C. F. Moss, & M. Vater (Eds.), *Echolocation in bats and dolphins* (pp. 404–410). Chicago: University of Chicago Press.

Hoch, E. (2000). Olfaction in whales: Evidence from a young odontocete of the Late Oligocene North Sea. *Historical Biology, 14*(1–2), 67–89.

Hughes, A. (1976). A supplement to the cat schematic eye. *Vision Research, 16*(2), 149–154.

Hughes, A. (1977). The topography of vision in mammals of contrasting life style. In F. Crescitelli (Ed.), *The visual system in vertebrates* (Vol. VII/5, pp. 613–756). Berlin: Springer.

Hui, C. A. (1994). Lack of association between magnetic patterns and the distribution of free-ranging dolphins. *Journal of Mammalogy, 75*(2), 399–405.

Jacobs, M. S., Morgane, P. J., & McFarland, W. L. (1971). The anatomy of the brain of the bottlenose dolphin (*Tursiops truncatus*). Rhinic lobe (Rhinencephalon). I. The paleocortex. *Journal of Comparative Neurology, 141*(2), 205–272.

Kamiya, T., & Pirlot, P. (1974). Brain morphogenesis in *Stenella coeruleoalba*. *Scientific Reports of the Whale Research Institute Tokyo, 26*, 245–253.

Kellogg, R. (1928). The history of whales: Their adaptation to life in the water (concluded). *Quarterly Review of Biology, 3*(2), 174–208.

Ketten, D. R. (2000). Cetacean ears. In W. W. L. Au, A. N. Popper, & R. R. Fay (Eds.), *Hearing by whales and dolphins* (pp. 43–108). New York: Springer.

Kim, S., Amundin, M., & Laska, M. (2013). Olfactory discrimination ability of South African fur seals (*Arctocephalus pusillus*) for enantiomers. *Journal of Comparative Physiology A: Neuroethology, Sensory, Neural, and Behavioral Physiology, 199*(6), 535–544.

Kirschvink, J. L. (1989). Magnetite biomineralization and geomagnetic sensitivity in higher animals: An update and recommendations for future study. *Bioelectromagnetics, 10*(3), 239–259.

Kirschvink, J. L. (1990). Geomagnetic sensitivity in cetaceans: An update with live stranding records in the United States. In J. A. Thomas & R. A. Kastelein (Eds.), *Sensory abilities of cetaceans: Laboratory and field evidence* (Vol. 196, pp. 639–649). New York: Plenum.

Kirschvink, J. L., Dizon, A. E., & Westphal, J. A. (1986). Evidence from strandings for geomagnetic sensitivity in cetaceans. *Journal of Experimental Biology, 120,* 1–24.

Kirschvink, J. L., & Gould, J. L. (1981). Biogenic magnetite as a basis for magnetic field detection in animals. *BioSystems, 13*(3), 181–201.

Kirschvink, J. L., Winklhofer, M., & Walker, M. M. (2010). Biophysics of magnetic orientation: Strengthening the interface between theory and experimental design. *Journal of the Royal Society, Interface, 7,* S179–S191.

Kishida, T., Kubota, S., Shirayama, Y., & Fukami, H. (2007). The olfactory receptor gene repertoires in secondary-adapted marine vertebrates: Evidence for reduction of the functional proportions in cetaceans. *Biology Letters, 3*(4), 428–430.

Kishida, T., & Thewissen, J. G. M. (2012). Evolutionary changes of the importance of olfaction in cetaceans based on the olfactory marker protein gene. *Gene, 492*(2), 349–353.

Klinowska, M. (1985). Cetacean live stranding sites relate to geomagnetic topography. *Aquatic Mammals, 11*(1), 27–32.

Klinowska, M. (1986). Cetacean live stranding dates relate to geomagnetic disturbances. *Aquatic Mammals, 11*(3), 109–119.

Kolchin, S. P., & Belkovich, V. M. (1973). Tactile sensitivity in *Delphinus delphis. Zoologicheskii Zhurnal, 52*(4), 620–622.

Komatsu, S., & Yamasaki, F. (1980). Formation of the pits with taste buds at the lingual root in the striped dolphin, *Stenella coeruleoalba. Journal of Morphology, 164*(2), 107–119.

Kowalewsky, S., Dambach, M., Mauck, B., & Dehnhardt, G. (2006). High olfactory sensitivity for dimethyl sulphide in harbour seals. *Biology Letters, 2*(1), 106–109.

Kröger, R. H. H., & Kirschfeld, K. (1989). Visual accommodation in cetaceans. In P. G. H. Evans & C. Smenk (Eds.), *European research on cetaceans* (pp. 38–40). Leiden: European Cetacean Society.

Kruger, L. (1959). The thalamus of the dolphin (*Tursiops truncatus*) and comparison with other mammals. *Journal of Comparative Neurology, 111*(1), 133–194.

Kuznetsov, V. B. (1990). Chemical sense of dolphins: Quasi-olfaction. In J. A. Thomas & R. A. Kastelein (Eds.), *Sensory abilities of cetaceans: Laboratory and field evidence* (Vol. 196, pp. 481–503). New York: Plenum.

Kuznetsov, V. B. (1992). Quasi-olfaction of dolphins. In R. L. Doty & D. Müller-Schwarze (Eds.), *Chemical signals in vertebrates* (Vol. 6, pp. 543–549). New York: Plenum.

Kuznetsov, V. B. (1999). Vegetative responses of dolphin to changes in permanent magnetic field. *Biophysics, 44*(3), 488–494.

Laska, M., Lord, E., Selin, S., & Amundin, M. (2010). Olfactory discrimination of aliphatic odorants in South African fur seals (*Arctocephalus pusillus*). *Journal of Comparative Psychology, 124*(2), 187–193.

Laska, M., Svelander, M., & Amundin, M. (2008). Successful acquisition of an olfactory discrimination paradigm by South African fur seals, *Arctocephalus pusillus*. *Physiology and Behavior, 93*(4–5), 1033–1038.

Layne, J. N., & Caldwell, D. K. (1964). Behavior of the Amazon dolphin, *Inia geoffrensis* (Blainville), in captivity. *Zoologica, 49*, 81–108.

LeDuc, R. (2009). Delphinids: Overview. In W. F. Perrin, B. Würsig, & J. G. M. Thewissen (Eds.), *Encyclopedia of marine mammals* (pp. 298–302). Amsterdam: Academic Press.

Ling, J. K. (1974). The integument of marine mammals. In R. J. Harrison (Ed.), *Functional anatomy of marine mammals* (Vol. 2, pp. 1–44). London: Academic Press.

Ling, J. K. (1977). Vibrissae of marine mammals. In R. J. Harrison (Ed.), *Functional anatomy of marine mammals* (Vol. 3, pp. 387–415). London: Academic Press.

Litwiler, T. L., & Cronin, T. W. (2001). No evidence of accommodation in the eyes of the bottlenose dolphin, *Tursiops truncatus*. *Marine Mammal Science, 17*(3), 508–525.

Manger, P. R. (2013). Questioning the interpretations of behavioral observations of cetaceans: Is there really support for a special intellectual status for this mammalian order? *Neuroscience, 250*, 664–696.

Manger, P. R., & Pettigrew, J. D. (1996). Ultrastructure, number, distribution and innervation of electroreceptors and mechanoreceptors in the bill skin of the platypus, *Ornithorhynchus anatinus*. *Brain, Behavior, and Evolution, 48*(1), 27–54.

Mann, J., Sargeant, B. L., Watson-Capps, J. J., Gibson, Q. A., Heithaus, M. R., Connor, R. C., et al. (2008). Why do dolphins carry sponges? *PLoS ONE, 3*(12), e3868.

Mass, A. M., & Supin, A. Y. (1989). Distribution of ganglion cells in the retina of an Amazon river dolphin *Inia geoffrensis*. *Aquatic Mammals, 15*(2), 49–56.

Mass, A. M., & Supin, A. Y. (1995). Ganglion cell topography of the retina in the bottlenosed dolphin, *Tursiops truncatus*. *Brain, Behavior, and Evolution, 45*(5), 257–265.

Mass, A. M., & Supin, A. Y. (1999). Retinal topography and visual acuity in the riverine tucuxi (*Sotalia fluviatilis*). *Marine Mammal Science, 15*(2), 351–365.

Mass, A. M., & Supin, A. Y. (2007). Adaptive features of aquatic mammals' eye. *Anatomical Record (Hoboken, NJ), 290*(6), 701–715.

Mass, A. M., & Supin, A. Y. (2009). Vision. In W. F. Perrin, B. Würsig, & J. G. M. Thewissen (Eds.), *Encyclopedia of marine mammals* (pp. 1200–1211). Amsterdam: Academic Press.

Mass, A. M., Supin, A. Y., Abramov, A. V., Mukhametov, L. M., & Rozanova, E. I. (2013). Ocular anatomy, ganglion cell distribution and retinal resolution of a killer whale (*Orcinus orca*). *Brain, Behavior, and Evolution, 81*(1), 1–11.

Mass, A. M., Supin, A. Y., & Popov, V. V. (1988). Topographic organization of ganglionic layer of retina of Amazon river dolphin *Inia geoffrensis. Doklady Akademii Nauk SSSR, 303*(1), 219–222.

McGowen, M. R., Clark, C., & Gatesy, J. (2008). The vestigial olfactory receptor subgenome of odontocete whales: Phylogenetic congruence between gene-tree reconciliation and supermatrix methods. *Systematic Biology, 57*(4), 574–590.

Messenger, S. L., & McGuire, J. A. (1998). Morphology, molecules, and the phylogenetics of cetaceans. *Systematic Biology, 47*(1), 90–124.

Morgane, P. J., & Jacobs, M. S. (1972). Comparative anatomy of the cetacean nervous system. In R. Harrison (Ed.), *Functional anatomy of marine mammals* (Vol. 1, pp. 117–244). London: Academic Press.

Murayama, T., & Somiya, H. (1998). Distribution of ganglion cells and object localizing ability in the retina of three cetaceans. *Fisheries Science, 64*(1), 27–30.

Murayama, T., Somiya, H., Aoki, I., & Ishii, T. (1995). Retinal ganglion cell size and distribution predict visual capabilities of Dall's porpoise. *Marine Mammal Science, 11*(2), 136–149.

Nachtigall, P. E. (1980). Odontocete echolocation performance on object size, shape and material. In R. G. Busnel & J. F. Fish (Eds.), *Animal sonar systems* (pp. 71–95). New York: Plenum.

Nachtigall, P. E. (1986). Vision, audition, and chemoreception in dolphins and other marine mammals. In R. J. Schusterman, J. A. Thomas, & F. G. Wood (Eds.), *Dolphin cognition and behavior: A comparative approach* (pp. 79–113). Hillsdale, NJ: Erlbaum.

Nachtigall, P. E., & Hall, R. W. (1984). Taste reception in the bottlenosed dolphin. *Acta Zoologica Fennica, 172 (Suppl.)*, 147–148.

Nachtigall, P. E., Lemonds, D. W., & Roitblat, H. L. (2000). Psychoacoustic studies of dolphin and whale hearing. In W. W. L. Au, A. N. Popper, & R. R. Fay (Eds.), *Hearing by whales and dolphins* (pp. 330–363). New York: Springer.

Nachtigall, P. E., Murchinson, A. E., & Au, W. W. L. (1980). Cylinder and cube shape discrimination by an echolocating blindfolded bottlenosed dolphin. In R. G. Busnel & J. F. Fish (Eds.), *Animal sonar systems* (pp. 945–947). New York: Plenum.

Niesterok, B., & Hanke, W. (2013). Hydrodynamic patterns from fast-starts in teleost fish and their possible relevance to predator–prey interactions. *Journal of Comparative Physiology A: Neuroethology, Sensory, Neural, and Behavioral Physiology, 199*(2), 139–149.

Nummela, S., Thewissen, J. G. M., Bajpai, S., Hussain, T., & Kumar, K. (2007). Sound transmission in archaic and modern whales: Anatomical adaptations for underwater hearing. *Anatomical Record, 290*(6), 716–733.

Oelschläger, H. A., & Buhl, E. H. (1985a). Development and rudimentation of the peripheral olfactory system in the harbor porpoise *Phocoena phocoena* (Mammalia: Cetacea). *Journal of Morphology*, *184*(3), 351–360.

Oelschläger, H. A., & Buhl, E. H. (1985b). Occurrence of an olfactory bulb in the early development of the harbor porpoise (*Phocoena phocoena* L.). *Progress in Zoology*, *30*, 695–698.

Palmer, E., & Weddell, G. (1964). The relationship between structure, innervation and function of the skin of the bottlenose dolphin (*Tursiops truncatus*). *Proceedings of the Zoological Society of London*, *143*(4), 553–558.

Paulos, R. D., Dudzinski, K. M., & Kuczaj, S. A., II. (2008). The role of touch in select social interactions of Atlantic spotted dolphin (*Stenella frontalis*) and Indo-Pacific bottlenose dolphin (*Tursiops aduncus*). *Journal of Ethology*, *26*(1), 153–164.

Peters, R. C., Eeuwes, L. B. M., & Bretschneider, F. (2007). On the electrodetection threshold of aquatic vertebrates with ampullary or mucous gland electroreceptor organs. *Biological Reviews of the Cambridge Philosophical Society*, *82*(3), 361–373.

Pryor, K. W. (1990). Non-acoustic communication in small cetaceans: Glance, touch, position, gesture, and bubbles. In J. A. Thomas & R. A. Kastelein (Eds.), *Sensory abilities of cetaceans: Laboratory and field evidence* (Vol. 196, pp. 537–544). New York: Plenum.

Reep, R. L., Gaspard, J. C., III, Sarko, D., Rice, F. L., Mann, D. A., & Bauer, G. B. (2011). Manatee vibrissae: Evidence for a "lateral line" function. *Annals of the New York Academy of Sciences*, *1225*(1), 101–109.

Reep, R. L., Marshall, C. D., & Stoll, M. L. (2002). Tactile hairs on the postcranial body in Florida manatees: A mammalian lateral line? *Brain, Behavior, and Evolution*, *59*(3), 141–154.

Reiss, D., & Marino, L. (2001). Mirror self-recognition in the bottlenose dolphin: A case of cognitive convergence. *Proceedings of the National Academy of Sciences of the United States of America*, *98*(10), 5937–5942.

Ridgway, S. H., & Carder, D. A. (1990). Tactile sensitivity, somatosensory responses, skin vibrations, and the skin surface ridges of the bottlenose dolphin, *Tursiops truncatus*. In J. A. Thomas & R. A. Kastelein (Eds.), *Sensory abilities of cetaceans: Laboratory and field evidence* (Vol. 196, pp. 181–193). New York: Plenum.

Rossbach, K. A., & Herzing, D. L. (1997). Underwater observations of benthic-feeding bottlenose dolphins (*Tursiops truncatus*) near Grand Bahama Island, Bahamas. *Marine Mammal Science*, *13*(3), 498–504.

Schusterman, R. J., Thomas, J. A., & Wood, F. G. (Eds.). (1986). *Dolphin cognition and behavior: A comparative approach*. Hillsdale, NJ: Erlbaum.

Shirihai, H., & Jarret, B. (2006). *Whales, dolphins, and other marine mammals of the world*. Princeton: Princeton University Press.

Sinclair, J. G. (1966). The olfactory complex of dolphin embryos. *Texas Reports on Biology and Medicine*, *24*(3), 426–431.

Smolker, R., Richards, A., Connor, R., Mann, J., & Berggren, P. (1997). Sponge carrying by dolphins (Delphinidae, *Tursiops* sp.): A foraging specialization involving tool use? *Ethology*, *103*(6), 454–465.

Suchowskaja, L. I. (1972). The morphology of the taste organs in dolphins. *Investigations on Cetacea*, *4*, 201–204.

Supin, A. Y., & Nachtigall, P. E. (2013). Gain control in the sonar of odontocetes. *Journal of Comparative Physiology A: Neuroethology, Sensory, Neural, and Behavioral Physiology*, *199*(6), 471–478.

Supin, A. Y., Popov, V. V., & Mass, A. M. (2001). *The sensory physiology of aquatic mammals*. Boston: Kluwer.

Tarpley, R. J., Gelderd, J. B., Bauserman, S., & Ridgway, S. H. (1994). Dolphin peripheral visual pathway in chronic unilateral ocular atropy: Complete decussation apparent. *Journal of Morphology*, *222*(1), 91–102.

Thewissen, J. G. M., George, J., Rosa, C., & Kishida, T. (2011). Olfaction and brain size in the bowhead whale (*Balaena mysticetus*). *Marine Mammal Science*, *27*(2), 282–294.

Thewissen, J. G. M., & Nummela, S. (2008). *Sensory evolution on the threshold: Adaptations in secondarily aquatic vertebrates*. Berkeley: University of California Press.

Thomas, J. A., & Kastelein, R. A. (1990). *Sensory abilities of cetaceans: Laboratory and field evidence* (Vol. 196). New York: Plenum.

Thomas, J. A., Moss, C. F., & Vater, M. (2004). *Echolocation in bats and dolphins*. Chicago: University of Chicago Press.

Treiber, C. D., Salzer, M. C., Riegler, J., Edelman, N., Sugar, C., Breuss, M., et al. (2012). Clusters of iron-rich cells in the upper beak of pigeons are macrophages not magnetosensitive neurons. *Nature*, *484*(7394), 367–370.

Trevarthen, C. B. (1968). Two mechanisms of vision in primates. *Psychologische Forschung*, *31*(4), 299–337.

Turner, W. (1890). The convolutions of the brain: A study in comparative anatomy. *Journal of Anatomy and Physiology*, *25*(Pt. 1), 105–153.

van der Willigen, R. F., Frost, B. J., & Wagner, H. (1998). Stereoscopic depth perception in the owl. *Neuroreport*, *9*(6), 1233–1237.

Walcott, C., Gould, J. L., & Kirschvink, J. L. (1979). Pigeons have magnets. *Science*, *205*(4410), 1027–1029.

Walker, M. M. (1997). Magnetic orientation and the magnetic sense in arthropods. In M. Lehrer (Ed.), *Orientation and communication in arthropods* (Vol. 84, pp. 187–213). Berlin: Birkhäuser.

Walker, M. M., Kirschvink, J. L., Ahmed, G., & Dizon, A. E. (1992). Evidence that fin whales respond to the geomagnetic field during migration. *Journal of Experimental Biology*, *171*, 67–78.

Walls, G. L. (1963). *The vertebrate eye and its adaptive radiation*. New York: Hafner Press.

Weiffen, M., Möller, B., Mauck, B., & Dehnhardt, G. (2006). Effect of water turbidity on the visual acuity of harbor seals (*Phoca vitulina*). *Vision Research*, *46*, 1777–1783.

West, J. A., Sivak, J. G., Murphy, C. J., & Kovacs, K. M. (1991). A comparative study of the anatomy of the iris and ciliary body in aquatic mammals. *Canadian Journal of Zoology*, *69*(10), 2594–2607.

White, D., Cameron, N., Spong, P., & Bradford, J. (1971). Visual acuity of the killer whale (*Orcinus orca*). *Experimental Neurology*, *32*(2), 230–236.

Wieskotten, S., Mauck, B., Miersch, L., Dehnhardt, G., & Hanke, W. (2011). Hydrodynamic discrimination of wakes caused by objects of different size or shape in a harbour seal (*Phoca vitulina*). *Journal of Experimental Biology*, *214*(11), 1922–1930.

Yablokov, A. V. (1961). The "sense of smell" in marine mammals. *Akademii Nauk SSSR*, *12*, 87–93.

Yablokov, A. V., Belkovich, V. M., & Borisow, V. I. (1972). *Whales and Dolphins Part II*. Moscow: Nauka.

Yamasaki, F., Komatsu, S., & Kamiya, T. (1978a). Papillary projections at the lingual margin in the striped dolphin, *Stenella coeruleoalba*, with special reference to their development and regression. *Journal of Morphology*, *157*(1), 33–47.

Yamasaki, F., Komatsu, S., & Kamiya, T. (1978b). Taste buds in the pits at the posterior dorsum of the tongue of *Stenella coeruleoalba*. *Scientific Reports of the Whales Research Institute Tokyo*, *30*, 285–290.

Yamasaki, F., Satomi, H., & Kamiya, T. (1976). An observation on the papillary projections at the lingual margin in the striped dolphin. *Scientific Reports of the Whales Research Institute Tokyo, No. 28*, 137–140.

Yu, L., Jin, W., Wang, J. X., Zhang, X., Chen, M. M., Zhu, Z. H., et al. (2010). Characterization of TRPC$_2$, an essential genetic component of VNS chemoreception, provides insights into the evolution of pheromonal olfaction in secondary-adapted marine mammals. *Molecular Biology and Evolution*, *27*(7), 1467–1477.

Zoeger, J., Dunn, J. R., & Fuller, M. (1981). Magnetic material in the head of the common pacific dolphin. *Science*, *213*(4510), 892–894.

II Dolphin Communication

4 Dolphin Societies: Structure and Function

Bernd Würsig and Heidi C. Pearson

Introduction

Delphinids are social creatures, and a lone dolphin is rare (Johnson & Norris, 1986). Even when "lone" saltwater dolphins occur, without the immediate presence of conspecifics, they tend to associate with other cetacean species or with humans (e.g., Wilke, Bossley, & Doak, 2005), presumably as an outgrowth of their intrinsic sociality. While this has been understood for quite some time, we do not fully understand dolphins' social nature. This includes all aspects of their lives—resting, traveling, foraging, feeding, and social and sexual behavior—within the prevalent mating system of a dolphin species or community. Comparisons of behaviors and mating systems, among species and even communities that live in different ways in similar or different habitats, can help shed light on social needs relative to environment.

Dolphins in riverine and nearshore areas tend to occur in small groups of a few to several dozen, while many open-ocean groups occur in hundreds to thousands. Norris and Dohl (1980a), Wells, Irvine, and Scott (1980), and Johnson and Norris (1986) related group sizes to proximity from shore and depth of water (the latter being probable proxies for habitat structure), with pelagic species forming larger groups. Gowans, Würsig, and Karczmarski (2008) expanded this idea to build a framework of societal structure that takes into account what the habitat supplies in the form of food, exposure to or shelter from predators, and resources for the care and teaching of vulnerable young. The habitat also affects opportunities for and manner of communication, sociality, and cognitive abilities (Barrett & Würsig, 2014).

In the simplest sense, and here we follow the definition of Norris and Dohl (1980a), a school or group (also termed pod) of dolphins comprises those that regularly swim together. This definition ignores the possibility that dolphins some distance apart may be in acoustic contact with one another for as much as hundreds of meters. Nevertheless we accept this general definition by proximity, as dolphins interacting at some greater distance by sound cannot be associating in important other ways. That is, only while in proximity can they communicate by subtle signals of body motions, exercise caregiving by mother of young, cooperatively surround or herd their prey, or engage in precopulatory behavior or sex.

Beyond assessing how many animals are in a group, researchers must also determine who is staying with whom and for how long, what kinds of interactions they have while together, and what kinds of specialized subgroups such as mating units and mother care ("nursery") units might exist. For this, it is important to recognize individuals, generally by photo-identification. For example, photos or videos of the nicks and notches on the trailing edge of dorsal fins are now used to identify individuals of almost all delphinid species (Würsig & Würsig, 1977), and saddle patch patterns and other marks are often used for killer whales, *Orcinus orca* (Bigg, Olesiuk, Ellis, Ford, & Balcomb, 1990), as are the spot patterns that identify individuals and age classes of spotted dolphins (*Stenella frontalis* and *S. attenuata*; Perrin, 1975; Herzing, 1997) (fig. 4.1). Beyond recognizing who is associating with whom (e.g., Herzing & Brunnick, 1997; Elliser & Herzing, 2014; Smolker, Richards, Connor, & Pepper, 1992; Wells, 2003), descriptions of behavior are also needed. While most of these have traditionally been

Figure 4.1
Photo-identification is used to identify individuals according to distinctive natural markings such as saddle patch patterns and dorsal fin scars on killer whales (*Orcinus orca*; left) and nicks and notches on the trailing edge of a dusky dolphin (*Lagenorhynchus obscurus*) dorsal fin (right). Killer whale photo courtesy Dan Olson.

made above water, the best cases have come from below water, a situation that necessitates (unfortunately rare) water clarity (Dudzinski, Clark, & Würsig, 1995; Dudzinski, 1998; Herzing, 1996; Herzing & Brunnick, 1997; Marten, Herzing, Poole, & Newman-Allman, 2001; Vaughn, Würsig, & Packard, 2010).

Group Sizes and Social Structures in Different Habitats

In an extraordinary book chapter published over thirty years ago, Wells, Irvine, and Scott (1980) laid out a blueprint for linking society structure with habitat. Many more data on society structure, social behaviors, and habitats have been gathered since then (for a review, see Gowans et al., 2008). The basic blueprint they developed not only survives but has been reinforced by this new information. We will integrate some of Wells et al. (1980) here, along with early thoughts by Norris and Dohl (1980a), to generate our own syntheses.

In the open ocean, the often large schools of small-bodied dolphins (such as of the genera *Delphinus*, *Lagenodelphis*, and *Stenella*) may travel with only one "wall"—the surface—for life-giving air. As they move in pelagic waters (i.e., deep and offshore), at times too deep even to sense the bottom by echolocation, they may never come in contact with islands or continents. Their entire lives are spent within the school envelope, and as we delve into subschool structure, we realize that there are sex- and age-mixed units, nursery units, mating units, and perhaps subadult units as well (Norris & Dohl, 1980a). While the large school is the overall stable unit, there is much fission–fusion and individual interchange among subunits, so that one nursery unit may not be composed of exactly the same mothers and calves from one day to the next, and even from hour to hour (Norris & Dohl, 1980b; Würsig, Wells, Norris, & Würsig, 1994).

It makes sense that these rather small-bodied delphinids gather in large schools, for overall enhancement of sensory awareness, faster response time to large predators such as sharks and killer whales, perhaps a confusion effect when attacked, and shared vigilance, such as being able to rest while others are awake and wary. (Partridge, 1982, provides an early primer on grouping for defense, but only for schools of fish; see also Norris & Schilt, 1988.) Delphinid schools in the open ocean can form such large groups only if they have exceptionally large prey reserves available. Most of them move over large distances and feed on deep scattering layer (DSL) organisms that come closer to the surface at night (Norris & Dohl, 1980a, 1980b; Benoit-Bird & Au, 2009a). Moving over large distances may function to allow prey in a given area to replenish and may also dissuade predators from "waiting" in a particular area for the dolphins.

When multiple species with similar swimming capabilities coexist in the pelagic zone, they can potentially enhance group benefits such as vigilance against predators by forming mixed-species groups. Such groups have been well described, for example, in pantropical spotted (*Stenella attenuata*) and spinner (*S. longirostris*) dolphins, with other tropical species at times mixed in (Psarakos, Herzing, & Marten, 2003). We surmise that their somewhat different sensory capabilities and activity patterns (such as feeding and resting at different times or feeding on slightly different mesopelagic prey) allow animals in mixed schools to enhance their own capabilities of vigilance, predator detection, and perhaps even food acquisition. Their co-occurrence may be a form of mutualism (Foster & Wenseleers, 2006). While some researchers may argue that such mutualism in large-brained social dolphins has led to reciprocal altruism (for general descriptions, see Connor & Norris, 1982; Trivers, 1971), it is also possible that the mixed school may simply serve to keep each individual safer from predation from large sharks, owing to the dilution effect of larger numbers.

Like schools of fish, dolphins pay attention to each other while they coordinate travel and other movements, such as foraging or predator evasion. However, unlike most fish, dolphins must also sort out their partnerships and perhaps subtle dominance/subservience relationships, take care of their young, and presumably engage in learning from each other and possibly even in teaching (e.g., Connor, Mann, Tyack, & Whitehead, 1998; Johnson, 2010; Pearson & Shelton, 2010; Wells, 2003). In some respects, dolphin schools can be compared to herds of terrestrial grazers, such as ungulates, that are constantly on the move, probably for reasons such as those discussed earlier of food regeneration and outtraveling their more stationary predators (fig. 4.2). Dolphins (and ungulates) have other social communication needs, as they pay attention to each other and make overall decisions of where to go (see Lusseau et al., 2003). However, terrestrial ungulates may not have (as far as we know) the capabilities for sophisticated long-term memories and for the individual and culturally mediated learning that we are just beginning to understand in dolphins (e.g., Mann & Sargeant, 2003; Rendell & Whitehead, 2001; Whitehead, Rendell, Osborne, & Würsig, 2004).

In the open ocean and with the necessity of sharing prey resources, there is no economic defensibility of resources such as space or food. Since such resource defensibility is generally considered necessary for the development of monogamy (Emlen & Oring, 1977), we would not expect to find monogamous mating systems in these animals. Similarly, polygyny, the common mammalian system of one male attempting to father multiple offspring of multiple females, is also uncommon in open-ocean dolphins (and in most other delphinids, with notable exceptions to be mentioned later). Instead, most pelagic dolphins exhibit polygynandry, or multimate breeding during

Figure 4.2
Like their terrestrial ungulate ancestors, dolphins may occur in large, often fast-moving herds. A herd of wildebeest (*Connochaetes taurinus*) on the savanna of Amboseli National Park, Tanzania (top). Dusky dolphins (*Lagenorhynchus obscurus*) travel in a large pod off Kaikoura, New Zealand (bottom).

the same estrus cycle. That is, males show polygynous mating attempts while females show polyandrous ones, possibly with female choice at both behavioral and physiological (cryptic choice) levels (Eberhard, 1996; Orbach, Packard & Würsig, 2014). Most offshore delphinids tend to be remarkably monomorphic, with only subtle differences in body morphology beyond the genital slit differences to discern them as males and females (although we assume that they can tell each other's sex much better than we can). While such monomorphism is common in monogamy, it is also consistent with polygynandry.

There are likely to be exceptions to polygynandry among a few pelagic small-bodied dolphins, such as the eastern spinner dolphin (*S. longirostris orientalis*). Males in this subspecies have a huge postanal keel and strongly backward-canted dorsal fin, and it is assumed that polygyny is the norm in this open-ocean society (Norris, Würsig, Wells, & Würsig, 1994; Perrin & Mesnick, 2003). However, whether mating is mediated by male dominance relationships, female choice, or both is as yet unknown for this subspecies.

But what about the larger-bodied delphinids that also occur in the pelagic zone, such as killer, false killer (*Pseudorca crassidens*), and pilot (*Globicephala* spp.) whales or the Risso's dolphins (*Grampus griseus*)? They occur in the open ocean in quite variably sized groups from just a few members to several hundred. We assume that they are large and "fierce" enough, especially the killer whale, to be relatively free from predation and therefore do not have the same needs for a large school for safety against predators. Instead, school size is probably largely determined by the most efficient size for taking prey (Baird, 2000) and may vary according to mating strategy as well. Thus killer whales, for example, aggregate together in "super pods" lasting only hours to days, apparently for assortative mating among members of different pods. Pilot whale schools are extremely variable in size, and especially large schools appear to occur in areas of high potential killer whale predation on their young, as in the northern parts of the North Atlantic. In contrast, pilot whale schools are much smaller in areas with few killer whales, such as the Gulf of Mexico or the tropics. In either case, these larger-bodied animals tend to occur in mixed-age and mixed-sex groupings, even when in smaller groups. In this way, within both the small and the large schools, all potential functions of communication, courtship, mating, rearing of young, and long-term learning are possible.

Killer whales and pilot whales (and possibly Risso's dolphins) tend toward social matriarchies (for killer whales, see Baird, 2000; Amos, Schlotterer, & Tautz, 1993; Heimlich-Boran, 1993; for pilot whales, see Kasuya & Marsh, 1984). That is, mothers and partial sisters (from one mother but usually from different fathers) tend to stay together, so that there are close female bonds within at least a subunit of the larger school. All three species are sexually dimorphic, with the extreme in the large male killer whale, with its high and erect dorsal fin (fig. 4.1). While some element of polygynandry may certainly exist in matriarchies, it is likely that these systems tend more toward polygyny, with one male attaining more than one paternity during a given period of estrus. In such situations, female choice is probably extremely important, but it has not been thoroughly investigated (Mesnick & Ralls, 2009). Pilot whale and killer whale females also show reproductive senescence (menopause), with at least pilot whales continuing to nurse for some time after senescence (Amos et al., 1993; Kasuya & Marsh, 1984), a trait perhaps related to matriarchy. Reproductive senescence appears to be valuable in allowing mothers and grandmothers to continue to be productive in helping and teaching members of the society without the direct cost of pregnancy, and continued nursing may help foster generational bonds (for a discussion of the potential utility of menopause, see Sherman, 1998), one important component of culture (Rendell & Whitehead, 2001).

Before addressing nearshore dolphins, it is instructive to consider delphinids that switch between pelagic and nearshore zones on a daily and seasonal basis. If we can discern general, and perhaps more specific, changes in group size and structure per habitat use, we can more closely align the reasons for their school structure with habitat, and behavior in that habitat. Two good examples are island- and atoll-living spinner dolphins that change environments on a diel basis, and dusky dolphins (*Lagenorhynchus obscurus*) that do so on both diel and seasonal bases. We will first take each species separately and then compare them.

As noted earlier, spinner dolphins in the open ocean live in large societies, at times even as multispecies groupings (Psarakos et al., 2003). But spinner dolphins (we will call them "spinners") show variable uses of habitats, with some staying close to shore and others exhibiting a semipelagic lifestyle, switching between nearshore shallow and deepwater pelagic zones on a diel basis. We know this especially well for the Kona Coast, Hawaii (Norris et al., 1994; Norris & Dohl, 1980b), where spinners rest close to shore during the day in small schools of approximately 20 to 100 or more individuals (table 4.1).

The predominant daytime activity of these animals is resting in bays over sandy bottom, being careful to avoid rocky outcroppings or coral where they might be surprised by a shark from below. In the late afternoon, spinners become more active, begin a zig-zag pattern of more rapid swimming, and vocalize more as they become more alert and social. After leaving the protected bays, they move alongshore to pick up pod members in other nearshore areas, leading school sizes to swell to hundreds of animals (upper limit unknown, but estimated from at least 200 to 400 at any one time). This is all in

84 — Bernd Würsig and Heidi C. Pearson

Table 4.1

Comparison of habitat, food type and distribution, predation risk, group size, and fission–fusion dynamics for spinner dolphins (*Stenella longirostris*) off the Kona Coast and Midway Island, Hawaii; and for dusky dolphins (*Lagenorhynchus obscurus*) in Golfo San José, Argentina; and Kaikoura and Admiralty Bay, New Zealand.

	Spinner Dolphins[1]		Dusky Dolphins[2]		
	Kona Coast	Midway Atoll	Golfo San José	Kaikoura	Admiralty Bay
Habitat	Nearshore bays and offshore pelagic waters	Atoll fringed by continuous emergent reef and surrounded by > 2000 m deep pelagic waters	Large bay	Coastal waters and open, deepwater canyon	Small, shallow enclosed bay
Primary food type, distribution, and foraging mode	More predictable mesopelagic fish, squid, and shrimp in the DSL; possible cooperation in pairs	More predictable mesopelagic fish, squid, and shrimp in the DSL; foraging mode unknown	Patchily distributed anchovy; cooperative "bait-balling"	More predictable nonschooling fish and squid associated with the DSL; individual or small group	Patchily distributed pilchard; cooperative "bait-balling"
Relative[3] predation risk	High	Moderate[4]	Moderate	High	Low

Table 4.1 (continued)

	Spinner Dolphins[1]		Dusky Dolphins[2]		
	Kona Coast	Midway Atoll	Golfo San José	Kaikoura	Admiralty Bay
Typical group size	20–100 (nearshore), 200–400 (offshore)	211	10–12[5] 150–300[7]	10–12[6] 250–1000[8]	7
Relative fission–fusion fluidity[9]	Medium	Low	High	Low	Medium

[1] References: Norris, Würsig, Wells, & Würsig, 1994; Karczmarski, Würsig, Gailey, Larson, & Vanderlip, 2005; Benoit-Bird & Au, 2003; Benoit-Bird & Au, 2004.

[2] References: Golfo San José: Würsig & Würsig, 1980. Kaikoura: Benoit-Bird, Würsig, & McFadden, 2004; Markowitz, 2004; Srinivasan & Markowitz, 2010; Würsig, Duprey, & Weir, 2007. Admiralty Bay: Benoit-Bird, Würsig, & McFadden, 2004; Pearson, 2009; Vaughn, Würsig, & Packard, 2010.

[3] Relative to other populations of the same species.

[4] Low predation risk in the immediate vicinity of the atoll, but high predation risk in nearby surrounding pelagic waters.

[5] Small group "fission."

[6] Small nursery and other satellite.

[7] Large group "fusion" feeding/social.

[8] Large regular daytime.

[9] The degree of group fission and fusion in terms of changes in group size and composition; high fluidity indicates extreme and oftentimes frequent changes in group size and composition.

preparation for offshore travel to meet the DSL, where they forage (often cooperatively; Benoit-Bird & Au, 2009a) on mesopelagic squid and lantern fishes. As described for other species earlier, the large groups no doubt offer the dolphins protection in these deeper, shark-dangerous waters. Sound recordings at night (Benoit-Bird & Au, 2009b) also indicate that spinners are highly vocal during this period, apparently doing much socializing at night. Thus the large nighttime schools may also be important for mating activities. After the DSL has descended again in the early morning, the large spinner school splits up (Benoit-Bird & Au, 2003, 2004), and smaller schools fit into the bays. "Fit into" appears to be an accurate term, as larger bays will tend to have larger schools than smaller ones (Würsig et al., 1994). Thus spinners exhibit a diel fission–fusion society.

While there remain long-term associations between mothers and calves, as well as some "friendship bonds" that involve older individuals that stay together for days to years, the composition of a school in Kealakekua Bay, Hawaii, still varies from day to day. The lability of interanimal affiliation in this fission–fusion society suggests to us that many of the dolphins of the greater nighttime feeding society may know each other reasonably well, and differences in affiliations are useful in maintaining and enhancing bonds. Such bonds may be particularly useful for dolphins that need to communicate efficiently during nighttime, not only to detect and avoid predators but perhaps also to coordinate cooperative feeding and to form sexual partnerships (Würsig et al., 1994).

In contrast to the Kona Coast, this diel pattern of group fission and fusion is not present for spinner dolphins that occur around lone atolls such as Midway Island, Hawaii, where other atolls are great distances (e.g., ≥ 50 km) away (table 4.1; Karczmarski, Würsig, Gailey, Larson, & Vanderlip, 2005). Instead a particular group of animals will use the same atoll for daytime rest, day after day, implying that the same animals travel together to open waters to feed at night. Presumably this closed society arose because other atolls are simply too far away for the efficient transfer of individuals. Additional research is needed to determine how this absence of diel fission–fusion affects other aspects of social organization and communication in these groups.

Karczmarski et al. (2005) described an incidence of "macro fission–fusion" in these animals, when a subset of dolphins from one atoll moved to another atoll. At first, individuals in this subset rested in the new atoll separately from the original group, but eventually they became physically and (apparently) socially integrated with the resident group. Such emigration from one atoll to another with dolphins already present may represent occasional, accidental wanderings. But it may also result from societal exclusion from one area or the need to find new mating partners. Indeed, if a resident group of about 100 dolphins at one atoll stayed sexually closed for several generations, unacceptable levels of inbreeding would occur. As it is, atoll genetic diversity is known to be low (Andrews et al., 2006), and comparisons of gene flow among those atolls indicates that interatoll matings take place, presumably as a result of these occasional and sporadic "macro fission–fusion" events (see also Andrews et al., 2010).

Dusky dolphins are another semipelagic species, but unlike spinners, they seldom occur in pelagic waters (Gaskin, 1968). However, duskies occur both over extensive coastal shallows, where the continental slope can be as far as 200 km away, and near coasts with precipitous drop-offs into oceanic waters. In the former habitat type, off southern Argentina and in the Marlborough Sounds, New Zealand, duskies feed on schooling fishes in the daytime and rest in shallow nearshore waters at night (for a summary, see Würsig, Würsig, & Cipriano, 1989). In Golfo San José, Argentina, duskies form a strong fission–fusion society, separating into small subunits of only about ten dolphins while resting in shallows at night (table 4.1). Arguably, this is a strategy against shark and killer whale predation, since the subunits are less conspicuous than the larger school (i.e., reduced encounter effect). We do not know how duskies react to shark predation, but when killer whales approach, duskies close enough to shore for evasive action attempt to hide in the surf zone (Würsig & Würsig, 1980). During the morning, groups coalesce during foraging on southern anchovy (*Engraulis anchoita*), and it appears that large groups (up to 300 or so individuals) are more efficient than smaller ones at herding fish into tight "baitballs" at the surface. After feeding, duskies engage in high levels of social and sexual activities in the large coalesced school before splitting into small subunits for the night once again. Other than mother-calf pairs, the subunit composition is unlikely to be the same night after night (Würsig & Würsig, 1980). In short, the fission–fusion events for both duskies off Argentina and spinners off the Kona Coast are related to feeding, but with day- and nighttime resting and feeding reversed between the two species.

Duskies that live near shore but near the precipitous drop-off at the edge of the Kaikoura Canyon on the South Island, New Zealand, exhibit very different grouping and foraging patterns. Off Kaikoura, they rest and socialize in open-water shallows throughout the day, generally staying within the large school envelope of approximately 50 to 1,000 individuals (table 4.1; Würsig, Duprey, & Weir, 2007). At night they move only one to several kilometers away to feed in the deep Kaikoura Canyon on DSL fishes and squid, just as do spinners. Our observations from active-acoustic surveys that detect both the DSL and dolphins down to a depth of about 150 m suggest that foraging occurs individually, with the possibility of some coordinated subgroups of two to five individuals (Benoit-Bird, Würsig, & McFadden, 2004).

Instead of going into deep rest for most of the day as spinners do, duskies alternate rest and social activities in daytime. Within the large school, varying levels of alertness may be present at any one time, yielding a collectively useful sensory awareness by the entire group for vigilance against predation. All activities—rest, socializing, mating, and nursing—occur during an overall slow level of traveling and occur within the large school envelope. However, intensively mating units (Markowitz, Markowitz, & Morton, 2010; Orbach, Packard, & Würsig, 2014) and nursery groups (Deutsch, Pearson, & Würsig, 2014; Weir, Deutsch, & Pearson, 2010) often segregate from the large school. Nursery groups in particular move into very shallow water close to shore, presumably to minimize predation risk (Srinivasan & Markowitz, 2010; Srinivasan, Grant, Swannack, & Ragan, 2010), but also likely to avoid the boisterous sociosexual activities and the constant traveling that occur in the main school. Although further research is needed on the nighttime activities of nursery groups, evidence indicates that they do not always join the main group to feed in deeper waters at night and may instead feed in the shallows during the day (Weir, 2008).

Thus we can observe diel variations in behavior and habitat use patterns both on the extensive shelf of Argentina and in the nearshore deep waters of New Zealand. New Zealand is unique, however, as there is also a marked seasonal difference for a subset of the overall population. In winter and early spring (about May–October), several hundred or more duskies, of the general Kaikoura area, travel approximately 275 kilometers north to Admiralty Bay, a shallow (≤ 105 m) bay in the Marlborough Sounds. Here they radically change their feeding habits and exhibit behaviors similar to those of the Argentina duskies (table 4.1). In the Marlborough Sounds, duskies corral baitballs of schooling fishes (e.g., pilchard, *Sardinops neopilchardus*) into tight schools at the surface. This cooperative herding behavior is generally carried out in small schools of ten to several dozen individuals (Vaughn, Shelton, Timm, Watson, & Würsig, 2007; Vaughn, Würsig, Shelton, Timm, & Watson, 2008; Vaughn, Degrati, & McFadden, 2010; Vaughn, Würsig, & Packard, 2010). Although group size is quite labile in Admiralty Bay, group sizes are smaller overall (averaging seven individuals; Pearson, 2009), and there is not the marked fission–fusion of the society seen off Argentina.

Potential explanations for these differences are the low predation risk in Admiralty Bay (Pearson, 2009) and the fact that not as many duskies can fit into this small bay as in the larger bay off Argentina. Reduced prey availability, compared to Argentina, is another potential reason for the more muted fission–fusion grouping patterns of Admiralty Bay. Some of the same duskies annually make the summer-winter trek between the deep waters off Kaikoura and the shallow waters of Admiralty Bay (Markowitz, 2004; Shelton, Harlin-Cognato, Honeycutt, & Markowitz, 2010). This behavior, by only a subset of the population, hints at the possibility of cultural differences between those that feed year-round off Kaikoura and those that change foraging strategies and habitat according to season (Whitehead et al., 2004; Würsig & Pearson, 2014).

Many populations of dolphins—including common (*Tursiops truncatus*) and Indo-Pacific (*T. aduncus*) bottlenose, humpback (*Sousa* sp.), Irrawaddy River (*Orcaella brevirostris*), Australian snubfin (*Orcaella heinsohni*), Costero (*Sotalia guianensis*), and three of the four *Cephalorhynchus* species—spend their entire lives in shallow coastal waters. However, a few populations, including common bottlenose dolphins, can be found in shelf and deep ocean waters as well. While not all of these species have been studied intensively, most occur in small groups (generally no more than one dozen at one time) and exhibit both lone and cooperative foraging techniques (for a primer, see Pryor & Norris, 1991).

Much of what we know about bottlenose dolphin (*Tursiops* spp.) societies comes from long-term studies off Sarasota, Florida, and in Shark Bay, Australia. In these relatively shallow (< 15 m) nearshore habitats, decades of behavioral observation and photo-identification have provided tremendous insight into not only grouping patterns but also the nature of social bonds (e.g., Connor, Wells, Mann, & Read, 2000; Connor, Heithaus, & Barre, 2001; Gibson & Mann, 2008; Mann, 2006; Wells, 2003; Wells et al., 1980). Both species of bottlenose dolphin that have been studied intensively live in highly dynamic fission–fusion societies where group structure (i.e., both size and composition) may change on timescales of hours to minutes (Connor et al., 2000). Groups typically contain between five and seven individuals, although nursery groups tend to be larger for increased predator protection (Mann, Connor, Barre, & Heithaus, 2000; Smolker et al., 1992; Wells, 2003). Upon closer examination of the social bonds formed within these groups, some striking differences are apparent between the sexes. Bonds between males tend to be stronger than bonds between females, likely due to the cooperative alliance relationships formed between males to gain access to estrous females (Connor, Smolker, & Richards, 1992; Connor et al., 2001). In contrast, females tend to form weaker bonds with one another but have wider social networks, as their social bonds change according to reproductive status (i.e., mother/nonmother, cycling/noncycling; as summarized in Pearson, 2011). Except when cycling, females rarely associate with males, likely to avoid male harassment and sexual coercion.

In general, when feeding on nonschooling fish (Barros & Wells, 1998; Patterson & Mann, 2011), individual foraging strategies prevail, while cooperative strategies prevail when feeding on schooling fish. For example, cooperative foraging on schooling fish has been observed for bottlenose dolphins in the Black Sea (Bel'kovich, Ivanova, Yefremenkova, Kozarovitsky, & Kharitonov, 1991), Argentina (Würsig & Würsig, 1979),

South Africa (Tayler & Saayman, 1972), the Gulf of Mexico (Leatherwood, 1975), Baja California (Leatherwood, 1975; Würsig, 1986), and the Florida Keys (Gazda, Connor, Edgar, & Cox, 2005). Bottlenose dolphins are even known to cooperatively forage with humans. In Laguna, Brazil, dolphins have been observed driving schools of mullet (*Mugil cephalus*) into fishermen's nets and then capturing fish for themselves as some escape from the nets (Pryor, Lindbergh, & Milano,1990).

Interactions of Habitat, Foraging Types, Social Behaviors, and Society Structure

As described in the previous section, foraging strategies vary according to prey type and availability, which in turn vary by habitat. These differences in foraging strategies may then parallel differences in social behaviors and social structures.

Killer whales are the most cosmopolitan delphinid, and several distinct ecotypes (i.e., populations that have overlapping ranges but are socially and reproductively isolated and also exhibit differences in body size and coloration) have been identified (summarized in Morin et al., 2010). Much of what we know about killer whale foraging and social structure comes from studies of the Northeast Pacific, where resident, transient, and offshore ecotypes coexist. At the broadest level, these three ecotypes are differentiated by their foraging strategies, as residents and offshores feed on fish while transients feed on marine mammals (Dahlheim et al., 2008; Ford et al., 1998; Saulitis, Matkin, Barrett-Lennard, Eise, & Ellis, 2000).

Resident killer whales are the only known mammals in which neither sex emigrates from the natal pod (Baird, 2000). The social structure of transients is less well known, but pod membership appears to be more fluid than in residents, with both sexes dispersing from the natal pod in response to mating and hunting opportunities (Baird, 2000; Bigg et al., 1990; Saulitis, Matkin, & Fay, 2005). Pod size in transients (1–4) is smaller than in residents (3–59) due to transients' reliance on more patchily distributed large prey items that require stealth to hunt and result in food sharing (Baird, 2000; Baird & Dill, 1996). In contrast, residents benefit by foraging in larger groups and sharing echolocation information when searching for smaller prey items such as salmon (Barrett-Lennard, Ford, & Heise, 1996). The optimal strategy for transients, then, is for one or both sexes to disperse from the natal pod. Typically, all but firstborn transient males disperse, although a female may return to her natal pod later in life if she is unable to reproduce (Baird, 2000). Little is known about offshore social structure, making this a rich area for future research (Dahlheim et al., 2008).

Culture offers one parsimonious explanation for why these stark differences among killer whale ecotypes may occur. Killer whale societies are aptly described as "multilevel" and "multicultural," as cultural variation is present between most societal levels (Bigg et al., 1990). Resident killer whales, for example, form stable matrilines containing an older female and up to three generations of offspring. Closely related matrilines form pods, each with a unique vocal dialect or set of calls. Pods that share portions of their vocal dialect then form vocal clans. It is likely that these distinct calls enable individuals to identify their close relatives, thereby avoiding inbreeding, so that mating then occurs between vocal clans, at the community level (Baird, 2000; Ford, 1991).

Genetic factors may also be involved in ecotype differences, since members of different ecotypes do not interact or mate. This leads to reproductive isolation, genetic differentiation, and the persistence of the mammal-eating versus fish-eating strategies that distinguish transient from resident ecotypes (e.g., Baird, 2000; Hoelzel, Dahlheim, & Stern, 1998; Pilot, Dahlheim, & Hoelzel, 2010). Simply put, it is likely that, at some point in their evolutionary history, some killer whales in the Northeast Pacific began to specialize on fish, and others began to specialize on marine mammals. These foraging specializations likely persisted as they were "shared" along familial lines, within and between generations, eventually leading to the ecotypes we see today (Hoelzel et al., 1998, 2007). In fact, emerging genetic evidence indicates that residents and transients, and some of the Southern Ocean ecotypes, may indeed be separate species (Morin et al., 2010).

Delphinids have slow life histories, characterized by long life spans and prolonged periods of gestation, infant dependency, and juvenility. A slow life history provides time for an individual to develop the cognitive adaptations necessary to respond to socioecological demands (van Schaik & Deaner, 2003). As has been documented for other animals with slow life histories (e.g., primates, elephants), a prolonged period of infant dependency in delphinids provides extra time for learning, especially between mother and offspring, but also between conspecifics (e.g., Mann, 2006; Mann & Sargeant, 2003). Bottlenose dolphins and killer whales offer two excellent examples of the importance of calf learning in the development of foraging strategies. It is also likely that older females, even when themselves no longer reproductive, help serve to teach and care for younger animals and thus provide a long-term reservoir of "wisdom" from having lived a long life of social learning (Norris & Pryor, 1991; Foster et al., 2012).

Bottlenose dolphins in Shark Bay are dependent on their mothers for an average of four years, providing a long period of time for learning foraging and social strategies (Mann et al., 2000; Mann & Sargeant, 2003). Some individuals (about 5 percent) in this population specialize in sponging, or carrying a sponge on the rostrum for protection while probing in the substrate during benthic foraging (Mann & Patterson, 2013).

This foraging strategy allows individuals to exploit lipid-rich fish that lack swim bladders (e.g., barred sandperch, *Parapercis nebulosa*) and would otherwise be undetectable via echolocation (Patterson & Mann, 2011). Since this tool-using behavior was first observed in 1984 (Smolker, Richards, Connor, Mann, & Berggren, 1997), studies have revealed that the majority of individuals that carry sponges are females, only a small subset of females sponge, and calves learn this behavior from their mothers during their second or third year of life (Mann & Sargeant, 2003; Mann et al., 2008). This is a vertically transmitted cultural behavior, as individuals (mostly daughters) will only sponge if their mother also sponged (Mann et al., 2008; Mann, Stanton, Patterson, Bienenstock, & Singh, 2012).

Sponging females spend more time alone or only with their calf than do other females, likely due to the increased amount of time required to perform this foraging technique (Mann et al., 2008; Smolker et al., 1997). The relatively asocial nature of sponging dolphins is related to the sex bias in this behavior, which can be explained by intersexual differences in reproductive success. In societies such as these where males do not provide parental care, female reproductive success depends largely on gaining access to food and avoiding predation, whereas male reproductive success depends largely on gaining and maintaining access to estrous females (Krebs & Davies, 1993; Wrangham, 1979). Therefore, while it is beneficial for female bottlenose dolphins to invest time in sponging at the possible expense of social partners, this behavior is not advantageous for males, as their reproductive success is increased through integration into the social network and the formation of mating alliances (Connor et al., 1998). Ultimately, these intersexual differences in reproductive success affect social structure, as male bottlenose dolphins are typically more gregarious than females (summarized in Pearson, 2011). In short, the extended period of bottlenose dolphin calf dependency provides time for females to learn specialized foraging techniques and males time to learn about future alliance partners. Similar intersexual differences in gregariousness and the development of tool-based foraging strategies are seen in chimpanzees (*Pan troglodytes*), where mothers that spend more time termite fishing spend more time alone (Lonsdorf, 2006), and female offspring spend more time termite fishing than do male offspring (Lonsdorf, Pusey, & Eberly, 2004).

Delphinid calves may also be taught foraging skills. Briefly, teaching is a behavior performed at a cost by an "actor," which only occurs in the presence of a "naive observer" and results in acquisition of a skill or knowledge by that observer (Caro & Hauser, 1992). This is difficult to quantify in the wild, but multiple studies suggest that teaching may occur in cetaceans (Bender, Herzing, & Bjorklund, 2009; Guinet, 1991; Guinet & Bouvier, 1995; Hoelzel, 1991; Lopez & Lopez, 1985).

For example, in Patagonia (Lopez & Lopez, 1985) and the Crozet archipelago (Guinet, 1991; Guinet & Bouvier, 1995), killer whales intentionally strand themselves to capture pinniped pups (both southern sea lions, *Otaria flavescens*, and southern elephant seals, *Mirounga leonina*). Adults beach themselves in the presence of juveniles but will then incur an energetic cost, by either not attempting to capture the pup or capturing a pup but then relinquishing it to a juvenile. In turn, juveniles benefit by copying the adult's stranding behavior or receiving the prey item. Calves practice this risky foraging strategy for several years before becoming proficient (Guinet, 1991; Guinet & Bouvier, 1995). The prolonged period of calf dependency, high maternal investment, and stable social structures in killer whales facilitate the development of this hunting strategy. Furthermore, the presence of several related adults in one pod indicates that this may be a kin-selected behavior that facilitates alloparental teaching (Guinet, 1991).

Understanding Dolphin Society Structure and Function to Inform Conservation Actions

Most dolphin societies are characterized by fission–fusion dynamics, where group size and composition fluctuate according to the shifting balance of costs and benefits associated with prey availability, mating opportunities, and predation risk (Gowans et al., 2008). Grouping patterns thus reflect strategies to optimize survival and should be considered in conservation actions. The New Zealand dusky dolphin population provides a case study for how knowledge of grouping and social behavior has been applied in conservation and management strategies. In Kaikoura, dusky dolphin ecotourism operations have existed since the late 1980s (Buurman, 2010). Studies during the same period have revealed that duskies occur in large, generally fast-moving mixed-sex groups in addition to satellite mating groups and nursery groups, all of which have lower levels of activity at midday (summarized in Markowitz et al., 2010). As tour vessels focus on the large groups and adhere to a mandated midday rest period, this is generally considered to be a sustainable industry with no significant long-term effects detected (summarized in Childerhouse & Baxter, 2010; Markowitz et al., 2010). Additional studies of this population in the wintertime foraging habitat of Admiralty Bay (see earlier) have revealed that cooperative foraging strategies are hindered by mussel farms (Markowitz, Harlin, Würsig, & McFadden, 2004). These findings were presented to the New Zealand Environmental Court and to date have been successful in halting, or at least delaying, expansion of mussel farms in Admiralty Bay (Childerhouse & Baxter, 2010). Data show a decline in the dolphins' use of Admiralty Bay in recent years, perhaps due to decreased prey availability, bottom-up effects from mussel farms, and/or climate change effects (Pearson, Vaughn-Hirshorn, Srinivasan, & Würsig, 2012; Srinivasan, Pearson, Vaughn-Hirshorn, Würsig, & Murtugudde, 2012). Continued monitoring of this system is necessary to determine the cause for this shift in use of this seasonal habitat, and a continuing dialogue between researchers and managers is necessary for determining sustainable or multiple uses of this habitat.

Avenues for Comparative Research

Dolphins, like all cetaceans, have a unique evolutionary history, as they have been separated from their closest terrestrial relatives, the ungulates, for over 50 million years (Fordyce, 2009). The comparative approach may therefore be used to understand convergent versus divergent influences on behavior. Shared ancestry may help to explain some similarities between dolphins and ungulates, such as the production of few to single offspring, at times prolonged periods of offspring dependency, offspring following behavior, and large-group formation for protection against predators (Jarman, 1974; Mann & Smuts, 1999; Norris & Dohl, 1980a; Würsig, 1989). However, dolphins also exhibit convergence with less closely related taxa, such as social carnivores, elephants, and primates (fig. 4.3). Commonalities that dolphins share with these latter taxa include a large relative brain size with complex cognition and sophisticated social strategies (summarized in Pearson & Shelton, 2010; Johnson, 2010).

In particular, dolphins and great apes exhibit convergent cognitive abilities, social structures, and behaviors. These taxa are evolutionarily separated by 95 million years (Bromham, Phillips, & Penny, 1999) and thus offer a powerful approach for testing hypotheses of cognitive and social evolution while minimizing "phylogenetic complication." One such hypothesis is the social brain hypothesis, which posits that large relative brain size (specifically a large relative neocortex size) and complex cognition evolved to solve problems associated with monitoring changing group properties, social hierarchies, and networks of relationships (Dunbar, 2003; Humphrey, 1976; Jolly, 1966). According to this hypothesis, relative neocortex size should increase with group size. While this is supported by data for anthropoid primates (Dunbar & Shultz, 2007) and odontocetes (Marino, 1996), we need more data on the nature and kind of relationships beyond group size. We emphasize that "dolphins are not aquatic apes" (Barrett & Würsig, 2014), and while there has been much convergent evolution, especially of social capabilities, there are also many differences, such as in morphology, physiological capabilities and constraints, and generally extreme habitat differences.

Figure 4.3
Dolphins exhibit convergence in social structure and cognitive capabilities with the terrestrial mammals shown here. Clockwise from top right: Indo-Pacific bottlenose dolphin (*Tursiops aduncus*), short-beaked common dolphin (*Delphinus delphis*), dusky dolphin (*Lagenorhynchus obscurus*), African elephant (*Loxodonta africana*), chimpanzee (*Pan troglodytes*), lion (*Panthera leo*), and spotted hyena (*Crocuta crocuta*). All photos except dusky dolphin and chimpanzee courtesy Chris Pearson.

Present and Future Research Directions

Much of the behavioral field research on delphinids has been by "bread-and-butter" basic observation. This includes work from shore, often with the help of theodolite tracking (e.g., Lundquist, Gemmell, & Würsig, 2012), from small vessels using focal follows (Mann, 1999; Pearson, 2009), and from underwater with ever-more-sophisticated sound acquisition and video techniques (Dudzinski et al., 1995; Herzing, 1996, 2000; Vaughn, Muzi, Richardson, & Würsig, 2011). Some limited work has also been carried out with conventional radio tracking, satellite tracking, and data tags, especially on larger delphinids such as pilot whales (Baird, Borsani, Hanson, & Tyack, 2002). Ever-smaller, more streamlined, and well-designed tags are now making such remote sensing even more tractable, with minimal disturbance to the tagged animals.

While it is not possible to predict the development of all the new sophisticated research techniques for behavioral studies, we are particularly excited by developments in the underwater localization of dolphin vocalizations (Schotten, Lammers, Sexton, & Au, 2005) and the rapid development of aerial observation techniques by unmanned vehicles, such as minicopters equipped with video and still cameras (e.g., Nowacek, 2002). We envision that very soon we will be able to have a small vessel slowly moving with a social group of dolphins, describing focal animal and focal group behavior (including the use of recognition photography), while an underwater video camera slaved to a directionalizing set of hydrophones gathers interindividual behaviors and individually discriminated sounds, and a minicopter "eye in the sky" gathers better interindividual distance and interactional data. A theodolite operator based onshore could also describe the speed of the group, its orientations and reorientations, measurements of meandering, and other group movement behaviors. These devices have all now been individually tested, but we believe it is the combination of old and new techniques, preferably with as little sound and vessel movement intrusion on the lives of the animals as possible, that will help us unravel ever-more-sophisticated aspects of the social behavior of dolphins.

References

Amos, B., Schlotterer, C., & Tautz, D. (1993). Social structure of pilot whales revealed by analytical DNA profiling. *Science, 260,* 670–672.

Andrews, K. R., Karczmarski, L., Au, W. W. L., Rickards, S. H., Vanderlip, C. A., Bowen, B. W., et al. (2010). Rolling stones and stable homes: Social structure, habitat diversity, and population genetics of the Hawaiian spinner dolphin (*Stenella longirostris*). *Molecular Ecology, 19,* 732–748.

Andrews, K. R., Karczmarski, L., Au, W. W. L., Rickards, S. H., Vanderlip, C. A., & Toonen, R. J. (2006). Patterns of genetic diversity of the Hawaiian spinner dolphin (*Stenella longirostris*). *Atoll Research Bulletin, 543,* 65–73.

Baird, R. W. (2000). The killer whale: Foraging specialties and group hunting. In J. Mann, R. C. Connor, P. L. Tyack, & H. Whitehead (Eds.), *Cetacean societies: Field studies of dolphins and whales.* Chicago: University of Chicago Press.

Baird, R. W., Borsani, J. F., Hanson, M. B., & Tyack, P. L. (2002). Diving and night-time behavior of long-finned pilot whales in the Ligurian Sea. *Marine Ecology Progress Series, 237,* 301–305.

Baird, R. W., & Dill, L. M. (1996). Ecological and social determinants of group size in transient killer whales. *Behavioral Ecology, 7,* 408–416.

Barrett, L., & Würsig, B. (2014). Why dolphins are not aquatic apes. *Animal Behavior and Cognition, 1*, 1–18.

Barrett-Lennard, L. G., Ford, J. K. B., & Heise, K. A. (1996). The mixed blessing of echolocation: Differences in sonar use by fish-eating and mammal-eating killer whales. *Animal Behaviour, 51*, 553–565.

Barros, N. B., & Wells, R. S. (1998). Prey and feeding patterns of resident bottlenose dolphins (*Tursiops truncatus*) in Sarasota Bay, Florida. *Journal of Mammalogy, 79*, 1045–1059.

Bel'kovich, V. M., Ivanova, E. E., Yefremenkova, O. V., Kozarovitsky, L. B., & Kharitonov, S. P. (1991). Searching and hunting behavior in the bottlenose dolphin (*Tursiops truncatus*) in the Black Sea. In K. Pryor & K. S. Norris (Eds.), *Dolphin societies: Discoveries and puzzles*. Berkeley: University of California Press.

Bender, C. E., Herzing, D. L., & Bjorklund, D. F. (2009). Evidence of teaching in Atlantic spotted dolphins (*Stenella frontalis*) by mother dolphins foraging in the presence of their calves. *Animal Cognition, 12*, 43–53.

Benoit-Bird, K. J., & Au, W. W. L. (2003). Prey dynamics affect foraging by a pelagic predator (*Stenella longirostris*) over a range of spatial and temporal scales. *Behavioral Ecology and Sociobiology, 53*, 364–373.

Benoit-Bird, K. J., & Au, W. W. L. (2004). Diel migration dynamics of an island-associated sound-scattering layer. *Deep-Sea Research Part I, 51*, 707–719.

Benoit-Bird, K. J., & Au, W. W. L. (2009a). Cooperative prey herding by a pelagic dolphin, *Stenella longirostris*. *Journal of the Acoustical Society of America, 125*, 539–546.

Benoit-Bird, K. J., & Au, W. W. L. (2009b). Phonation behavior of cooperatively foraging spinner dolphins. *Journal of the Acoustical Society of America, 125*, 125–137.

Benoit-Bird, K. J., Würsig, B., & McFadden, C. J. (2004). Dusky dolphin (*Lagenorhynchus obscurus*) foraging in two different habitats: Active acoustic detection of dolphins and their prey. *Marine Mammal Science, 20*, 215–231.

Bigg, M. A., Olesiuk, P. F., Ellis, G. M., Ford, J. K. B., & Balcomb, K. C. (1990). Social organization and genealogy of resident killer whales (*Orcinus orca*) in the coastal waters of British Columbia and Washington State. *Report of the International Whaling Commission, Special Issue, 12*, 383–405.

Bromham, L., Phillips, M. J., & Penny, D. (1999). Growing up with dinosaurs: Molecular dates and the mammalian radiation. *Trends in Ecology and Evolution, 13*, 113–118.

Buurman, D. (2010). Dolphin swimming and watching: One tourism operator's perspective. In B. Würsig & M. Würsig (Eds.), *The dusky dolphin: Master acrobat off different shores*. Amsterdam: Elsevier/Academic Press.

Caro, T. M., & Hauser, M. D. (1992). Is there teaching in nonhuman animals? *Quarterly Review of Biology, 67*, 151–174.

Childerhouse, S., & Baxter, A. (2010). Human interactions with dusky dolphins: A management perspective. In B. Würsig & M. Würsig (Eds.), *The dusky dolphin: Master acrobat off different shores.* Amsterdam: Elsevier/Academic Press.

Connor, R. C., Heithaus, M. R., & Barre, L. M. (2001). Complex social structure, alliance stability and mating access in a bottlenose dolphin "super-alliance." *Proceedings of the Royal Society of London: Biological Sciences, 268,* 263–267.

Connor, R. C., Mann, J., Tyack, P. L., & Whitehead, H. (1998). Social evolution in toothed whales. *Trends in Ecology & Evolution, 13,* 228–232.

Connor, R. C., & Norris, K. S. (1982). Are dolphins reciprocal altruists? *American Naturalist, 119,* 358–374.

Connor, R. C., Smolker, R. A., & Richards, A. F. (1992). Two levels of alliance formation among male bottlenose dolphins (*Tursiops* sp.). *Proceedings of the National Academy of Sciences of the United States of America, 89,* 987–990.

Connor, R. C., Wells, R. S., Mann, J., & Read, A. J. (2000). The bottlenose dolphin: Social relationships in a fission–fusion society. In J. Mann, R.C. Connor, P.L. Tyack, and H. Whitehead (Eds.), *Cetacean societies: Field studies of dolphins and whales.* Chicago: University of Chicago Press.

Dahlheim, M. E., Shulman-Janiger, A., Black, N., Ternullo, R., Ellifrit, E., & Balcomb, K. C., III. (2008). Eastern temperate North Pacific offshore killer whales (*Orcinus orca*): Occurrence, movements, and insights into feeding ecology. *Marine Mammal Science, 24,* 719–729.

Deutsch, S., Pearson, H., & Würsig, B. (2014). Development of leaps in dusky dolphin (*Lagenorhynchus obscurus*) calves. *Behaviour, 151,* 1555–1577.

Dudzinski, K. M. (1998). Contact behavior and signal exchange among Atlantic spotted dolphins (*Stenella frontalis*). *Aquatic Mammals, 24,* 129–142.

Dudzinski, K. M., Clark, C. W., & Würsig, B. (1995). A mobile video/acoustic system for simultaneously recording dolphin behavior and vocalizations underwater. *Aquatic Mammals, 21,* 187–193.

Dunbar, R. I. (2003). The social brain: Mind, language, and society in evolutionary perspective. *Annual Review of Anthropology, 32,* 163–181.

Dunbar, R. I. M., & Shultz, S. (2007). Understanding primate brain evolution. *Philosophical Transactions of the Royal Society of London: Biological Sciences, 362,* 649–658.

Eberhard, W. G. (1996). *Female control: Sexual selection by cryptic female choice.* Princeton: Princeton University Press.

Elliser, C. R., & Herzing, D. L. (2014). Long-term social structure of a resident community of Atlantic spotted dolphins, *Stenella frontalis*, in the Bahamas, 1991–2002. *Marine Mammal Science, 30,* 308–328.

Emlen, S. T., & Oring, L. W. (1977). Ecology, sexual selection, and the evolution of mating systems. *Science, 197,* 215–223.

Ford, J. K. B. (1991). Vocal traditions among resident killer whales (*Orcinus orca*) in coastal waters of British Columbia. *Canadian Journal of Zoology, 69,* 1454–1483.

Ford, J. K. B., Ellis, G. M., Barrett-Lennard, L. G., Morton, A. B., Palm, R. S., & Balcomb, K. C., III. (1998). Dietary specialization in two sympatric populations of killer whales (*Orcinus orca*) in coastal British Columbia and adjacent waters. *Canadian Journal of Zoology, 76,* 1456–1471.

Fordyce, R. E. (2009). Cetacean evolution. In W. F. Perrin, B. Würsig, & J. G. M. Thewissen (Eds.), *Encyclopedia of marine mammals* (2nd ed., pp. 201–207). Amsterdam: Elsevier/Amsterdam Press.

Foster, E. A., Franks, D. W., Mazzi, S., Darden, S. K., Balcomb, K. C., Ford, J. K. B., et al. (2012). Adaptive prolonged postreproductive life span in killer whales. *Science, 337,* 1313.

Foster, K. R., & Wenseleers, T. (2006). A general model for the evolution of mutualism. *Journal of Evolutionary Biology, 19,* 1283–1293.

Gaskin, D. E. (1968). Distribution of Delphinidae (Cetacea) in relation to sea surface temperatures off eastern and southern New Zealand. *New Zealand Journal of Marine and Freshwater Research, 2,* 527–534.

Gazda, S. K., Connor, R. C., Edgar, R. K., & Cox, F. (2005). A division of labour with role specialization in group-hunting bottlenose dolphins (*Tursiops truncatus*) off Cedar Key, Florida. *Proceedings of the Royal Society of London: Biological Sciences, 272,* 135–140.

Gibson, Q. A., & Mann, J. (2008). The size, composition and function of wild bottlenose dolphin (*Tursiops* sp.) mother-calf groups in Shark Bay, Australia. *Animal Behaviour, 76,* 389–405.

Gowans, S., Würsig, B., & Karczmarski, L. (2008). The social structure and strategies of delphinids: Predictions based on an ecological framework. *Advances in Marine Biology, 53,* 195–294.

Guinet, C. (1991). Intentional stranding apprenticeship and social play in killer whales (*Orcinus orca*). *Canadian Journal of Zoology, 69,* 2712–2716.

Guinet, C., & Bouvier, J. (1995). Development of intentional stranding hunting techniques in killer whale (*Orcinus orca*) calves at Crozet Archipelago. *Canadian Journal of Zoology, 73,* 27–33.

Heimlich-Boran, J. R. (1993). *Social organisation of the short-finned pilot whale, Globicephala macrorhynchus, with special reference to the comparative social ecology of Delphinids.* Doctoral dissertation, University of Cambridge, Cambridge.

Herzing, D. L. (1996). Vocalizations and associated underwater behavior of free-ranging Atlantic spotted dolphins, *Stenella frontalis*, and bottlenose dolphins, *Tursiops truncatus. Aquatic Mammals, 22,* 61–79.

Herzing, D. L. (1997). The life history of free-ranging Atlantic spotted dolphins (*Stenella frontalis*): Age classes, color phases, and female reproduction. *Marine Mammal Science, 13,* 576–595.

Herzing, D. L. (2000). Acoustics and social behavior of wild dolphins: Implications for a sound society. In W. W. L. Au, A. N. Popper, & R. R. Fay (Eds.), *Hearing by whales and dolphins*. New York: Springer.

Herzing, D. L., & Brunnik, B. J. (1997). Coefficients of association of reproductively active female Atlantic spotted dolphins, *Stenella frontalis*. *Aquatic Mammals, 23*, 155–162.

Hoelzel, A. R. (1991). Killer whale predation on marine mammals at Punta Norte, Argentina: Food sharing, provisioning and foraging strategy. *Behavioral Ecology and Sociobiology, 29*, 197–204.

Hoelzel, A. R., Hey, J., Dahlheim, M. E., Nicholson, C., Burkanov, V., & Black, N. (2007). Evolution of population structure in a highly social top predator, the killer whale. *Molecular Biology and Evolution, 24*, 1407–1415.

Hoelzel, A. R., Dahlheim, M., & Stern, S. J. (1998). Low genetic variation among killer whales (*Orcinus orca*) in the Eastern North Pacific and genetic differentiation between foraging specialists. *Journal of Heredity, 89*, 121–128.

Humphrey, N. K. (1976). The social function of intellect. In P. P. G. Bateson & R. A. Hinde (Eds.), *Growing points in ethology: Based on a conference sponsored by St. John's College and King's College*. Cambridge: Cambridge University Press.

Jarman, P. J. (1974). The social organisation of antelope in relation to their ecology. *Behaviour, 48*, 215–267.

Johnson, C. M. (2010). Observing cognitive complexity in primates and cetaceans. *International Journal of Comparative Psychology, 23*, 587–624.

Johnson, C. M., & Norris, K. S. (1986). Delphinid social organization and social behavior. In R. J. Schusterman, J. A.Thomas, & F. G. Wood (Eds.), *Dolphin cognition and behavior: A comparative approach*. London: Lawrence Erlbaum Associates.

Jolly, A. (1966). Lemur social behavior and primate intelligence. *Science, 153*, 501–506.

Karczmarski, L., Würsig, B., Gailey, G., Larson, K. W., & Vanderlip, C. (2005). Spinner dolphins in a remote Hawaiian atoll: Social grouping and population structure. *Behavioral Ecology, 16*, 675–685.

Kasuya, T., & Marsh, H. (1984). Life history and reproductive biology of the short-finned pilot whale, *Globicephala macrorhynchus*, off the Pacific coast of Japan. *Report of the International Whaling Commission. Special Issue, 6*, 259–310.

Krebs, J. R., & Davies, N. B. (1993). *An introduction to behavioral ecology*. Oxford: Blackwell Scientific Publications.

Leatherwood, S. (1975). Some observations of feeding behavior of bottle-nosed dolphins (*Tursiops truncatus*) in the Northern Gulf of Mexico and (*Tursiops* cf *T. gilli*) off Southern California, Baja California, and Nayarit, Mexico. *Marine Fisheries Review, 37*, 10–16.

Lonsdorf, E. V. (2006). What is the role of mothers in the acquisition of termite-fishing behaviors in wild chimpanzees (*Pan troglodytes schweinfurthii*)? *Animal Cognition, 9*, 36–46.

Lonsdorf, E. V., Pusey, A. E., & Eberly, L. (2004). Sex differences in learning in chimpanzees. *Nature, 428*, 715–716.

Lopez, J. C., & Lopez, D. (1985). Killer whales (*Orcinus orca*) of Patagonia, and their behavior of intentional stranding while hunting nearshore. *Journal of Mammalogy, 66*, 181–183.

Lundquist, D., Gemmell, N., & Würsig, B. (2012). Behavioural responses of dusky dolphin (*Lagenorhynchus obscurus*) groups to tour vessels off Kaikoura, New Zealand. *PLoS ONE, 7*, e41969.

Lusseau, D., Schneider, K., Boisseau, O. J., Haase, P., Slooten, E., & Dawson, S. M. (2003). The bottlenose dolphin community of Doubtful Sound features a large proportion of long-lasting associations. *Behavioral Ecology and Sociobiology, 54*, 396–405.

Mann, J. (1999). Behavioral sampling methods for cetaceans: A review and critique. *Marine Mammal Science, 15*, 102–122.

Mann, J. (2006). Establishing trust: Socio-sexual behaviour and the development of male-male bonds among Indian Ocean bottlenose dolphins. In V. Sommer & P. L. Vassey (Eds.), *Homosexual behaviour in animals: An evolutionary perspective*. Cambridge: Cambridge University Press.

Mann, J., Connor, R. C., Barre, L. M., & Heithaus, M. R. (2000). Female reproductive success in bottlenose dolphins (*Tursiops* sp.): Life history, habitat, provisioning, and group-size effects. *Behavioral Ecology, 11*, 210–219.

Mann, J., & Patterson, E. M. (2013). Tool use by aquatic animals. *Proceedings of the Royal Society of London: Biological Sciences, 368*.

Mann, J., & Sargeant, B. (2003). Like mother, like calf: The ontogeny of foraging traditions in wild Indian Ocean bottlenose dolphins (*Tursiops* sp.). In D. Fragaszy & S. Perry (Eds.), *The biology of traditions*. Cambridge: Cambridge University Press.

Mann, J., Sargeant, B. L., Watson-Capps, J. J., Gibson, Q. A., Heithaus, M. R., Connor, R. C., et al. (2008). Why do dolphins carry sponges? *PLoS ONE, 3*, e3868.

Mann, J., & Smuts, B. (1999). Behavioral development in wild bottlenose dolphin newborns (*Tursiops* sp.). *Behaviour, 136*, 529–566.

Mann, J., Stanton, M. A., Patterson, E. M., Bienenstock, E. J., & Singh, L. O. (2012). Social networks reveal cultural behavior in tool-using dolphins. *Nature Communications, 3*, 980. doi:10.1038/ncomms1983.

Marino, L. (1996). What can dolphins tell us about primate evolution? *Evolutionary Anthropology, 5*, 81–86.

Markowitz, T. M. (2004). *Social organization of the New Zealand dusky dolphin*. Doctoral dissertation, Texas A&M University, College Station.

Markowitz, T. M., Harlin, A. D., Würsig, B., & McFadden, C. J. (2004). Dusky dolphin foraging habitat: Overlap with aquaculture in New Zealand. *Aquatic Conservation: Marine and Freshwater Ecosystems, 14,* 133–149.

Markowitz, T. M., Markowitz, W., & Morton, L. M. (2010). Mating habits of New Zealand dusky dolphins. In B. Würsig & M. Würsig (Eds.), *The dusky dolphin: Master acrobat off different shores.* Amsterdam: Elsevier/Academic Press.

Marten, K., Herzing, D. L., Poole, M., & Newman-Allman, K. (2001). The acoustic predation hypothesis: Linking underwater observations and recordings during odontocete predation and observing the effects of loud impulsive sounds on fish. *Aquatic Mammals, 27,* 56–66.

Mesnick, S. L., & Ralls, K. (2009). Mating systems. In W. F. Perrin, B. Würsig, & J. G. M. Thewissen (Eds.), *The encyclopedia of marine mammals* (2nd ed.). Amsterdam: Elsevier/Academic Press.

Morin, P. A., Archer, F. I., Foote, A. D., Vilstrup, J., Allen, E. E., Wade, P., et al. (2010). Complete mitochondrial genome indicates multiple species. *Genome Research, 20,* 908–916.

Norris, K. S., & Dohl, T. P. (1980a). The structure and function of cetacean schools. In L. M. Herman (Ed.), *Cetacean behavior: Mechanisms and functions.* New York: Wiley Interscience.

Norris, K. S., & Dohl, T. P. (1980b). Behavior of the Hawaiian spinner dolphin, *Stenella longirostris. Fish Bulletin, 77,* 821–849.

Norris, K. S., & Pryor, K. (1991). Essay: Some thoughts on grandmothers. In K. Pryor & K. S. Norris (Eds.), *Dolphin societies: Discoveries and puzzles.* Berkeley: University of California Press.

Norris, K. S., & Schilt, C. R. (1988). Cooperative societies in three-dimensional space: On the origins of aggregations, flocks, and schools, with special reference to dolphins and fish. *Ethology and Sociobiology, 9,* 149–179.

Norris, K. S., Würsig, B., Wells, R. S., & Würsig, M. (1994). *The spinner dolphin.* Berkeley: University of California Press.

Nowacek, D. (2002). Sequential foraging behaviour of bottlenose dolphins, *Tursiops truncatus,* in Sarasota Bay, FL. *Behaviour, 139*(9), 1125–1145.

Orbach, D., Packard, J., & Würsig, B. (2014). Mating group size in dusky dolphins (*Lagenorhynchus obscurus*): Costs and benefits of scramble competition. *Ethology, 120,* 804–815.

Partridge, B. L. (1982). Structure and function of fish schools. *Scientific American, 246,* 114–123.

Patterson, E. M., & Mann, J. (2011). The ecological conditions that favor tool use and innovation in wild bottlenose dolphins (*Tursiops* sp.). *PLoS ONE, 6,* e22243.

Pearson, H. C. (2009). Influences on dusky dolphin fission–fusion dynamics in Admiralty Bay, New Zealand. *Behavioral Ecology and Sociobiology, 63,* 1437–1446.

Pearson, H. C. (2011). Sociability of female bottlenose dolphins (*Tursiops* sp.) and chimpanzees (*Pan troglodytes*): Understanding evolutionary pathways toward social convergence. *Evolutionary Anthropology, 20,* 85–95.

Pearson, H. C., & Shelton, D. E. (2010). *A large-brained social animal*. In B. Würsig & M. Würsig (Eds.), *The dusky dolphin: Master acrobat off different shores*. Amsterdam: Elsevier/Academic Press.

Pearson, H. C., Vaughn-Hirshorn, R. L., Srinivasan, M., & Würsig, B. (2012). Avoidance of mussel farms by dusky dolphins (*Lagenorhynchus obscurus*) in New Zealand. *New Zealand Journal of Marine and Freshwater Research, 46*, 567–574.

Perrin, W. F. (1975). Variation of spotted and spinner porpoise (genus *Stenella*) in the eastern tropical Pacific and Hawaii. *Bulletin of the Scripps Institute of Oceanography, 21*.

Perrin, W. F., & Mesnick, S. L. (2003). Sexual ecology of the spinner dolphin, *Stenella longirostris*: Geographic variation in mating system. *Marine Mammal Science, 19*, 462–483.

Pilot, M., Dahlheim, M. E., & Hoelzel, A. R. (2010). Social cohesion among kin, gene flow without dispersal and the evolution of population genetic structure in the killer whale (*Orcinus orca*). *Journal of Evolutionary Biology, 23*, 20–31.

Pryor, K., Lindbergh, J., & Milano, R. (1990). A dolphin-human fishing cooperative in Brazil. *Marine Mammal Science, 6*, 77–82.

Pryor, K., & Norris, K. S. (1991). *Dolphin societies: Discoveries and puzzles*. Berkeley: University of California Press.

Psarakos, S., Herzing, D. L., & Marten, K. (2003). Mixed species associations between pantropical spotted dolphins (*Stenella attenuata*) and Hawaiian spinner dolphins (*Stenella longirostris*) off Oahu, Hawaii. *Aquatic Mammals, 29*, 390–395.

Rendell, L., & Whitehead, H. (2001). Culture in whales and dolphins. *Behavioral and Brain Sciences, 24*, 309–382.

Sargeant, B. L., Mann, J., Berggren, P., & Krutzen, M. (2005). Specialization and development of beach hunting, a rare foraging behavior, by wild bottlenose dolphins (*Tursiops* sp.). *Canadian Journal of Zoology, 83*, 1400–1410.

Saulitis, E., Matkin, C., Barrett-Lennard, L., Eise, K., & Ellis, G. (2000). Foraging strategies of sympatric killer whale (*Orcinus orca*) populations in Prince William Sound, Alaska. *Marine Mammal Science, 16*, 74–107.

Saulitis, E. L., Matkin, C. O., & Fay, F. H. (2005). Vocal repertoire and acoustic behavior of the isolated AT1 killer whale subpopulation in southern Alaska. *Canadian Journal of Zoology, 83*, 1015–1029.

Schotten, M., Lammers, M. O., Sexton, K., & Au, W. W. L. (2005). Application of a diver-operated 4-channel acoustic/video recording device to study wild dolphin echolocation and communication. *Journal of the Acoustical Society of America, 117*, 2552.

Shelton, D. E., Harlin-Cognato, A. D., Honeycutt, R. L., & Markowitz, T. M. (2010). Sexual segregation and genetic relatedness in New Zealand. In B. Würsig & M. Würsig (Eds.), *The dusky dolphin: Master acrobat off different shores*. Amsterdam: Elsevier/Academic Press.

Sherman, P. W. (1998). The evolution of menopause. *Nature, 392*, 759–761.

Smolker, R., Richards, A., Connor, R., Mann, J., & Berggren, P. (1997). Sponge carrying by dolphins (Delphinidae, *Tursiops* sp.): A foraging specialization involving tool use? *Ethology, 103*, 454–465.

Smolker, R. A., Richards, A. F., Connor, R. C., & Pepper, J. W. (1992). Sex differences in patterns of association among Indian Ocean bottlenose dolphins. *Behaviour, 123*, 38–69.

Srinivasan, M., Grant, W. E., Swannack, M., & Ragan, J. (2010). Behavioral games involving a clever prey avoiding a clever predator: An individual-based model of dusky dolphins and killer whales. *Ecological Modelling, 22*, 2687–2698.

Srinivasan, M., & Markowitz, T. M. (2010). Predator threats and dusky dolphin survival strategies. In B. Würsig & M. Würsig (Eds.), *The dusky dolphin: Master acrobat off different shores*. Amsterdam: Elsevier/Academic Press.

Srinivasan, M., Pearson, H. C., Vaughn-Hirshorn, R. L., Würsig, B., & Murtugudde, R. (2012). Using climate downscaling to hypothesize impacts on apex predator marine ecosystem dynamics. *New Zealand Journal of Marine and Freshwater Research, 46*, 575–584.

Tayler, C. K., & Saayman, G. S. (1972). The social organisation and behaviour of dolphins (*Tursiops aduncus*) and baboons (*Papio ursinus*): Some comparisons and assessments. *Annals of the Cape Provincial Museums: Natural History, 9*, 11–49.

Trivers, R. L. (1971). The evolution of reciprocal altruism. *Quarterly Review of Biology, 46*, 35–57.

van Schaik, C. P., & Deaner, R. O. (2003). Life history and cognitive evolution in primates. In F. B. M. de Waal & P. L. Tyack (Eds.), *Animal social complexity: Intelligence, culture, and individualized societies*. Cambridge, MA: Harvard University Press.

Vaughn, R. L., Degrati, M., & McFadden, C. J. (2010). Dusky dolphins foraging in daylight. In B. Würsig & M. Würsig (Eds.), *The dusky dolphin: Master acrobat off different shores*. Amsterdam: Elsevier/Academic Press.

Vaughn, R. L., Muzi, E., Richardson, J. L., & Würsig, B. (2011). Dolphin bait-balling behaviors in relation to prey ball escape behaviors. *Ethology, 117*, 859–871.

Vaughn, R. L., Shelton, D. E., Timm, L. L., Watson, L. A., & Würsig, B. (2007). Dusky dolphin (*Lagenorhynchus obscurus*) feeding tactics and multi-species associations. *New Zealand Journal of Marine and Freshwater Research, 41*, 391–400.

Vaughn, R., Würsig, B., & Packard, J. (2010). Dolphin prey herding: Prey ball mobility relative to dolphin group and prey ball sizes, multispecies associates, and feeding duration. *Marine Mammal Science, 26*, 213–225.

Vaughn, R. L., Würsig, B., Shelton, D. S., Timm, L. L., & Watson, L. A. (2008). Dusky dolphins influence prey accessibility for seabirds in Admiralty Bay, New Zealand. *Journal of Mammalogy, 89*, 1051–1058.

Weir, J. S. (2008). *Dusky dolphin nursery groups off Kaikoura, New Zealand.* Master's thesis, Texas A&M University, College Station.

Weir, J., Deutsch, S., & Pearson, H. C. (2010). Dusky dolphin calf rearing. In B. Würsig & M. Würsig (Eds.), *The dusky dolphin: Master acrobat off different shores.* Amsterdam: Elsevier/Academic Press.

Wells, R. S. (2003). Dolphin social complexity: Lessons from long-term study and life history. In F. B. M. de Waal & P. L. Tyack (Eds.), *Animal social complexity: Intelligence, culture, and individualized societies.* Cambridge, MA: Harvard University Press.

Wells, R. S., Irvine, A. B., & Scott, M. D. (1980). The social ecology of inshore odontocetes. In L. M. Herman (Ed.), *Cetacean behavior: Mechanisms and functions.* New York: Wiley-Interscience.

Whitehead, H., Rendell, L., Osborne, R. W., & Würsig, B. (2004). Culture and conservation of non-humans with reference to whales and dolphins: Review and new directions. *Biological Conservation, 120,* 427–437.

Wilke, M., Bossley, M., & Doak, W. (2005). Managing human interactions with solitary dolphins. *Aquatic Mammals, 31,* 427–433.

Wrangham, R. W. (1979). On the evolution of ape social systems. *Social Sciences Information, 18,* 335–368.

Würsig, B. (1986). Delphinid foraging strategies. In R. J. Schusterman, J. A. Thomas, & F. G. Wood (Eds.), *Dolphin cognition and behavior: A comparative approach.* Hillsdale: Lawrence Erlbaum Publishers.

Würsig, B. (1989). Cetaceans. *Science, 244,* 1550–1557.

Würsig, B., Duprey, N., & Weir, J. (2007). Dusky dolphins (*Lagenorhynchus obscurus*) in New Zealand waters: Present knowledge and research goals. *DOC Research and Development Series, 270,* 1–28.

Würsig, B., & Pearson, H. C. (2014). Dusky dolphins: Flexibility in foraging and social strategies. In J. Yamagiwa & L. Karczmarski (Eds.), *Primates and cetaceans: Field research and conservation of complex mammalian societies.* New York: Springer.

Würsig, B., Wells, R. S., Norris, K. S., & Würsig, M. (1994). A spinner dolphin's day. In K. S. Norris, B. Würsig, R. S. Wells, & M. Würsig (Eds.), *The Hawaiian spinner dolphin.* Berkeley: University of California Press.

Würsig, B., & Würsig, M. (1977). The photographic determination of group size, composition, and stability of coastal porpoises (*Tursiops truncatus*). *Science, 198,* 755–756.

Würsig, B., & Würsig, M. (1979). Behavior and ecology of the bottlenose dolphin, *Tursiops truncatus,* in the South Atlantic. *U.S. Fisheries Bulletin, 77,* 399–412.

Würsig, B., & Würsig, M. (1980). Behavior and ecology of the dusky dolphin, *Lagenorhynchus obscurus,* in the South Atlantic. *U.S. Fisheries Bulletin, 77,* 871–890.

Würsig, B., Würsig, M., & Cipriano, F. (1989). Dolphins in different worlds. *Oceanus, 32,* 71–75.

5 Analyzing the Acoustic Communication of Dolphins

Marc O. Lammers and Julie N. Oswald

Introduction

The sea can be a vast place in which to live, both horizontally, with tens to thousands of kilometers separating physical boundaries, and vertically, with depths ranging from a few meters to almost eleven kilometers. The sea is also largely dark and opaque. Light from the surface reaches to a depth of only about 200 meters (in clear waters), and depending on turbidity, the range of visibility is restricted to between centimeters and tens of meters. Such large spatial scales and limits on visual perception present a sensory and communicative challenge for aquatic animals such as dolphins, which disperse widely to find food, mates, and protection from predators but must also stay in contact with other individuals to share information, synchronize activities, and coordinate behaviors.

In contrast to vision, the use of sound for signaling and sensing is quite effective in the aquatic environment. Due to water's higher density, sound travels about 4.5 times faster in water than in air, making the acoustic modality well suited for signaling over long ranges. However, sound transmission in the aquatic world is also complex. Changes in pressure, temperature, and salinity each affect the propagation of sound, and these characteristics vary with depth, latitude, and local oceanographic features (e.g., currents, eddies, upwelling fronts). As a result, even without physical barriers or obstructions, regions form with variable sound propagation characteristics, including channels of enhanced or diminished sound transmission.

In this chapter, we explore the use of sound by dolphins for communication. We do so mindful that the acoustic signaling of these species is adapted to both exploit and overcome the advantages and complexities of the aquatic world. We explore both the history and the state of the art of research on this complex topic and consider some of the challenges that we, as terrestrial mammals, face in our investigations, as well as the solutions that are available to us. Finally, we look toward the future and project where we might go next in our search for answers.

A Complex Problem

We face a host of challenges in our studies of dolphin acoustic signaling, not least of which is our own sensory bias toward air-adapted, terrestrial living. Simply put, as humans we are remarkably limited in our ability to hear, observe, and move in the aquatic environment compared to our study subjects. We lose approximately 30 decibels (dB) (about a 32-fold reduction) of hearing sensitivity when we enter the water, and the morphology and spacing of our ears relative to the wavelengths of aquatic sounds make it nearly impossible for us to localize sound underwater unaided by technology. These handicaps make it extremely difficult to examine dolphin sounds in the context of their associated behaviors, as we might for terrestrial mammals, birds, or insects. Thus direct observation and inference of the relationships among sounds, individuals, and their actions, which form the basis for learning about animal communication, present major challenges with respect to studying dolphins and other aquatic animals.

Another hurdle we must overcome is the disparity in the frequency sensitivities in the hearing of dolphins and humans. The auditory range of most human adults ranges from approximately 20 Hz to 15 kHz, with a maximal sensitivity between 1 and 6 kHz (Sivian & White, 1933). In comparison, a healthy bottlenose dolphin (*Tursiops truncatus*) can typically hear sounds from 75 Hz to 150 kHz, with the best sensitivity between 15 and 110 kHz (Johnson, 1967). This is also the frequency range of the majority of the signals produced by most dolphin species, so many of their signals lie well above the range of human hearing. Therefore, in addition to having poor sensitivity underwater, human hearing is tuned to very different frequencies than both the hearing and the signaling of most dolphins.

Finally, less quantifiable, but perhaps most challenging, are the hurdles we must overcome related to the ecological, behavioral, and cognitive differences that exist between humans and dolphins. Although we may share certain similarities as a result of being social mammals, humans and dolphins occupy very different ecological niches and live in distinct sensory worlds. Therefore we have to assume that substantial differences exist in the way humans and dolphins perceive and communicate about their world. Effectively studying dolphin communication requires that we also conceptually understand their *umwelt*, or cognitive frame of reference, which is no easy task.

Early Investigations and Dolphin "Language"

Modern scientific interest in dolphin acoustics began in the 1940s, when the first qualitative reports of dolphin sounds were published (McBride & Hebb, 1948; Schevill & Lawrence, 1949). In the 1950s, researchers hypothesized (Kellogg, Kohler, & Morris, 1953; McBride, 1956; Schevill & Lawrence, 1956) and later experimentally demonstrated that dolphins use echolocation for sensing and navigating (Norris, Prescott, Asa-Dorian, & Perkins, 1961; Kellogg, 1958). Around this time, the first descriptions of dolphin communication signals and associated behaviors were published (Essapian, 1953; Wood, 1953), followed by acoustic communication studies in the 1960s (Dreher, 1961; Lang & Smith, 1965; Lilly & Miller, 1961a, 1961b). These early researchers recognized that dolphins produce stereotyped tonal sounds, or whistles, when alarmed or under stress (Dreher & Evans, 1964; Lilly, 1963). Subsequent experiments showed that isolated animals produced individually distinctive whistles, leading to the "signature whistle" hypothesis (Caldwell & Caldwell, 1965). However, the large brains of these animals and their complex acoustic repertoire also led to much further speculation, and by the late 1960s, a scientific debate was under way about the communicative function of dolphin whistles, including whether a dolphin "language" existed, equivalent to that of humans (Caldwell & Caldwell, 1968; Dreher, 1966; Lilly, 1967); if dolphins could be taught to speak English with sufficient training (Lilly, 1967); and whether two-way communication was possible with humans (Lilly, 1961). These types of questions continued to receive attention for many years, with most discussions centering primarily on two issues: whether the system of communication used by dolphins has the attributes of "true" language (Herman & Tavolga, 1980), and whether dolphins have the cognitive capacity to learn and apply an artificial or human-mediated language (Batteau & Markey, 1967; Herman, Richards, & Wolz, 1984; Sigurdson, 1993). The question of whether dolphins possess or are capable of acquiring language persists to this day, both scientifically and in popular culture. Perhaps because science is poorly suited to prove a negative, or because methodological hurdles continue to hinder researchers investigating this question, the debate is likely to continue for some time to come.

The Role of Technology in Driving Advances

The methodological challenges associated with studying dolphin communication previously outlined require the application of cutting-edge technological tools. Where human senses fall short, high-bandwidth hydrophones and recorders must be used to capture the sounds produced by dolphins. Until the 1990s and the onset of the digital age, this was accomplished by using large tape recorders that could spin at rates high enough to record ultrasonic sounds. However, these recorders were expensive and not well suited for making field recordings because they lacked portability and were limited to only a few minutes of recording at a time. In addition, before computers became widely available, methods for analyzing signals were often convoluted. Recordings had to be slowed down to bring sounds into the human audible range, viewed on oscilloscope traces, or transcribed onto paper sonograms using spectral analyzers (e.g.,

Caldwell & Caldwell, 1968; Lilly & Miller, 1961a; Watkins, 1966). Not surprisingly, most early efforts to study dolphin communication were conducted on captive animals (Batteau & Markey, 1967; Caldwell & Caldwell, 1972; Lang & Smith, 1965; Lilly & Miller, 1961b) and were limited to descriptions of signal characteristics (Evans & Prescott, 1962; Powell, 1966; Watkins & Schevill, 1971). However, there were exceptions, including some efforts to overcome methodological challenges in the field. For example, to match sounds with the behavior of free-ranging animals, Watkins and Schevill (1974) used a three-dimensional hydrophone array to acoustically localize spinner dolphins in Kealakekua Bay, Hawaii. This effort resulted in a description of "burst pulse" exchanges between dolphins (discussed further hereafter) but also highlighted the challenge of matching dolphin sounds and behaviors in the field.

As the importance of whistles in dolphin communication became clear, and partly because of the challenge of obtaining broadband recordings in the field, many researchers opted for the more widely available (and affordable) audio recorders designed for popular use. These were typically restricted in bandwidth to between 15 and 22 kHz, but they allowed longer recordings to be made, were easily taken into the field, and were sufficient for capturing the fundamental components of dolphin whistles. As a result, fuller descriptions of dolphin signaling patterns began to emerge in the in the 1980s and 1990s, albeit focused largely on whistles and sounds occurring in the human-audible frequency range (e.g., Herzing, 1996; Overstrom, 1983; Sayigh, Tyack, Wells, & Scott, 1990; Steiner, 1981; Tyack, 1986), with some exceptions (e.g., Dawson, 1991; Evans, Awbrey, & Hackbarth, 1988). Also, technological advances in computing made new experimental paradigms possible, allowing researchers to explore the boundaries of dolphins' signal production and mimicking abilities (Reiss & McCowan, 1993; Richards, Wolz, & Herman, 1984; Sigurdson, 1993).

Around the same time, efforts were also made to address other challenges, such as resolving the identity of signaling animals (Tyack, 1985). This had been (and continues to be) a persistent problem for many researchers investigating the exchanges of signals among individuals, leading to ambiguity in the interpretation of signaling patterns. Tyack's (1985) "vocalight," which was placed on the animal's head and would light up when the animal produced a sound, allowed the identification of whistling dolphins in captivity and helped provide further evidence for the role of whistles as signature signals (Tyack, 1986). The technology also led to more sophisticated microcontroller-based data loggers that could be placed on wild dolphins (Tyack & Recchia, 1991), which became the prototypes of modern-day acoustic tags that are placed on animals to record their movement and acoustic signaling (e.g., the D-tag; Johnson and Tyack, 2003).

The beginning of the digital age in the 1990s ushered in both dramatic advances in technological solutions and more sophisticated ways of analyzing acoustic data. Bandwidth limitations on field recordings of dolphin signals were overcome as computer-controlled analog-to-digital converters quickly evolved to become faster and smaller (Au, Lammers, & Aubauer, 1999). These led to fuller descriptions of the signaling repertoire of free-ranging dolphins (Lammers, Au, & Herzing, 2003) and the ability to routinely use hydrophone arrays for localizing signaling animals (Janik, 2000; Janik, Van Parijs, & Thompson, 2000; Lammers & Au, 2003; Lammers, Schotten, & Au, 2006; Miller, 2002; Miller & Tyack, 1998; Rankin, Oswald, & Barlow, 2008). Faster computing speeds also paved the way to more complex analyses of signals and signaling patterns, including signal warping (Buck & Tyack, 1993), the use of cluster analyses (McCowan, 1995), neural networks (Murray, Mercado, & Roitblat, 1998), and information theory (McCowan, Hanser, & Doyle, 1999), to name only a few.

Today research on dolphin communication is less constrained by technological limitations and more by other factors, including access to funding and facilities able to support the work needed to resolve important questions. Nevertheless researchers continue to push forward and apply novel solutions to familiar challenges.

Dolphin Whistles

The most commonly investigated communication sounds produced by dolphins are whistles. Whistles are continuous, narrow-band, frequency-modulated signals produced by many, but not all, delphinid species (May-Collado, Agnarsson, & Wartzok, 2007a). Whistles can range in duration from tens of milliseconds to several seconds (Tyack & Clark, 2000). They are composed of a fundamental frequency contour and often one or more harmonics. Some whistles are very simple, with few or no inflections, while others are more complex, with multiple inflections, breaks, and steps in the contour (fig. 5.1). Most dolphin species produce whistles with fundamental frequencies between 2 and 20 kHz, but whistles extending beyond this range have been reported for several species, including spinner dolphins (*Stenella longirostris*), Atlantic spotted dolphins (*S. frontalis*), striped dolphins (*S. coeruleoalba*), white-beaked dolphins (*Lagenorhynchus albirostris*), Guiana dolphins (*Sotalia guianensis*), and killer whales (*Orcinus orca*) (Lammers et al., 2003; Rasmussen & Miller, 2002; Oswald, Rankin, & Barlow, 2004; Samarra et al., 2010; May-Collado & Wartzok, 2009).

The shape, frequency content, and duration of whistles can be highly variable both within species and within the repertoire of individuals. While some species produce whistles that are distinctive, such as the simple, low-frequency whistles produced by

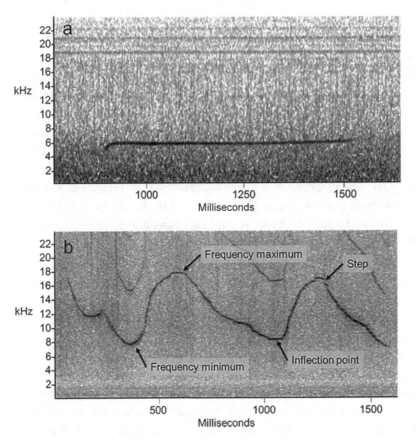

Figure 5.1
Example of a simple whistle (a) produced by a false killer whale, and a complex whistle (b) produced by a bottlenose dolphin. Several features commonly measured from whistles are labeled in (b).

false killer whales (*Pseudorca crassidens*), many other species produce highly variable whistles that often overlap in time and frequency, making them difficult to distinguish from those of other species (Oswald, Rankin, Barlow, & Lammers, 2007). There is a correlation between the average body size of species and the fundamental frequency of their whistles, with larger dolphins generally producing lower-frequency whistles (Matthews, Rendell, Gordon, & Macdonald, 1999; May-Collado, Agnarsson, & Wartzok, 2007b; Wang, Würsig, & Evans, 1995). It has also been suggested that whistle production may be related to the degree of gregariousness in odontocete species (Herman & Tavolga, 1980; Oswald, Rankin, & Barlow, 2008). However, while correlations have been observed between increased whistle modulation and both the average group size and the social structure of different species, not all odontocete species that are social produce whistles (May-Collado et al., 2007a; Oswald et al., 2008).

Analyzing Whistles

Most efforts to understand the role of whistles in dolphin communication have focused on quantifying the characteristics of the fundamental frequency contour. One common method for describing whistles is to qualitatively classify them based on their contour shape. Common shape categories include upsweep, downsweep, convex, concave, and sinusoid (e.g., Acevedo & Van Sluys, 2005; Bazua-Duran & Au, 2002; Lopez, 2011). Such whistle categories can be defined broadly, accepting a large amount of variation within each, or narrowly, with new subcategories defined to encompass differences in similar contours. The amount of variation accepted in categories is often subjective and can differ considerably between researchers. For example, an upswept whistle may have an inflection at the beginning or end. This whistle might be lumped into a broad upsweep category or split into a more narrowly defined group based on the location and degree of inflection. Figure 5.2 illustrates the ambiguity that can be encountered. The whistles shown span a continuum of contour shapes ranging from concave to upsweep to convex with intermediary forms between categories. Ultimately the decision of whether to lump or split categories depends on the question being examined and on the tendencies of the researcher. Universally accepted guidelines for manually classifying or characterizing whistles do not presently exist, often making it difficult to compare results among studies.

A less subjective method for describing whistles is to make quantitative measurements of the whistle contour. Among the variables that are commonly measured from whistle contours are the whistle's frequency minima and maxima, beginning and ending frequencies, duration, the number of steps, and number of inflection points (see fig. 5.1; Lopez, 2011; Matthews et al., 1999; Oswald et al., 2007; Rendell, Matthews, Gill, Gordon, & Macdonald, 1999). Several automated methods have been used to characterize both whistle contours (e.g., Buck & Tyack, 1993; Roch et al., 2011) and whistle repertoires based on measurements of the contour shape (Brown & Miller, 2007; Deecke & Janik, 2006; Janik, 1999; McCowan, 1995). Various statistical methods have also been applied for grouping whistles based on measured variables (e.g., classification tree analysis, hidden Markov models, artificial neural networks, discriminant function analysis, to name only a few) (Brown & Miller, 2007; Deecke & Janik, 2006; Fristrup & Watkins, 1993; Oswald, Barlow, & Norris, 2003; Roch, Soldevilla, Burtenshaw, Henderson, & Hildebrand, 2007). However, while automating the measurement and categorization process does remove subjectivity, a certain amount of ambiguity

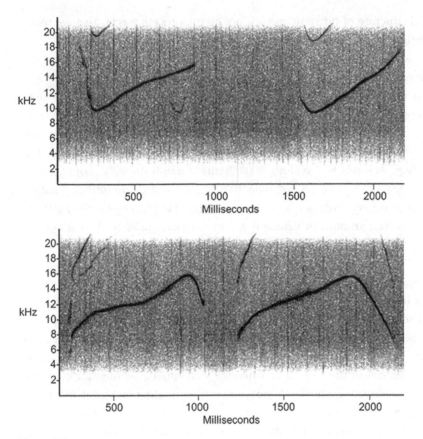

Figure 5.2
Whistles recorded from Hawaiian spinner dolphins. These whistles illustrate a continuum of contour shapes ranging from concave (top left) to upsweep (top right) to convex (bottom right).

usually remains regarding whether the resulting classifications are biologically meaningful (Janik, 1999). In addition, ambiguity can occur because the contours of whistles recorded may be affected by signal directionality and propagation loss. Therefore signal categorization, whether manual or automated, can be influenced by factors related to the recording context (e.g., the distance and orientation of animals relative to the hydrophone) that determine the quality of the recordings obtained.

Measuring and categorizing contours are important steps in describing whistles and studying their occurrence. However, determining the signal features used by dolphins as communication cues is a difficult task because humans and dolphins perceive sounds differently. As a result, characteristics of whistles that seem important to human analysts may not be important to the animals and vice versa. One approach is to combine detailed observations of behavior with acoustic recordings. For example, Lopez (2011) found that among free-ranging bottlenose dolphins in the Mediterranean, foraging behavior was associated with multilooped whistles that had long durations and high peak frequencies, while upswept or "rise" whistles were associated with social behaviors. Hawkins and Gartside (2010) found that in a population of Indo-Pacific bottlenose dolphins (*Tursiops aduncus*), flat and upswept whistles were associated with social behaviors, and "sine" whistles were used commonly during travel. These findings suggest that at least some whistle parameters are tied to behavior and therefore hold important clues about communication. However, while researchers have made some progress in understanding which features of whistles are important to dolphins (e.g., Dudzinski, Clark, & Würsig, 1995; Harley, 2008; Herzing, 1996; Ralston & Herman, 1995; Watwood, Owen, Tyack, & Wells, 2005), the challenges inherent to associating specific whistles with individuals and observing the full repertoire of their underwater behavior have made advancements slow.

Directionality and Harmonics

It is generally assumed that whistles are relatively omnidirectional, so few studies have focused on the directional properties of these signals and how they might be relevant to communication. However, Lammers and Au (2003) found that whistles produced by free-ranging spinner dolphins swimming toward a hydrophone array were on average higher in amplitude and richer in harmonic structure than those recorded from dolphins moving away from the array. They also observed that the lower-frequency components of whistles (the fundamental frequencies) had less directivity than the higher-frequency harmonic components. Evans et al. (1964) found a similar trend when they propagated sound through the heads of *Stenella* and *Tursiops* cadavers, and Branstetter et al. (2012) reported the same pattern based on experiments with captive *Tursiops*. Similarly, Miller (2002) found that some killer whale calls are made up of a directional high-frequency component and a more omnidirectional low-frequency component. Signals that contain both directional and nondirectional elements are said to have "mixed directionality" and are thought to play an important function in insect and bird communication (Laresen & Dabelsteen, 1990).

The harmonics of whistles occur at integer multiples of the fundamental frequency and usually extend into the ultrasonic (> 20 kHz) range. Harmonics occur when the whistle's waveform deviates from a pure sinusoid (Au & Hastings, 2008). The greatest amount of energy in a whistle is generally found in the fundamental frequency, while the energy contained in harmonics is usually lower and variable (Branstetter, Moore, Finneran, Tormey, & Aihara, 2012; Lammers et al., 2003). It is not clear yet whether dolphins exert some control over the harmonic content of their whistles, or whether harmonics are simply involuntary by-products of signal production. However, it has been suggested that harmonics could play a role in promoting group cohesion by providing a cue about the orientation of the whistling dolphin (Branstetter, Black, & Bakhtiari, 2013; Lammers & Au, 2003; Miller, 2002). Given that the harmonic content of whistles is correlated with the orientation of an animal, some researchers have proposed that dolphins may use this information to infer the direction of movement of conspecifics and detect changes in their orientation (Branstetter et al., 2013; Lammers & Au, 2003; Miller, 2002). Yuen et al. (2007) demonstrated that a captive false killer whale (*Pseudorca crassidens*) could in fact discriminate between a pure tone and complex tones with up to five harmonics, and discrimination thresholds existed for different harmonic combinations. Branstetter et al. (2013) similarly found that captive bottlenose dolphins could discriminate whistles based on their harmonic content and were able to analyze the spectral profile of whistles independently of the overall amplitude of the signals. Thus, acoustically and perceptually, the evidence strongly suggests that the harmonic components of whistles may play an important communicative role. However, whether dolphins actually use the mixed-directionality cue present in whistles is contingent on a learned association between a specific harmonic pattern and a signaler's orientation, which to date has not yet been demonstrated.

Amplitude Modulation and Graded Signaling

Whistles are typically thought of as frequency-modulated signals, but they also exhibit a degree of amplitude modulation (AM). Some whistles, or segments of whistles, have a particularly high level of AM, which makes them sound less like pure tones and gives them a more "raspy" quality (Lammers et al., 2003). These whistles take on pulse-like properties (fig. 5.3) and may be a vocalization type that is intermediate between whistles and burst pulses. Murray et al. (1998) used self-organizing neural networks to objectively identify categories of false killer whale phonations based on waveform measurements taken from whistles and click trains. Based on these measurements, Murray et al. (1998) suggested that the acoustic repertoire of false killer whales is not made up of discrete categories but rather is a continuum of vocalizations with whistles lying at one end and click trains at the other end. AM whistles have also been reported in Atlantic spotted dolphins and spinner dolphins (Lammers et al., 2003). It is presently not clear to what degree dolphins voluntarily control the AM of their whistles and whether, or how, this might factor into communication. However, because AM in a tone decreases with increasing distance from the source (Urick, 1983), some researchers have hypothesized that amplitude variations in whistles may primarily be relevant

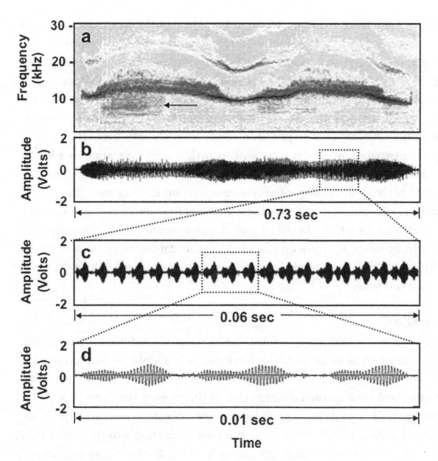

Figure 5.3
Amplitude-modulated Atlantic spotted dolphin whistle showing the frequency contour (a), the waveform pattern (b), and a progressive magnification of the pulselike structure of the signal [(c) and (d)]. The arrow points to a weak concurrent burst pulse. Reproduced with permission from Lammers et al. (2003). Copyright 2003 by the Acoustical Society of America.

for close-range communication (Lammers et al., 2003). Whistles with little or no AM may be more typical of signals used for longer-range communication, such as those for maintaining contact among dispersed individuals.

Graded signaling has also been proposed by Taruski (1979), who used a graded model to describe the whistles produced by North Atlantic pilot whales (*Globicephala melaena*) and concluded that their whistles could be arranged as a continuum of signals rather than discrete categories. In other words, each whistle contour in the repertoire is related to every other whistle through a series of intermediates. Moore and Ridgway

(1995) came to a similar conclusion after examining the whistle repertoire of common dolphins (*Delphinus delphis*) off Southern California, and a graded call repertoire has also been proposed for beluga whales (*Delphinapterus leucas*) (Karlsen, Bisther, Lydersen, Haug, & Kovacs, 2002; Sjare & Smith, 1986). Therefore it may be that for certain species, or perhaps when examining whistles at the level of populations rather than individuals, a graded model may be a more appropriate method of characterizing the repertoire than identifying discrete phonation categories. This also raises the possibility that graded signaling might form, at least in part, a basis for communication in whistles. Graded signals are often employed in other taxa when a continuous or relative condition is communicated in a favorable propagation environment (Bradbury & Vehrencamp, 1998). Among dolphins, signaling among individuals in close proximity might be a situation in which the use of graded signals could play a communicative role. Presently, however, we have no empirical data to determine whether dolphins perceive whistles along a continuum or categorically, so it is not possible to draw any conclusions at this time.

Signature Whistles

Many delphinid species that produce whistles are gregarious and live in large groups, and their whistles are believed to play an important role in social communication (Herzing, 2000; Janik, 2000; Janik & Slater, 1998). How whistles are used in communication is a much-discussed topic in the literature. One of the primary functions proposed is the identification of individuals, also known as the "signature whistle" hypothesis. Caldwell and Caldwell (1965) first coined the term "signature whistle" when they observed captive bottlenose dolphins producing individually distinctive and, in certain acoustic features, stereotyped whistles. The signature whistle hypothesis proposes that dolphins use these stereotyped whistles to broadcast their identity and location to other members of their social group (Caldwell, Caldwell, & Tyack, 1990). Bottlenose dolphins have been observed to mimic the signature whistles of other dolphins within a group (Caldwell & Caldwell, 1972; King, Sayigh, Wells, Fellner, & Janik, 2013; Tyack, 1986), leading to the hypothesis that these whistles may be used to establish and maintain contact between individuals, particularly mother-calf pairs (Sayigh et al., 1990; Smolker, Mann, & Smuts, 1993). The observed production of signature whistles during periods of separation between members of a group further supports a cohesion call function (Janik & Slater, 1998). Vocal learning appears to play an important role in the development of individually distinctive whistles (Fripp et al., 2005; Miksis, Tyack, & Buck, 2002; Tyack, 1997). Potential signature whistles have been recorded from bottlenose dolphins, *Stenella* species, *Lagenorhynchus* species, long-finned pilot whales, tucuxi

(*Sotalia fluviatilis*), and Amazon river dolphins (*Inia geoffrensis*) (Caldwell & Caldwell, 1965, 1971; Caldwell, Caldwell, & Miller, 1973; Caldwell et al., 1990; Steiner, 1981; Wang et al., 1995).

Some researchers have questioned the validity of the signature whistle hypothesis, raising doubts about whether original accounts of signature whistle use, which reportedly made up from 74 to 95 percent of a dolphin's whistle repertoire, were representative of typical dolphin acoustic behavior (McCowan & Reiss, 1995). These researchers have instead proposed that, in the context of isolation, dolphins produce a shared predominant whistle type that contains individual variability, not individually distinct whistle contours (McCowan & Reiss, 2001). However, although the predominance of signature whistles occurring in natural situations has been a past topic of debate (McCowan & Reiss, 1995, 2001), follow-on efforts have convincingly demonstrated that signature whistles are indeed used by dolphins and make up at least a portion of the whistle repertoire of bottlenose dolphins (Janik & Sayigh, 2013; Janik, King, Sayigh, & Wells, 2013; Watwood et al., 2005). Moreover, recent experimental efforts have shown that wild bottlenose dolphins respond to hearing a copy of their own signature whistle by calling back, providing evidence that a dolphin's learned signature whistle is used as a label when addressing conspecifics (King & Janik, 2013).

Pulsed Signals

In addition to producing whistles, delphinids (and all other odontocetes) also produce pulsed sounds used for sensing the environment and for communication (Popper, 1980). These pulses or "clicks" are short, lasting only tens to hundreds of microseconds (μs), are produced in "trains" with interclick intervals (ICIs) between a few and hundreds of milliseconds (ms), and typically have broadband energy mostly or exclusively at frequencies above the human hearing range (i.e., >15 kHz). These signals are generally differentiated into click trains used for echolocation and "burst pulses" associated with communication (fig. 5.4). Historically, burst pulses have been given many onomatopoeic names, including "squawks" and "squeaks" (Caldwell & Caldwell, 1971), "yelps" (Puente & Dewsbury, 1976), and "cries" (Dawson, 1991), to name only a few. This is because human listeners aurally perceive the audible components of burst pulses as continuous sounds rather than as series of discrete clicks, owing to the human auditory system's limited temporal acuity, which is between 22 ms and 3.2 ms, depending on frequency (Shailer & Moore, 1983). By contrast, the bottlenose dolphin is able to resolve clicks separated by about 260 μs (Au, 1993), a temporal resolution that is approximately twentyfold more acute than that of humans.

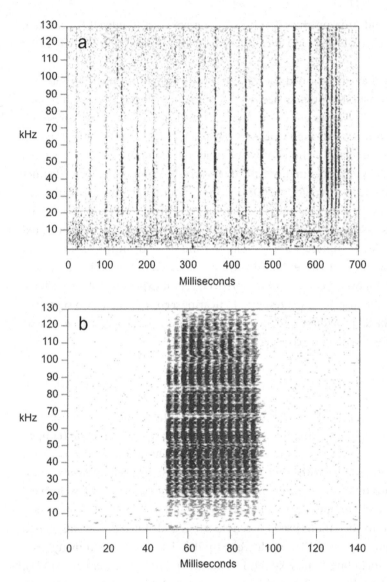

Figure 5.4
Spinner dolphin echolocation click train (a) with mean ICI = 32 ms. Spinner dolphin burst pulse
(b) with mean ICI = 3.4 ms. From Lammers et al. (2004). © 2004 by the University of Chicago. All
rights reserved.

Distinguishing between Echolocation and Burst Pulses

Although generally considered separately, neither the acoustic nor the functional distinction between echolocation and burst pulse click trains is well defined. Researchers studying click train production usually consider clicks with ICIs of tens or hundreds of milliseconds to be echolocation signals and those with ICIs of only a few milliseconds as burst pulses used for communication (Frankel & Yin, 2010; Lammers et al., 2003; Murray et al., 1998; Simard, Mann, & Gowans, 2008). This distinction is primarily made because experimental evidence has revealed that dolphins performing echolocation tasks space their clicks to account for the two-way travel time to the target plus a "processing time" of between 19 and 45 ms (Au, 1993). As a result, click trains with very short ICIs (i.e., burst pulses) are not considered functional in echolocation (as we presently understand it) because they lack the processing period. In addition, the mean ICIs of click trains produced by free-ranging spinner dolphins have been shown to exhibit a bimodal distribution, with peaks in occurrence at 3.5 ms and 80.0 ms and a gap at 10 ms (Lammers, Au, Aubauer, & Nachtigall, 2004), supporting a functional differentiation of click trains on the basis of ICIs.

An ambiguity in the distinction between echolocation and burst pulse click trains results from the fact that free-swimming, echolocating dolphins produce "terminal buzzes" during the final stages of approach to a target or prey. These terminal buzzes have very short ICIs of only a few milliseconds (i.e., they are burst pulse–like) but are continuous with echolocation click trains (Johnson, Madsen, Zimmer, De Soto, & Tyack, 2006; Morozov, Akopian, Zaytseva, & Sokovykh, 1972). Figure 5.5 illustrates an example from an Atlantic spotted dolphin recorded in the Bahamas (Lammers, unpublished data). In this sequence, click trains can be observed with ICIs consistent with echolocation, burst pulse signaling, and elements of both. What role, if any, terminal buzzes and other click trains with ICIs intermediary between categories play in echolocation is still unclear, but it brings into question whether very short ICIs are necessarily diagnostic of click trains involved in communication, or whether these also play a role in echolocation, perhaps with echoes processed differently than in click trains with longer ICIs. It is also not yet clear whether echolocation clicks might also be involved in information sharing among individuals. Experimental and field evidence suggests that dolphins have the ability to "eavesdrop" on one another's echolocation click trains (Xitco & Roitblat, 1996). In addition, nearby animals can probably infer some things about an echolocating individual, including the animal's location and orientation, as a result of the directionality of clicks (Branstetter et al., 2012). Therefore it may be that under certain circumstances, echolocation signals may also mediate some information flow among animals.

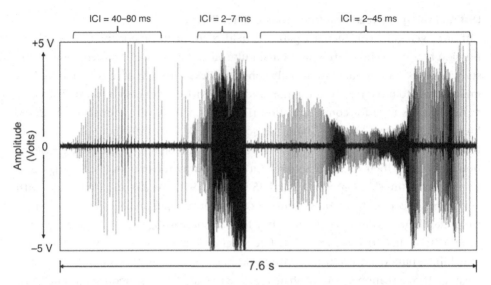

Figure 5.5
Click train sequences recorded from an Atlantic spotted dolphin (*S. frontalis*), exhibiting clicks
with relatively long ICIs consistent with echolocation (first sequence), very short ICIs representative of burst pulse signaling (second sequence), and ICIs varying between the two categories (third
sequence).

Pulsed Signaling and Communication

Despite some degree of ambiguity in the functional distinction between echolocation
and burst pulses click trains, ample evidence links echolocation with sensory perception
(for a detailed review, see Au, 1993) and burst pulses with a communicative role. Burst
pulses are documented to occur in agonistic encounters between animals both in the
field and in captivity (Blomqvist & Amundin, 2004; Herzing, 1988; Overstrom, 1983)
and also in conjunction with mating and courtship behaviors (Herzing, 1996). However,
it is likely that burst pulses are also involved in many additional communicative functions that are not yet evident to us. Lammers et al. (2003) showed that 39.2 percent of
spinner dolphin and 60.3 percent of spotted dolphin burst pulses recorded in the field
had less than 10 percent of their total energy below 20 kHz. Therefore, while some burst
pulses are clearly audible and prominent at sonic frequencies, many are either barely
detectable aurally or visually on a sonogram or are completely devoid of energy at those
frequencies. In other words, burst pulse signaling is probably a much more common
occurrence among dolphins than has been captured during most band-limited studies.

In addition, the importance of pulsed signaling in dolphin communication is
highlighted by the fact that several odontocete species (e.g., *Phocoena phocoena,*

Cephalorhynchus commersonii, C. hectori) that are behaviorally and socially similar to other whistle-producing delphinids do not produce whistles but mediate all their acoustic communication using clicks alone (Dawson, 1991; Morisaka & Connor, 2007). Similarly, dusky dolphins off New Zealand and Argentina and northern right whale dolphins (*Lissodelphis borealis*) in the north Pacific both often occur in large pods of several hundred animals, forage cooperatively and rely on group coordination and cohesion for protection from predators, and produce almost no whistles, but emit a large variety of click trains and burst pulses (Au, Lammers, & Yin, 2010; Rankin, Oswald, Barlow, & Lammers, 2007). This indicates that burst pulses and other click trains play communicative functions that have not yet been documented.

How burst pulses might be used to transmit information is presently unknown. Click trains can be produced in nearly infinite combinations of click numbers, ICIs, peak frequencies, and amplitudes. The potential exists for information to be encoded in these variations, but there is no evidence to suggest that dolphins discriminate or attend to these variations, largely because experiments to study these questions have not yet been conducted. However, some evidence does exist from field studies suggesting that burst pulses are used for relatively close-range signaling between animals. Lammers et al. (2006) examined the spatial context of signaling in free-ranging spinner dolphins and found that animals producing burst pulse signals were separated by a median distance of 14 meters (maximum = 72 meters), whereas whistling animals were separated by a median distance of 23 meters (maximum = 113 meters). Watkins and Schevill (1974) similarly found that burst pulses from one spinner dolphin were often answered within a half second by burst pulses from nearby dolphins but appeared to be ignored by more distant animals. Thus it would appear that whereas tonal signaling may be more important in communication among dispersed individuals, burst pulse signals probably play an important role in communication among animals in close proximity to one another.

Other Sounds

Whistles and burst pulses do not always occur independently. Several authors have reported the simultaneous production of whistles and burst pulses in experimental situations and in various behavioral contexts (Caldwell & Caldwell, 1967; Herzing, 1996; Lilly, 1962; Reiss, 1988). These sounds have sometimes been labeled "whistle/squawks." They often occur in situations of high excitement or distress and are therefore thought to be emotive expressions.

In addition, dolphins also produce sounds that do not fall into the categories of whistle, click, or burst pulse. For example, McCowan and Reiss (1995) reported low-frequency, wideband sounds called "thunks" that were recorded from captive female bottlenose dolphins. The authors hypothesized that these sounds function as an aggressive contact vocalization directed at infant calves. In addition, low-frequency pulses called "pops" have been recorded from free-ranging bottlenose dolphins in Shark Bay, Australia, and Sarasota, Florida (Connor & Smolker, 1996; Nowacek, 2005). These sounds have been described as a hybrid between an echolocation click and a burst pulse sound (Nowacek, 2005). In Shark Bay, pops were recorded only from males, and based on their observations, Connor and Smolker (1996) suggested that pops are threat vocalizations intended to induce females to remain close to the popping male during courtships. Nowacek (2005) also recorded pops from both lone bottlenose dolphins and individuals in groups in Sarasota and suggested that these sounds play a role in foraging.

Although little research exists on the topic, dolphins may also use nonvocal sounds for communication. Tail slaps and jaw claps have been recorded from several dolphin species in the context of aggressive interactions (Gazda, Connor, Edgar, & Cox, 2005; Herzing, 1988; Herzing, 1996; Overstrom, 1983). A specific type of tail slap, called a "kerplunk," has been described for bottlenose dolphins in Shark Bay, Australia, and Sarasota, Florida (Connor, Heithaus, Berggren, & Miksis, 2000). This sound is thought to aid in the location or capture of fish by eliciting a startle response in hiding fish (Connor, Wells, Mann, & Read, 2000). In addition to tail slaps and jaw claps, sounds produced by breaches and leaps may also function in communication, perhaps providing location cues to pod mates (Dudzinski, 1998; Norris, Würsig, Wells, & Würsig, 1994).

Charting a Way Forward

We now consider some of the steps we might take moving forward. Broadly speaking, we are still fundamentally deficient in three major areas of understanding: How do dolphins use acoustic communication in different behavioral contexts? What are the acoustic perceptual and conceptual boundaries of dolphins? How have social and ecological factors shaped the acoustic communication system of dolphins?

Understanding the Context of Dolphin Acoustic Communication
The challenges associated with examining dolphin signals within their communicative contexts are by now clear. The physics of underwater sound and the observational

limitations we face remain our greatest hurdles, so solutions will have to come from technological advances and basic human ingenuity. The primary area that must be improved on is our ability to reliably match behavioral and acoustic observations. One promising area is the development of instrumented tags that can be placed on animals to measure a suite of variables, including those related to animal movement, physiological state, and production of and exposure to sounds. Instrumented tags can also measure environmental conditions experienced by tagged animals. Many of the recent advancements in our understanding of relatively cryptic species, such as beaked whales, have come from data obtained from such tags (e.g., Johnson et al., 2006; Tyack et al., 2011). Conceivably, tags will one day be developed that, if synchronized and spatially referenced, will allow multiple receivers to be placed within a pod to function together as a hydrophone and data-gathering array (e.g., Jensen and Tyack, 2013). Obtaining such animal-based perspectives on behavior and signaling is one way that we will dramatically evolve our understanding of the spatial, temporal, and behavioral contexts of signal production.

Another promising methodology is the application of autonomous underwater vehicles (AUVs), such as gliders. Presently, AUVs are most commonly employed in oceanographic and biological remote-sensing and monitoring applications (e.g., Eriksen et al., 2001; Moore, Howe, Stafford, & Boyd, 2007). As with instrumented tags, the use of AUVs is still limited by their size, capabilities, and cost. However, as smaller, faster, and more maneuverable and capable AUVs are developed, these could one day remotely place human observers in the midst of a dolphin pod for extended periods of time. Coupled with acoustic localization technology, such observations would provide an entirely new perspective on questions about who, where, when, and why dolphins produce signals.

Finally, ample opportunity exists for applying rapidly expanding technological innovations in more traditional field study settings. In several places in the world (e.g., the Bahamas, Western Australia), long-term observational studies of free-ranging dolphin populations provide remarkable opportunities for in-depth observations of natural behavior (see Elliser & Herzing, 2014; Mann, Stanton, Patterson, Bienenstock, & Singh, 2012). In locations such as these, new technologies offering broadband signal recording, real-time processing, and localization of signaling animals, coupled with extended viewing and video recording (e.g., fig. 5.6; Schotten et al., 2005), could offer game-changing opportunities for advancing our understanding of dolphin behavior and acoustic communication.

Figure 5.6
A diver recording an Atlantic spotted dolphin's behavior and acoustic signaling using a four-channel hydrophone array and video system. Photo courtesy of the Wild Dolphin Project and Michiel Schotten.

Assessing Perceptual and Conceptual Boundaries

We also need to continue to probe how acoustic signals are processed and discriminated by dolphins. The majority of efforts in this area have focused on echolocation abilities and hearing acuity, usually as it relates to ambient noise (for a review, see Au, Popper, & Fay, 2000). Considerably fewer studies have examined how dolphins perceive and classify signals associated with communication (for a review, see Harley, 2008). However, this is an important area of future research, because if we continue to rely solely on human-based perception and statistical metrics to distinguish and categorize dolphin signals, we will inevitably leave questions open regarding the biological appropriateness of such an approach.

Some efforts have been made to understand dolphins' abilities to discriminate whistles, and specifically signature whistles. For example, Ralston and Herman (1995) showed that bottlenose dolphins distinguish among whistle contour shapes rather than simply attending to their absolute frequency. These authors also demonstrated that dolphins have the ability to classify whistlelike sounds as being the same, despite frequency transpositions of a full octave. In addition, Harley (2008) found that a bottlenose dolphin could discriminate between signature whistles, associated them with surrogate producers (objects or places), and did not categorize partial-contour (i.e., fragmented) whistles together with intact-contour whistles, indicating that partial-contour whistles are not perceived as simply variants of an intact contour. These experimental results illustrate the type of empirical data needed to advance our understanding of how dolphins associate and discriminate the signals they employ. Presently no data exist to determine whether dolphins make distinctions between subtle variations of nonsignature whistles. As a result, it remains a matter of speculation how much importance to place on whistle variations and graded signaling and whether these variations encode any salient information.

Arguably, the greatest shortcoming in our understanding of dolphin signals is the role that burst pulses play in communication. Much remains unclear about the use of burst pulses in normal social signaling, and effectively we know nothing about how dolphins perceive and process these sounds. We currently have no information about whether dolphins attend to the number of clicks, the interclick intervals, and the spectral content of burst pulses. These and other variables might hold the best potential for information transfer in dolphin signals, yet we know next to nothing about whether or how dolphins discriminate among these signals. It will not be possible to draw definitive conclusions about the potential and capacity of dolphin communicative signals without thoroughly exploring the production and perception of burst pulses.

Social and Ecological Factors That Shape Dolphin Communication

The majority of studies examining dolphin acoustic communication have focused on relatively few species. The production of whistles by bottlenose dolphins has historically dominated the literature largely for two reasons: they are the most common species in captivity and are cosmopolitan in their occurrence in coastal habitats, and thus most accessible. Consequently most of what we know about dolphin communication comes from inferences made from studies of bottlenose dolphins and a handful of other species (such as Atlantic spotted dolphins, spinner dolphins, and pilot whales). Yet despite their ubiquity, bottlenose dolphins represent a fairly specific delphinid type. Most information about their social signaling comes from captive colonies or coastal populations, where animals occur in small groups (one to ten individuals) and often have limited interaction with adjacent populations. It is not yet clear to what extent the communication system of these animals can be generalized to species with different ecological and social characteristics, such as pelagic and highly gregarious species.

Future studies should attempt to link acoustic communication with the ecological and social factors influencing different species. These factors will often not be obvious, but establishing the variables that promote individual survival will provide important clues about the information that is likely being communicated. Ultimately, communication signals must help individuals survive and reproduce. Thus, whether in the case of bottlenose dolphins producing signature whistles or dusky dolphins emitting clicks and burst pulses, the signals of each species have undoubtedly been adapted to promote both survival and the exploitation of a particular ecological niche. While such differences might present challenges for generalizing communication traits across species, they also highlight the adaptability of dolphins for using acoustic communication across a wide range of social and ecological conditions.

References

Acevedo, A. F., & Van Sluys, M. (2005). Whistles of tucuxi dolphins (*Sotalia fluviatilis*) in Brazil: Comparisons among populations. *Journal of the Acoustical Society of America, 117*, 1456–1464.

Au, W. W. L. (1993). *The sonar of dolphins*. New York: Springer.

Au, W. W. L., & Hastings, M. C. (2008). *Principles of marine bioacoustics*. New York: Springer.

Au, W. W. L., Lammers, M. O., & Aubauer, R. (1999). A portable broadband data acquisition system for field studies in bioacoustics. *Marine Mammal Science, 15*, 526–531.

Au, W. W. L., Lammers, M. O., & Yin, S. (2010). Acoustics of dusky dolphins. In B. Würsig & M. Würsig (Eds.), *The dusky dolphin: Master acrobat off different shores* (pp. 75–97). San Diego: Academic Press.

Au, W. W. L., Popper, A. N., & Fay, R. R. (2000). *Hearing by whales and dolphins*. New York: Springer.

Batteau, D. W., & Markey, P. R. (1967). *Man/dolphin communication*. Final report, contract N00123-67-1103, December 15, 1966–December 13, 1967. China Lake, CA: US Naval Ordnance Test Station.

Bazua-Duran, C., & Au, W. W. L. (2002). The whistles of Hawaiian spinner dolphins. *Journal of the Acoustical Society of America, 112*, 3064–3072.

Blomqvist, C., & Amundin, M. (2004). High-frequency burst pulse sounds in agonistic/aggressive interactions in bottlenose dolphins, *Tursiops truncatus*. In J. A. Thomas, C. F. Moss, & M. M. Vater (Eds.), *Echolocation in bats and dolphins* (pp. 425–431). Chicago: University of Chicago Press.

Bradbury, J. W., & Vehrencamp, S. L. (1998). *Principles of animal communication*. Sunderland: Sinauer.

Branstetter, B. K., Black, A., & Bakhtiari, K. (2013). Discrimination of mixed-directional whistles by a bottlenose dolphin (*Tursiops truncatus*). *Journal of the Acoustical Society of America, 134*, 2274–2285.

Branstetter, B. K., Moore, P. W., Finneran, J. J., Tormey, M. N., & Aihara, H. (2012). Directional properties of bottlenose dolphin (*Tursiops truncatus*) clicks, burst-pulse, and whistle sounds. *Journal of the Acoustical Society of America, 134*, 1613–1621.

Brown, J. C., & Miller, P. J. O. (2007). Automatic classification of killer whale vocalizations using dynamic time warping. *Journal of the Acoustical Society of America, 122*, 1201–1207.

Buck, J. R., & Tyack, P. L. (1993). A quantitative measure of similarity for *Tursiops truncatus* signature whistles. *Journal of the Acoustical Society of America, 94*, 2497–2506.

Caldwell, M. C., & Caldwell, D. K. (1965). Individualized whistle contours in bottlenose dolphins (*Tursiops truncatus*). *Nature, 207*, 434–435.

Caldwell, M. C., & Caldwell, D. K. (1967). Intraspecific transfer of information via the pulsed sound in captive odontocete cetaceans. In R. G. Busnel (Ed.), *Animal sonar systems: Biology and bionics* (pp. 879–936). Jouy-en-Josas, France: Laboratoire de Physiologie Acoustic.

Caldwell, M. C., & Caldwell, D. K. (1968). Vocalization of naive captive dolphins in small groups. *Science, 159*, 1121–1123.

Caldwell, M. C., & Caldwell, D. K. (1971). Underwater pulsed sounds produced by captive spotted dolphins, *Stenella plagiodon*. *Cetology, 1*, 1–7.

Caldwell, M. C., & Caldwell, D. K. (1972). Vocal mimicry in the whistles made by an Atlantic bottlenose dolphin. *Cetology, 9*, 1–8.

Caldwell, M. C., Caldwell, D. K., & Miller, J. F. (1973). Statistical evidence for individual signature whistles in the spotted dolphin, *Stenella plagiodon*. *Cetology, 16*, 1–21.

Caldwell, M. C., Caldwell, D. K., & Tyack, P. (1990). Review of the signature whistle hypothesis for the Atlantic bottlenose dolphin. In S. Leatherwood & R. R. Reeves (Eds.), *The bottlenose dolphin* (pp. 199–234). San Diego: Academic Press.

Connor, R. C., Heithaus, M. R., Berggren, P., & Miksis, J. L. (2000). "Kerplunking": Surface fluke-splashes during shallow-water bottom foraging by bottlenose dolphins. *Marine Mammal Science, 16*, 646–653.

Connor, R. C., & Smolker, R. A. (1996). "Pop" goes the dolphin: A vocalization male bottlenose dolphins produce during consortships. *Behaviour, 133*, 643–662.

Connor, R. C., Wells, R. S., Mann, J., & Read, A. J. (2000). The bottlenose dolphin: Social relationships in a fission–fusion society. In J. Mann, R. C. Connor, P. L. Tyack, & H. Whitehead (Eds.), *Cetacean societies: Field studies of dolphins and whales* (pp. 91–126). Chicago: University of Chicago Press.

Dawson, S. M. (1991). Clicks and communication: The behavioral and social contexts of Hector's dolphin vocalizations. *Ethology, 88,* 265–276.

Deecke, V. B., & Janik, V. M. (2006). Automated categorization of bioacoustics signals: Avoiding perceptual pitfalls. *Journal of the Acoustical Society of America, 119,* 645–653.

Dreher, J. J. (1961). Linguistic considerations of porpoise sounds. *Journal of the Acoustical Society of America, 33,* 1799–1800.

Dreher, J. J. (1966). Cetacean communication: Small group experiment. In K. S. Norris (Ed.), *Whales, dolphins, and porpoises* (pp. 529–543). Berkeley: University of California.

Dreher, J. J., & Evans, W. E. (1964). Cetacean communication. *Marine Bioacoustics, 1,* 373–393.

Dudzinski, K. M. (1998). Contact behavior and signal exchange in Atlantic spotted dolphins (*Stenella frontalis*). *Aquatic Mammals, 24,* 129–142.

Dudzinski, K. M., Clark, C. W., & Würsig, B. (1995). Mobile video/acoustic system for simultaneous under-water recording of dolphin interactions. *Aquatic Mammals, 21,* 187–193.

Elliser, C. R., & Herzing, D. L. (2014). Long-term social structure of a resident community of Atlantic spotted dolphins, *Stenella frontalis,* in the Bahamas, 1991–2002. *Marine Mammal Science, 30,* 308–328.

Eriksen, C. C., Osse, T. J., Light, R. D., Wen, T., Lehman, T. W., Sabin, P. L., et al. (2001). Seaglider: A long-range autonomous underwater vehicle for oceanographic research. *IEEE Journal of Oceanic Engineering, 26,* 424–436.

Essapian, F. S. (1953). The birth and growth of a porpoise. *Natural History, 62,* 392–399.

Evans, W. E., Awbrey, F. T., & Hackbarth, H. (1988). High frequency pulses produced by free-ranging Commerson's dolphin (*Cephalorhynchus commersonii*) compared to those of phocoenids. In R. L. Brownell & G. P. Donovan (Eds.), *Biology of the genus Cephalorhynchus* (pp. 173–181). Cambridge: International Whaling Commission.

Evans, W. E., & Prescott, J. H. (1962). Observations of the sound production capabilities of the bottlenosed porpoise: A study of whistles and clicks. *Zoologica, 47,* 121–128.

Evans, W. E., Sutherland, W. W., & Beil, R. G. (1964). The directional characteristics of delphinid sounds. In W. N. Tavolga (Ed.), *Marine bioacoustics* (pp. 353–372). New York: Pergamon Press.

Frankel, A. S., & Yin, S. (2010). A description of sounds recorded from melon-headed whales (*Peponocephala electra*) off Hawai'i. *Journal of the Acoustical Society of America, 127,* 3248–3255.

Fripp, D., Owen, C., Quintana-Rizzo, E., Shapiro, A., Buckstaff, K., Jankowski, K., et al. (2005). Bottlenose dolphin (*Tursiops truncatus*) calves appear to model their signature whistles on the signature whistles of community members. *Animal Cognition, 8,* 17–26.

Fristrup, K. M., & Watkins, W. A. (1993). Marine animal sound classification. *Woods Hole Oceanographic Institution Technical Report WHOI-94-13.*

Gazda, S. K., Connor, R. C., Edgar, R. K., & Cox, F. (2005). A division of labour with role specialization in group-hunting bottlenose dolphins (*Tursiops truncatus*) off Cedar Key, Florida. *Proceedings of the Royal Society B: Biological Sciences, 272*, 135–140.

Harley, H. E. (2008). Whistle discrimination and categorization by the Atlantic bottlenose dolphin (*Tursiops truncatus*): A review of the signature whistle framework and a perceptual test. *Behavioural Processes, 77*, 243–268.

Hawkins, E. R., & Gartside, D. F. (2010). Whistle emissions of Indo-Pacific bottlenose dolphins (*Tursiops aduncus*) differ with group composition and surface behaviors. *Journal of the Acoustical Society of America, 127*, 2652–2663.

Herman, L. M., Richards, D. G., & Wolz, J. P. (1984). Comprehension of sentences by bottlenosed dolphins. *Cognition, 16*, 129–219.

Herman, L. M., & Tavolga, W. N. (1980). The communication systems of cetaceans. In L. M. Herman (Ed.), *Cetacean behavior: Mechanisms and functions* (pp. 149–209). New York: Wiley Interscience.

Herzing, D. L. (1988). *A quantitative description and behavioral associations of a burst-pulsed sound, the squawk, in captive bottlenose dolphins, Tursiops truncatus*. Master's thesis, San Francisco State University.

Herzing, D. L. (1996). Vocalizations and associated underwater behavior of free-ranging Atlantic spotted dolphins, *Stenella frontalis*, and bottlenose dolphins, *Tursiops truncatus*. *Aquatic Mammals, 22*, 61–79.

Herzing, D. L. (2000). Acoustics and social behavior of wild dolphins: Implications for a sound society. In W. W. L. Au, A. N. Popper, & R. R. Fay (Eds.), *Hearing by whales and dolphins* (pp. 225–272). New York: Springer.

Janik, V. M. (1999). Pitfalls in the categorization of behavior: A comparison of dolphin whistle classification methods. *Animal Behaviour, 57*, 133–143.

Janik, V. M. (2000). Whistle matching in wild bottlenose dolphins (*Tursiops truncatus*). *Science, 289*, 1355–1357.

Janik, V. M., King, S. L., Sayigh, L. S., & Wells, R. S. (2013). Identifying signature whistles from recordings of groups of unrestrained bottlenose dolphins (*Tursiops truncatus*). *Marine Mammal Science, 29*, 109–122.

Janik, V. M., & Sayigh, L. S. (2013). Communication in bottlenose dolphins: 50 years of signature whistle research. *Journal of Comparative Physiology A: Neuroethology, Sensory, Neural, and Behavioral Physiology, 199*, 479–489.

Janik, V. M., & Slater, P. J. B. (1998). Context-specific use suggests that bottlenose dolphin whistles are cohesion calls. *Animal Behaviour, 56*, 829–838.

Janik, V. M., Van Parijs, S. M., & Thompson, P. M. (2000). A two-dimensional acoustic localization system for marine mammals. *Marine Mammal Science, 16*, 437–447.

Jensen, F. H., & Tyack, P. L. (2013). Studying acoustic communication in pilot whale social groups. *Journal of the Acoustical Society of America*, *134*, 4006.

Johnson, C. S. (1967). Sound detection thresholds in marine mammals. In W. Tavolga (Ed.), *Marine bioacoustics* (pp. 247–260). New York: Pergamon.

Johnson, M., Madsen, P. T., Zimmer, W. M. X., De Soto, N. A., & Tyack, P. L. (2006). Foraging Blainville's beaked whales (*Mesoplodon densirostris*) produce distinct click types matched to different phases of echolocation. *Journal of Experimental Biology*, *209*, 5038–5050.

Johnson, M. P., & Tyack, P. L. (2003). A digital acoustic recording tag for measuring the response of wild marine mammals to sound. *IEEE Journal of Oceanic Engineering*, *28*, 3–12.

Karlsen, J., Bisther, A., Lydersen, C., Haug, T., & Kovacs, K. (2002). Summer vocalisations of adult male white whales (*Delphinapterus leucas*) in Svalbard, Norway. *Polar Biology*, *25*, 808–817.

Kellogg, W. N. (1958). Echo ranging in the porpoise. *Science*, *128*, 982–988.

Kellogg, W. N., Kohler, R., & Morris, H. N. (1953). Porpoise sounds as sonar signals. *Science*, *117*, 239–243.

King, S. L., & Janik, V. M. (2013). Bottlenose dolphins can use learned vocal labels to address each other. *Proceedings of the National Academy of Sciences of the United States of America*, *110*, 13216–13221.

King, S. L., Sayigh, L. S., Wells, R. S., Fellner, W., & Janik, V. M. (2013). Vocal copying of individually distinctive signature whistles in bottlenose dolphins. *Proceedings of the Royal Society of London, Series B: Biological Sciences*, *280*, 1–9.

Lammers, M. O., & Au, W. W. L. (2003). Directionality in the whistles of Hawaiian spinner dolphins (*Stenella longirostris*): A signal feature to cue direction of movement? *Marine Mammal Science*, *19*, 249–264.

Lammers, M. O., Au, W. W. L., Aubauer, R., & Nachtigall, P. E. (2004). A comparative analysis of echolocation and burst-pulse click trains in *Stenella longirostris*. In J. A. Thomas, C. F. Moss, & M. M. Vater (Eds.), *Echolocation in bats and dolphins* (pp. 414–419). Chicago: University of Chicago Press.

Lammers, M. O., Au, W. W. L., & Herzing, D. L. (2003). The broadband signaling behavior of spinner and spotted dolphins. *Journal of the Acoustical Society of America*, *114*, 1629–1639.

Lammers, M. O., Schotten, M., & Au, W. W. L. (2006). The spatial context of free-ranging Hawaiian spinner dolphins (*Stenella longirostris*) producing acoustic signals. *Journal of the Acoustical Society of America*, *119*, 1244–1250.

Lang, T. G., & Smith, H. A. P. (1965). Communication between dolphins in separate tanks by way of an electronic acoustic link. *Science*, *150*, 1839–1844.

Laresen, O. N., & Dabelsteen, T. (1990). Directionality of blackbird vocalization: Implications for vocal communication and its further study. *Ornis Scandinavica*, *21*, 37–45.

Lilly, J. C. (1961). *Man and dolphin: Adventures of a new scientific frontier.* New York: Doubleday.

Lilly, J. C. (1962). Vocal behavior of the bottlenose dolphin. *Proceedings of the American Philosophical Society, 106,* 520–529.

Lilly, J. C. (1963). Distress call of the bottlenose dolphin: Stimuli and evoked behavioral responses. *Science, 139,* 116–118.

Lilly, J. C. (1967). *The mind of the dolphin.* New York: Doubleday.

Lilly, J. C., & Miller, A. M. (1961a). Sounds emitted by the bottlenose dolphin. *Science, 133,* 1689–1693.

Lilly, J. C., & Miller, A. M. (1961b). Vocal exchanges between dolphins: Bottlenose dolphins "talk" to each other with whistles, clicks, and a variety of other noises. *Science, 134,* 1873–1876.

Lopez, B. D. (2011). Whistle characteristics in free-ranging bottlenose dolphins (*Tursiops truncatus*) in the Mediterranean Sea: Influence of behavior. *Mammalian Biology, 76,* 180–189.

Mann, J., Stanton, M. A., Patterson, E. M., Bienenstock, E. J., & Singh, L. O. (2012). Social networks reveal cultural behaviour in tool-using using dolphins. *Nature Communications, 3,* 1181. doi:10.1038/ncomms1983.

Matthews, J. N., Rendell, L. E., Gordon, J. C. D., & Macdonald, D. W. (1999). A review of frequency and time parameters of cetacean tonal calls. *Bioacoustics, 10,* 47–71.

May-Collado, L. J., Agnarsson, I., & Wartzok, D. (2007a). Phylogenetic review of tonal sound production in whales in relation to sociality. *BMC Evolutionary Biology, 7,* 136.

May-Collado, L. J., Agnarsson, I., & Wartzok, D. (2007b). Reexamining the relationship between body size and tonal signals frequency in whales: A comparative approach using a novel phylogeny. *Marine Mammal Science, 23,* 524–552.

May-Collado, L. J., & Wartzok, D. (2009). A characterization of Guyana dolphin (*Sotalia guianensis*) whistles from Costa Rica: The importance of broadband recording systems. *Journal of the Acoustical Society of America, 125,* 1202–1213.

McBride, A. F. (1956). Evidence for echolocation by cetaceans. *Deep-Sea Research, 3,* 153–154.

McBride, A. F., & Hebb, D. O. (1948). Behavior of the captive bottlenose dolphin, *Tursiops truncatus. Journal of Comparative and Physiological Psychology, 41,* 111.

McCowan, B. (1995). A new quantitative technique for categorizing whistles using simulated signals and whistles from captive bottlenose dolphins (*Tursiops truncatus*). *Ethology, 100,* 177–193.

McCowan, B., Hanser, S. F., & Doyle, L. R. (1999). Quantitative tools for comparing animal communication systems: Information theory applied to bottlenose dolphin whistle repertoires. *Animal Behaviour, 57,* 409–419.

McCowan, B., & Reiss, D. (1995). Maternal aggressive contact vocalizations in captive bottlenose dolphins (*Tursiops truncatus*): Wide-band, low frequency signals during mother/aunt-infant interactions. *Zoo Biology, 14,* 293–309.

McCowan, B., & Reiss, D. (2001). The fallacy of "signature whistles" in bottlenose dolphins: A comparative perspective of "signature information" in animal vocalizations. *Animal Behaviour, 62,* 1151–1162.

Miksis, J. L., Tyack, P. L., & Buck, J. R. (2002). Captive dolphins, *Tursiops truncatus,* develop signature whistles that match acoustic features of human-made model sounds. *Journal of the Acoustical Society of America, 112,* 728–739.

Miller, P. J. O. (2002). Mixed-directionality of killer whale stereotyped calls: A direction of movement cue? *Behavioral Ecology and Sociobiology, 52,* 262–270.

Miller, P. J., & Tyack, P. L. (1998). A small towed beamforming array to identify vocalizing resident killer whales (*Orcinus orca*) concurrent with focal behavioral observations. *Deep-Sea Research Part II: Topical Studies in Oceanography, 45,* 1389–1405.

Moore, S. E., Howe, B. M., Stafford, K. M., & Boyd, M. L. (2007). Including whale call detection in standard ocean measurements: Application of acoustic Seagliders. *Marine Technology Society Journal, 41,* 53–57.

Moore, S. E., & Ridgway, S. H. (1995). Whistles produced by common dolphins from the Southern California Bight. *Aquatic Mammals, 21,* 55–63.

Morisaka, T., & Connor, R. C. (2007). Predation by killer whales (*Orcinus orca*) and the evolution of whistle loss and narrow-band high frequency clicks in odontocetes. *Journal of Evolutionary Biology, 20,* 1439–1458.

Morozov, V. P., Akopian, A. I., Zaytseva, K. A., & Sokovykh, Y. A. (1972). Tracking frequency of the location signals of dolphins as a function of the distance to the target. *Biofizika, 17,* 139–143.

Murray, S. O., Mercado, E., & Roitblat, H. L. (1998). Characterizing the graded structure of false killer whale (*Pseudorca crassidens*) vocalizations. *Journal of the Acoustical Society of America, 104,* 1679–1688.

Norris, K. S., Prescott, J. H., Asa-Dorian, P. V., & Perkins, P. (1961). An experimental demonstration of echolocation behavior in the porpoise, *Tursiops truncatus* (Montagu). *Biological Bulletin, 120,* 163–176.

Norris, K. S., Würsig, B., Wells, R. S., & Würsig, M. (1994). *The Hawaiian spinner dolphin.* Berkeley: University of California Press.

Nowacek, D. P. (2005). Acoustic ecology of foraging bottlenose dolphins (*Tursiops truncatus*), habitat-specific use of three sound types. *Marine Mammal Science, 21,* 587–602.

Oswald, J. N., Barlow, J., & Norris, T. F. (2003). Acoustic identification of nine delphinid species in the eastern tropical Pacific Ocean. *Marine Mammal Science, 19,* 20–37.

Oswald, J. N., Rankin, S., & Barlow, J. (2004). The effect of recording and analysis bandwidth on acoustic identification of delphinid species. *Journal of the Acoustical Society of America, 116,* 3178–3185.

Oswald, J. N., Rankin, S., & Barlow, J. (2008). To whistle or not to whistle? Geographic variation in the whistling behavior of small odontocetes. *Aquatic Mammals, 34,* 288–302.

Oswald, J. N., Rankin, S., Barlow, J., & Lammers, M. O. (2007). A tool for real-time acoustic species identification of delphinid whistles. *Journal of the Acoustical Society of America, 122,* 587–595.

Overstrom, N. A. (1983). Association between burst-pulse sounds and aggressive behavior in captive Atlantic bottlenose dolphins, *Tursiops truncatus. Zoo Biology, 2,* 93–103.

Popper, A. N. (1980). Sound emission and detection by delphinids. In L. M. Herman (Ed.), *Cetacean behavior: Mechanisms and functions* (pp. 1–52). New York: Wiley Interscience.

Powell, B. A. (1966). Periodicity of vocal activity of captive Atlantic bottlenose dolphins, *Tursiops truncatus. Bulletin of the Southern California Academy of Sciences, 65,* 237–244.

Puente, A. E., & Dewsbury, D. A. (1976). Courtship and copulatory behavior of bottlenosed dolphins (*Tursiops truncatus*). *Cetology, 21,* 1–9.

Ralston, J. V., & Herman, L. M. (1995). Perception and generalization of frequency contours by a bottlenose dolphin (*Tursiops truncatus*). *Journal of Comparative Psychology, 109,* 268–277.

Rankin, S., Oswald, J. N., & Barlow, J. (2008). Acoustic behavior of dolphins in the Pacific Ocean: Implications for using passive acoustic methods for population studies. *Canadian Acoustics, 36,* 88–92.

Rankin, S., Oswald, J., Barlow, J., & Lammers, M. (2007). Patterned burst-pulse vocalizations of the northern right whale dolphin, *Lissodelphis borealis. Journal of the Acoustical Society of America, 121,* 1213–1218.

Rasmussen, M. H., & Miller, L. A. (2002). Whistles and clicks from white-beaked dolphins, *Lagenorhynchus albirostris,* recorded in Faxaflói Bay, Iceland. *Aquatic Mammals, 28,* 78–89.

Reiss, D. (1988). Observations on the development of echolocation in young bottlenose dolphins. In P. E. Nachtigall & P. W. B. Moore (Eds.), *Animal sonar* (pp. 121–127). New York: Plenum Press.

Reiss, D., & McCowan, B. (1993). Spontaneous vocal mimicry and production by bottlenose dolphins (*Tursiops truncatus*): Evidence for vocal learning. *Journal of Comparative Psychology, 107,* 301.

Rendell, L. E., Matthews, J. N., Gill, A., Gordon, J. C. D., & Macdonald, D. W. (1999). Quantitative analysis of tonal calls from five odontocete species, examining interspecific and intraspecific variation. *Journal of the Zoological Society, London, 249,* 403–410.

Richards, D. G., Wolz, J. P., & Herman, L. M. (1984). Vocal mimicry of computer-generated sounds and vocal labeling of objects by a bottlenosed dolphin, *Tursiops truncatus. Journal of Comparative Psychology, 98,* 10–28.

Roch, M. A., Brandes, T. S., Patel, B., Barkley, Y., Baumann-Pickering, S., & Soldevilla, M. S. (2011). Automated extraction of odontocete whistle contours. *Journal of the Acoustical Society of America, 130,* 2212–2223.

Roch, M. A., Soldevilla, M. S., Burtenshaw, J. C., Henderson, E. E., & Hildebrand, J. A. (2007). Gaussian mixture model classification of odontocetes in the Southern California Bight and the Gulf of California. *Journal of the Acoustical Society of America, 121*, 1737–1748.

Samarra, F. I., Deecke, V. B., Vinding, K., Rasmussen, M. H., Swift, R. J., & Miller, P. J. O. (2010). Killer whales (*Orcinus orca*) produce ultrasonic whistles. *Journal of the Acoustical Society of America Express Letters, 128*, EL205–EL210.

Sayigh, L. S., Tyack, P. L., Wells, R. S., & Scott, M. D. (1990). Signature whistles of free-ranging bottlenose dolphins, *Tursiops truncatus*: Stability and mother-offspring comparisons. *Behavioral Ecology and Sociobiology, 26*, 247–260.

Shailer, M. J., & Moore, B. C. J. (1983). Gap detection as a function of frequency, bandwidth, and level. *Journal of the Acoustical Society of America, 74*, 467–473.

Schevill, W. E., & Lawrence, B. (1949). Underwater listening to the white porpoise (*Delphinapterus leucas*). *Science, 109*, 143–144.

Schevill, W. E., & Lawrence, B. (1956). Food-finding by a captive porpoise (*Tursiops truncatus*). *Breviora, 53*, 1–15.

Schotten, M., Lammers, M. O., Sexton, K., & Au, W. W. L. (2005). Application of a diver-operated 4-channel acoustic/video recording device to study wild dolphin echolocation and communication. *Journal of the Acoustical Society of America, 117*, 2552.

Sigurdson, J. (1993). Frequency-modulated whistles as a medium for communication with the bottlenose dolphin (*Tursiops truncatus*). In H. L. Roitblat, L. M. Herman, & P. E. Nachtigall (Eds.), *Language and communication: Comparative perspectives* (pp. 153–173). Hillsdale, NJ: Erlbaum.

Simard, P., Mann, D. A., & Gowans, S. (2008). Burst-pulse sounds recorded from white-beaked dolphins (*Lagenorhynchus albirostris*). *Aquatic Mammals, 34*, 464–470.

Sivian, L. J., & White, S. D. (1933). On minimum audible sound fields. *Journal of the Acoustical Society of America, 4*, 288–321.

Sjare, B. L., & Smith, T. G. (1986). The vocal repertoire of white whales, *Delphinapterus leucas*, summering in Cunningham Inlet, Northwest Territories. *Canadian Journal of Zoology, 64*, 407–415.

Smolker, R. A., Mann, J., & Smuts, B. B. (1993). Use of signature whistles during separations and reunions by wild bottlenose dolphin mothers and infants. *Behavioral Ecology and Sociobiology, 33*, 393–402.

Steiner, W. W. (1981). Species-specific differences in pure tonal whistle vocalizations of five western North Atlantic dolphin species. *Behavioral Ecology and Sociobiology, 9*, 241–246.

Taruski, A. G. (1979). The whistle repertoire of the North Atlantic pilot whale (*Globicephala melaena*) and its relationship to behavior and environment. In H. E. Winn & B. L. Olla (Eds.), *Behavior of marine animals: Current perspectives on research: Vol. 3. Cetaceans* (pp. 345–368). New York: Plenum Press.

Tyack, P. (1985). An optical telemetry device to identify which dolphin produces a sound. *Journal of the Acoustical Society of America, 78,* 1892–1895.

Tyack, P. (1986). Whistle repertoires of two bottlenosed dolphins, *Tursiops truncatus:* Mimicry of signature whistles? *Behavioral Ecology and Sociobiology, 18,* 251–257.

Tyack, P. L. (1997). Development and social functions of signature whistles in bottlenose dolphins, *Tursiops truncatus. Bioacoustics, 8,* 21–46.

Tyack, P. L., & Clark, C. W. (2000). Communication and acoustic behavior of dolphins and whales. In W. W. L. Au, A. N. Popper, & R. R. Fay (Eds.), *Hearing by whales and dolphins* (pp. 156–224). New York: Springer.

Tyack, P. L., & Recchia, C. A. (1991). A datalogger to identify vocalizing dolphins. *Journal of the Acoustical Society of America, 90,* 1668–1671.

Tyack, P. L., Zimmer, W. M., Moretti, D., Southall, B. L., Claridge, D. E., Durban, J. W., et al. (2011). Beaked whales respond to simulated and actual navy sonar. *PLoS ONE, 6,* e17009.

Urick, R. J. (1983). *Principles of underwater sound* (3rd ed.). New York: McGraw-Hill.

Wang, D., Würsig, B., & Evans, W. (1995). Comparisons of whistles among seven odontocete species. In R. A. Kastelein & J. A. Thomas (Eds.), *Sensory systems of aquatic mammals* (pp. 283–290). Woerden, the Netherlands: De Spil.

Watkins, W. A. (1966). Listening to cetaceans. In K. S. Norris (Ed.), *Whales, dolphins, and porpoises* (pp. 471–476). Berkeley: University of California Press.

Watkins, W. A., & Schevill, W. E. (1971). Underwater sounds of *Monodon* (narwhal). *Journal of the Acoustical Society of America, 49,* 595–599.

Watkins, W. A., & Schevill, W. E. (1974). Listening to Hawaiian spinner porpoises, *Stenella cf. longirostris,* with a three-dimensional hydrophone array. *Journal of Mammalogy, 55,* 319–328.

Watwood, S. L., Owen, E. C. G., Tyack, P. L., & Wells, R. S. (2005). Signature whistle use by temporarily restrained and free-swimming bottlenose dolphins, *Tursiops truncatus. Animal Behaviour, 69,* 1373–1386.

Wood, F. G., Jr. (1953). Underwater sound production and concurrent behavior of captive porpoises, *Tursiops truncatus* and *Stenella plagiodon. Bulletin of Marine Science, 3,* 120–133.

Xitco, M. J., & Roitblat, H. L. (1996). Object recognition through eavesdropping: Passive echolocation in bottlenose dolphins. *Animal Learning and Behavior, 24,* 355–365.

Yuen, M. M. L., Nachtigall, P. E., Breese, M., & Vlachos, S. A. (2007). The perception of complex tones by a false killer whale (*Pseudorca crassidens*). *Journal of the Acoustical Society of America, 121,* 1768–1774.

6 Making Sense of It All: Multimodal Dolphin Communication

Denise L. Herzing

History of Sound/Behavior Correlates

Making sense of any communication system, including that of aquatic mammals, requires multimodal analysis of their signal use, not just single physical signal analysis. The total sensory envelope of potential information available for dolphins includes vocal as well as nonvocal signals including visual, tactile, kinesthetic, and chemoreceptive. The challenges of studying multimodal communication are vast and have evolved from early theoretical ideas (Johnson, 1993; Tavolga, 1983) to quantitative techniques and frameworks of analysis (Herzing, 2006; Oswald, Barlow, & Norris, 2003). Ironically, early work on delphinids was rich with observations of both behavior and sound correlates. Although most studies were descriptive in nature and focused primarily on bottlenose dolphins in captivity, they led to some of the baseline understanding of dolphin body postures and vocalization use.

Dolphin vocalizations are typically categorized into three types. Frequency-modulated whistles were historically, and continue to be, the most commonly studied and behaviorally correlated type of delphinid vocalization. Echolocation clicks have been studied heavily by the military to inform sonar design, and later for additional social explanations of delphinid behavior. Burst pulse sounds, typically clumps of clicks with varying repetition rates, have been described (Lammers & Oswald, chap. 5, this vol.) using many terms, including "squawks," "moans," and "groans," but remain difficult to categorize and correlate with social functions. Even the previously assumed categories of dolphin sounds (whistles, clicks, burst pulse) have come into question, as the possibility of their occurring along a continuum has challenged the idea of discrete categories (Madsen, Jensen, Carder, & Ridgway, 2012; Murray, Mercado, & Roitblat, 1998). Nonvocal acoustics—those not modulated by vocal mechanisms but produced percussively—were also generally described, but only recently have they been spectrally studied and behaviorally correlated. In this chapter, I review past and present attempts to describe and analyze multimodal signals, with an eye to integrating these modalities for future work.

1960–1990: Acoustic and Postural Communication Signals

In the early history of studying dolphin communication signals, most research occurred in captivity, and many reports were anecdotal. The best comprehensive books of this era are *Whales, Dolphins, and Porpoises* (Norris, 1966), *Cetacean Communication: Mechanisms and Functions* (Herman, 1980), and *Dolphin Cognition and Behavior: A Comparative Approach* (Schusterman, Thomas, & Wood, 1986).

Some of the early observations of bottlenose dolphin (*Tursiops truncatus*) mother-calf behavior led to acoustic observations of the signature whistle (Tavolga & Essapian, 1957). Caldwell and Caldwell (1965) first suggested and later revised (Caldwell, Caldwell, & Tyack, 1990) that stereotyped whistles, termed "signature" whistles, were individually specific and functioned as identifiers of individual dolphins. Signature whistles were later described (Caldwell, Caldwell, & Miller, 1973) for captive Atlantic spotted dolphins (*Stenella plagiodon/frontalis*). Caldwell and Caldwell (1967) also reported a "whistle-squawk" produced by bottlenose dolphins when removed from the water, and suggested that this type of sound was emitted when the dolphin was "protesting." Reiss (1988) later reported the presence of the "whistle-squawk" for two infant bottlenose dolphins and suggested that these sounds were produced during emotional contexts when animals were in an excited state.

During these early decades, general associations between whistles and behavior in the wild were reported. These included increased whistling rates during excitement and stress in a number of species, including spinner dolphins, *Stenella longirostris* (Norris, Würsig, Wells, & Würsig, 1994), during bow riding and feeding in the common dolphin, *Delphinus delphis* (Busnel & Dziedzic, 1966), during feeding in pilot whales, *Globicephala* sp. (Dreher & Evans, 1964), and during fleeing and cooperative behavior in bottlenose dolphins (Evans, 1966), pilot whales, *Globicephala melaena* (Taruski, 1979), and beluga whales, *Delphinapterus leucas* (Sjare & Smith, 1986). Whistles were originally thought to display relatively low directionality and serve as long-distance social communicative signals owing to their highly modulated frequency characteristics (Norris & Dohl, 1980). During these decades, whistles were already understood to function for the broadcast of individual identity and to maintain and initiate contact between individuals, although more detailed analysis did not occur until the 1990s, including the discovery of the directional nature of the harmonics found with whistles (Lammers, Au, & Herzing, 2003).

Echolocation click trains, termed a "buzz," are known to emanate forward from the melon in a cone of directional sound and are associated primarily with gathering environmental information and navigation. The use of echolocation signals in detecting and retrieving prey items has been well established. In addition, intense sound pressure levels of over 220 dB *re* 1 µPa (Au, Floyd, & Haun, 1978; reviewed by Au, 1993; Norris, Prescott, Asa-Dorian, & Perkins, 1961) emitted by bottlenose dolphins were also proposed to have a debilitating effect on prey species (Norris & Mohl, 1981), including during the herding of fish (Hult, 1982). Echolocation buzzes were first reported in threat situations between bottlenose dolphins (Lilly & Miller, 1961) and were subsequently described during general social interactions (Reiss, 1988). Various types of buzzes were also reported from captivity during interspecific encounters between dolphins and sharks (Wood, 1953; Wood, Caldwell, & Caldwell, 1970).

These early reports also suggested that burst pulse sounds play a significant role in dolphin social communication (Caldwell & Caldwell, 1966, 1967; Caldwell, Caldwell, & Evans, 1966; Gish, 1979; Herman & Tavolga, 1980). General associations between behaviors and burst pulse sounds were reported for captive bottlenose dolphins, including high-intensity broadband "cracks" or "pops" of alarm and fright (Caldwell & Caldwell, 1967) and squawks during head-to-head and open-mouth agonistic encounters (Caldwell & Caldwell, 1967; Gish, 1979; Herzing, 1988; Overstrom, 1983), as well as during play behavior (Caldwell & Caldwell, 1967).

Burst pulse vocalizations were also reported for a variety of wild odontocetes, including exchanges between spinner dolphins (Norris et al., 1994; Watkins & Schevill, 1974), pilot whales and harbor porpoise, *Phocoena phocoena* (Busnel & Dziedzic, 1966), and narwhals, *Monodon monoceros* (Ford & Fisher, 1978). In addition, the use of pulsed codas was thought to carry signature information in sperm whales, *Physeter macrocephalus* (Watkins & Schevill, 1977). These studies suggested that burst pulse sounds were an important but often overlooked part of the social vocal repertoire of many odontocetes.

Dolphins may also be able to convey and extract important information through the use of aerial behaviors and percussive sounds associated with body and tail movements both underwater and at the surface. Nonvocal impulse sounds, including sounds produced by slamming of body parts, cavitation, percussive thrashing during attempted hits, closure of the jaw, and various aerial behaviors, were described spectrally and in behavioral contexts for multiple species during this period (Caldwell & Caldwell, 1967; Marten, Norris, Moore, & Englund, 1988). Such signals typically have short durations and were thought to function as supplemental or exclamatory signals during activity, either from a distance or in close proximity. Nonvocal acoustics have been correlated with behavioral activity for captive bottlenose dolphins, including jaw claps in aggressive contexts (Caldwell & Caldwell, 1971; Herzing, 1988; Overstrom, 1983), interspecifically during the intimidation of a subordinate dolphin by dominant conspecifics (Wood, 1953), and during contexts of agitation and stress (Caldwell, Haugen, & Caldwell, 1962; Norris & Dohl, 1980). Other nonvocal acoustics and behavioral correlates were also described for free-ranging dolphins, including bubbles and bubble rings (tori) used during aggressive interactions of pantropical spotted dolphins, *Stenella attenuata* (Pryor & Kang, 1980).

Hints of the importance of the coordination and synchronization of behavior and vocalizations were first reported in the choruses of barks during interspecific chases in captivity by dolphins exposed to a tiger shark (Wood, 1953). Mammalian patterns of synchronized behavior during coordinated efforts of activity have been described during coordination activities of male baboon coalitions (Smuts, 1985) and in the escalating pep rallies of wild dogs as signals of readiness to fight (Estes & Goddard, 1967), but little work was done on dolphin synchrony until the 1990s.

1960–1990: Nonacoustic Communication Signals

Studies of other communicative modalities were, by necessity, restricted to captive or experimental environments, although attempts at understanding the evolutionary and functional significance of these senses were made in both captive and natural settings (see Hanke & Erdsack, chap. 3, this vol.). Delphinid vision, which is highly discriminative, especially for moving stimuli (Nachtigall, 1986), has been discussed as a possible mechanism for emotional communication (Herman & Tavolga, 1980). For instance, visual acuity and low light detection have been suggested to play a role in many visual displays of body posture in smaller dolphin species (Caldwell & Caldwell, 1972; Puente & Dewsbury, 1976; Tavolga, 1966).

Gustatory (taste) senses were also studied in the discrimination of prey (salt, quinine, acids) and for bodily functions such as sexual or excretory behavior (Bullock & Gurevich, 1979; Kuznetsov, 1974; Mountcastle, 1974; Sokolov & Volkova, 1973; reviewed in Nachtigall, 1986). However, during this period of dolphin science, pheromone communication was not examined in any detail. A few studies attempted to make behavioral correlations with chemical senses and ovulation by measuring serum estradiol in captive spinner dolphins (Wells, 1984), but the work was inconclusive.

The use of vibrissal crypts, the remnants of vibrissae follicles, and cutaneous skin receptors for the tactile sense (touch) were noted for tactile behavior during nursing and sex (Bullock & Gurevich, 1979; Palmer & Weddell, 1964; Yablokov, Bel'kovich, &

Borisov, 1972). General descriptions of touch were often included in the description of general social behavior among various delphinids (Bateson, 1965; Brown & Norris, 1956; Brown, Caldwell, & Caldwell, 1966; Essapian, 1953; McBride & Hebb, 1948; Puente & Dewsbury, 1976; Tavolga & Essapian, 1957).

In the wild, observations of tactile contact were noted for a variety of dolphin species (Indian Ocean bottlenose dolphin, *Tursiops aduncus*: Saayman, Tayler, & Bower, 1973; humpback dolphins, *Sousa chinensis*: Saayman & Tayler, 1979; various: Norris & Dohl, 1980; pantropical spotted dolphins: Pryor & Kang, 1980). However, observations on other senses continued to be limited and difficult to obtain in free-ranging dolphins.

Current Approaches to Multimodal Studies

The 1990s ushered in new technology, new frameworks, and new opportunities for long-term fieldwork. The emergence of multimodal thinking began with Karen Pryor's (1990) thoughtful chapter on the senses available to dolphins. Researchers also began approaching the science of dolphin communication with more standardized techniques used in animal behavior, including the development of systematic behavior ethograms and vocal repertoires. Even the conflict over harvesting tuna by encircling dolphins in nets provided opportunities for behavioral observations while trying to solve this fisheries management problem (Pryor & Kang-Shallenberger, 1991). Nonetheless, studies that allowed the simultaneous correlation of behavior and sound or other modalities were still limited.

Probably the biggest change in the 1990s was increased access to wild populations of dolphins for both surface and underwater observations of known individuals and family groups. Work at the surface tended to involve general behavioral categories such as travel, rest, socializing, and foraging, while underwater observations could include detailed ethograms of body postures, vocalizations, swimming activity, and sexual-social behavior. For other deep-water odontocetes, such as sperm whales and pilot whales, new technologies entailed the attachment of monitoring devices and included sophisticated data collection (e.g., of orientation of animal, velocity, and sound) that allowed researchers to visualize diving and foraging features (e.g., Johnson, Madsen, Zimmer, Aguilar de Soto, & Tyack, 2006). To date, this remains the only plausible way to gain information about species that go out of sight during their daily lives. This shift to focusing more on behavior in the wild enabled us to look at the complexities of dolphin communication in a more intimate, real-world, and detailed way.

1990 to Present: Acoustic and Postural Communication Signals

General reviews of wild dolphin behavior and sound correlation are summarized in Herzing (2000) and updated here for multimodalities (table 6.1). During these decades, ethograms and vocal repertoires were also described for a variety of odontocete species beyond small delphinids and matched with basic behavioral contexts (e.g., López, 2011; Scheer, Hofmann, & Behr, 2004; Xiao & Wang, 2005).

Whistles (primarily signature whistles) and their behavioral correlates continued to be studied in captive dolphins (Caldwell, Caldwell, & Tyack, 1990; Sayigh, Esch, Wells, & Janik, 2007; Tyack, 1993). The use of bubble streams both to identify the individual vocalizing and to verify the vocalizer was suggested by some researchers (McCowan & Reiss, 1995a) but later challenged by others working both in the wild (Fripp, 2006) and in captivity (Sayigh et al., 2007).

Many new studies in the wild supported earlier thoughts about signature whistle function, including their use during mother-calf reunions, alloparental care, and courtship (Indian Ocean bottlenose dolphins: Smolker, Mann, & Smuts, 1993; bottlenose dolphins: Tyack, 1993; Atlantic spotted dolphins: Herzing, 1996). These observations are supported by both the stability of whistle contours over the years and kin-related identifiers in whistles (bottlenose dolphins: Sayigh, Tyack, Wells, & Scott, 1990; Atlantic spotted dolphins: Bebus & Herzing, 2015; Burris, 2004). New work describes the exchange of signature whistles during greetings at sea between dolphin groups (Quick & Janik, 2012). Vocal copying of signature whistles between dolphins has also been documented, suggesting that dolphins modify their whistles when addressing each other (King, Sayigh, Wells, Fellner, & Janik, 2013). Much of this new work suggests that more detailed information is contained within a signature whistle than previously thought.

Evidence of group cohesion whistles by male Indian Ocean bottlenose dolphins (Smolker & Pepper, 1999) was also later verified in a second community of bottlenose dolphins (Janik & Slater, 1998; Watwood, Tyack, & Wells, 2004). Other types of whistles, including overlapping whistles (termed screams or chorusing), were also noted for a few delphinid species (Herzing, 1996, 2000; Janik, Simard, Sayigh, Mann, & Frankel, 2011).

The 1990s also ushered in new observations of dolphin vocalization other than whistles. Researchers began investigating burst pulse sounds in captive behavioral scenarios, including during play fights (Blomqvist, Mello, & Amundin, 2005) and aggression (Blomqvist & Amundin, 2004a). Burst pulse sounds (including those labeled "thunks") were also described during disciplinary behavior between bottlenose dolphin mothers and infants (McCowan & Reiss, 1995b; Schultz, Cato, Corkeron, & Bryden, 1995). Other burst pulse sounds, including the commonly reported "squawks," were observed for various species during aggressive activity in the wild (spinner dolphins: Driscoll, 1995; Atlantic spotted dolphins: Herzing, 1996; bottlenose dolphins: Robinson, 2014). The "whistle-squawk" has also been reported in the wild as the predominate vocalization produced during alarm and distress in Atlantic spotted dolphins (Herzing, 1996). Recently, Ridgway and colleagues (2014) also reported on a "victory" squeal correlated with success in fish catching.

Burst pulse sounds were also reported during aggression, including "pops" during the herding of females by male coalitions of Indian Ocean bottlenose dolphins (Connor & Smolker, 1996), "barks" during interspecific aggression between Atlantic spotted and bottlenose dolphins (Herzing, 1996), and "low-frequency" narrow-band sounds produced between conspecifics in bottlenose dolphins (Schultz, Cato, Corkeron, & Bryden, 1995) and Pacific humpback dolphins (van Parijs & Corkeron, 2001). "Brays" were originally described for bottlenose dolphins in Portugal (dos Santos, Caporin, Moreira, Ferreira, & Coelho, 1990; dos Santos, Ferreira, & Harzen, 1995) and were later correlated with surface observations of foraging behavior (Janik, 2000). However, when these calls were observed underwater, they were correlated with aggressive fights and courtship behavior (Herzing, 2015). Correlations between silence and certain dolphin behaviors were also noted for bottlenose dolphins in Portugal (dos Santos et al., 1990) and in the Bahamas, such as when dyads or small subgroups of Atlantic spotted dolphins lay motionless on the bottom when a shark passed by (Herzing, 1996).

Functional aspects of nonwhistles, including echolocation, have now been described during foraging and the stunning of prey for multiple free-ranging delphinids (Herzing, 1996, 2004; Marten, Herzing, Poole, & Newman-Allman, 2001). High-repetition-rate echolocation trains (buzzes) were described for Atlantic spotted dolphins during social behaviors, including tactile stimulation between conspecifics during courtship, interspecific dolphin aggression, discipline, and dolphin-shark interactions (Herzing, 1996, 2004). A correlation of various sound types was reported during foraging for bottlenose dolphins and Atlantic spotted dolphins, including razor buzzes, trills, and upswept whistles (dos Santos, Ferreira, & Harzen, 1995; Herzing, 1996, 2000; Rossbach & Herzing, 1997), and subsequent analysis showed that some of these foraging signals penetrated bottom sediment, enabling the recognition of objects (Roitblat, Au, Nachtigall, Shizumura, & Moons, 1995).

Correlations between general acoustic categories and surface behavior have also been described for a number of species, including Risso's dolphins, *Grampus griseus* (Corkeron & van Parijs, 2001), Hector's dolphins (Dawson, 1991; Slooten, 1994), common dolphins (Henderson, Hildebrand, Smith, & Falcone, 2012), bottlenose dolphins (Oehen, 1996), spinner dolphins (Ostman, 1994), killer whales (Simila & Ugarte, 1993), beluga whales (Sjare & Smith, 1986), pilot whales (Hofmann, Scheer, & Behr, 2004; Scheer et al., 2004), and multiple delphinids (Wang, Würsig, & Evans, 1995). A variety of killer whale projects involved in the long-term tracking of resident pods in the Pacific Northwest continued to report the correlation of surface behavior with sound (Ford, 1991; Grahman & Noonan, 2010; Rehn, Teichert, & Thomsen, 2007). For killer whales, burst pulse sounds have actually been studied more than whistles, reversing the emphasis on sound type analysis historically for the Delphinidae. For example, whereas many dolphin species use whistles to convey signature information, killer whales appear to use burst pulse calls as pod and community identifiers (Thomsen, Franck, & Ford, 2002).

Nonvocal acoustics, including visual and acoustic aspects of bubbles and bubble rings, have been correlated with play activity in captive bottlenose dolphins (Marten, Shariff, & Psarakos, 1996; McCowan, Marino, Vance, Walke, & Reiss, 2000), although for at least Atlantic spotted dolphins, this function is different in the wild (Herzing, 1996, 2004). A few studies in marine parks using underwater viewing windows began to emerge (bottlenose dolphins: Favaro, Gnome, & Pessani, 2013; Gnome, Moriconi, & Gambini, 2005; Tizzi & Accorsi, 2010; beluga whales: Horback, Friedman, & Johnson, 2010) that allowed greater clarity of various underwater postural and acoustic signals in this setting.

A variety of nonvocal acoustics in a variety of behavioral contexts were now reported in free-ranging Atlantic spotted dolphins (e.g., Herzing, 1996). These included the occurrence of tail slaps during annoyance or attention-seeking behavior, jaw claps during escalated aggression, and aerial splashing during both play and intra- and interspecific aggressive chases. A variety of bubble displays (bubble trails, full and half bubbles, and bubble rings or tori) were also observed during annoyance and aggressive contexts. In-air vocalizations, including the chuff (an explosive exhalation), were observed during annoyance, and the raspberry (a constricted exhalation) was reported during interspecies affiliative contexts for multiple species (pantropical spotted dolphins: Pryor & Kang-Shallenberger, 1991; spinner dolphins: Driscoll, 1995; Atlantic spotted dolphins and bottlenose dolphins: Herzing, 1996).

Researchers were also beginning to describe the visual information available in wild populations of dolphins, including body postures and touch (Herzing, 1996; Paulos, Dudzinski, & Kuczaj, 2008; Würsig, Kieckhefer, & Jefferson, 1990). While researchers first began to document these behaviors decades ago, studies of contact behaviors in captive dolphins have continued to emphasize flipper rubbing as a social behavior that varies with age and sex (Johnson & Moewe, 1999; Sakai, Hishii, Takeda, & Kohshima, 2006). Although Tamaki and colleagues (2006) determined that the direction of rubbing did not vary with the identity of the dolphin, they did find that flipper rubbing in bottlenose dolphins may function to restore friendly relationships after aggression or even reduce potential conflicts. In addition, measurements of the tactile pressures produced during vocalizations in common dolphins (Kolchin & Bel'kovich, 1973) suggest that calls like buzzes and pops may also have a tactile component. Applied to Atlantic spotted dolphins, this may be a feature of social behavior in which animals direct buzzes at specific parts of one another's bodies during play and aggression (Herzing, 2004), suggesting an important tactile component of acoustic signals.

Table 6.1
Species comparisons of sensory modalities studied in the wild and in captivity.

Species	Sensory Modalities	Captive or Wild	Behavioral Context	References
	Sound			
Bottlenose dolphin, *Tursiops truncatus*	Sound, various vocalization types, and nonacoustic sounds	Captive	Aggression, alarm, chases, distress, dominance, exploration, feeding, fear, fights, foraging, courtship, separation, sexual play	Blomqvist & Amundin, 2004a; Blomqvist, Mello, & Amundin, 2005; Caldwell & Caldwell, 1965, 1967; Caldwell et al., 1962, 1990; Favaro et al., 2013; Gish, 1979; Gnome et al., 2005; Herzing, 1988; Lilly & Miller, 1961; Marten et al., 1988, 1996; McCowan & Reiss, 1995a, 1995b; McCowan et al., 2000; Overstrom, 1983; Reiss, 1988; Sayigh et al., 2007; Tavolga & Essapian, 1957; Tizzi & Accorsi, 2010; Tyack, 1993; Wood, 1953; Wood et al., 1970
Bottlenose dolphin, *Tursiops truncatus*	Sound, various vocalization types	Wild	Aggression, arousal, cohesion, contact, foraging, play, reunions, scanning, sexual, social	Boisseau, 2005; dos Santos et al., 1990, 1995; Herzing, 1996, 2000, 2004, 2015; Hult, 1982; Janik, 1999, 2000; Janik & Slater, 1998; Janik et al., 2011; King & Janik, 2013; King et al., 2013; López, 2011; Marten et al., 2001; Norris & Mohl, 1981; Oehen, 1996; Quick & Janik, 2012; Wang et al., 1995; Robinson, 2014; Rossbach & Herzing, 1997; Schultz et al., 1995; Watwood et al., 2004

Table 6.1 (continued)

Species	Sensory Modalities	Captive or Wild	Behavioral Context	References
	Sound			
Indian Ocean bottlenose dolphin, *Tursiops aduncus*	Sound, various vocalization types	Wild	Consortship, reunions	Connor & Smolker, 1996; Smolker & Pepper, 1999; Smolker et al., 1993
Atlantic spotted dolphin, *Stenella frontalis*	Sound, various vocalization types, and nonacoustic sounds	Captive	Various behavioral states	Caldwell & Caldwell, 1966, 1971; Caldwell et al., 1973; Wood, 1953
Atlantic spotted dolphin, *Stenella frontalis*	Sound, various vocalization types, and nonacoustic sounds	Wild	Affiliative, aggression, alloparental, annoyance, chase, contact, courtship, discipline, distress, excitement, fights, intra/ interspecies, reunions, social play	Burris, 2004; Bebus & Herzing, 2015; Herzing, 1996, 2000, 2004, 2015; Marten et al., 1988
Spinner dolphin, *Stenella longirostris*	Sound, various vocalization types	Wild	Affiliative (interspecific), coordination, nighttime activity, travel	Driscoll, 1995; Lammers & Au, 1996; Lammers et al., 2006; Norris & Dohl, 1980; Norris et al., 1994; Ostman, 1994; Watkins & Schevill, 1974
Pantropical spotted dolphin, *Stenella attenuata*	Sound, various vocalization types, and nonacoustic sounds	Wild	Affiliative (intra/ interspecies), aggression, frustration, bubbles	Pryor & Kang, 1980; Pryor & Kang-Shallenberger, 1991
Hector's dolphin, *Cephalorhynchus hectori*	Sound, various vocalization types	Wild	Sexual, aggression, play, feeding, aerial	Dawson, 1991; Slooten, 1994

Table 6.1 (continued)

Species	Sensory Modalities	Captive or Wild	Behavioral Context	References
	Sound			
Common dolphin, *Delphinus* sp.	Sound, various vocalization types	Wild	Feeding, bow riding	Busnel & Dziedzic, 1966; Henderson et al., 2012; Papale et al., 2013
Humpback dolphin, *Sousa chinensis*	Sound, various vocalization types	Wild	Aggression, surface behavior	Corkeron & van Parijs, 2001; van Parijs & Corkeron, 2001
Killer whale, *Orcinus orca*	Sound, nonacoustic sounds	Captive	Body cavitations, jumps	Marten et al., 1988
Killer whale, *Orcinus orca*	Sound, various vocalization types	Wild	Feeding, social	Ford, 1991; Grahman & Noonan, 2010; Rehn et al., 2007; Simila & Ugarte, 1993; Thomsen et al., 2002
Yangtze finless porpoise, *Neophocaena phocaenoides*	Sound, various vocalization types	Captive	Variety of behavior	Xiao & Wang, 2005
Amazon river dolphin, *Inia geoffrensis*	Sound, various vocalization types, and nonacoustic sounds	Captive	Alarm, feeding, jaw claps, various behavioral states	Caldwell et al., 1966
Beluga whale, *Delphinapterus leucas*	Sound, various vocalization types	Captive	Aggression, S-posture	Horback et al., 2010
Beluga whale, *Delphinapterus leucas*	Sound, various vocalization types	Wild	Contact, coordination	Sjare & Smith, 1986
Narwhal, *Monodon monoceros*	Sound, various vocalization types	Wild	Social	Ford & Fisher, 1978
Sperm whale, *Physeter macrocephalus*	Sound, click codas	Wild	Social, diving, feeding	Watkins & Schevill, 1977; Whitehead, 1996

Table 6.1 (continued)

Species	Sensory Modalities	Captive or Wild	Behavioral Context	References
Touch				
Bottlenose dolphin, *Tursiops truncatus*	Touch	Captive	Nursing, sexual, social behavior	Bateson, 1965; Brown et al., 1966; Brown & Norris, 1956; Bullock & Gurevich, 1979; Essapian, 1953; Johnson & Moewe, 1999; McBride & Hebb, 1948; Mountcastle, 1974; Palmer & Weddell, 1964; Puente & Dewsbury, 1976; Tavolga & Essapian, 1957; Yablokov et al., 1972
Variety of delphinids	Touch	Wild	Nursing, sexual, social behavior	Herzing, 1996; Norris & Dohl, 1980; Papale et al., 2013; Paulos et al., 2008; Pryor & Kang, 1980; Saayman et al., 1973; Saayman & Tayler, 1979; Sakai et al., 2006; Würsig et al., 1990
Other Senses				
Variety of delphinids	Vision	Captive	Body postures	Caldwell & Caldwell, 1972; Nachtigall, 1986; Puente & Dewsbury, 1976; Tavolga, 1966; Wang et al., 1995
Bottlenose dolphin, *Tursiops truncatus*	Taste (gustatory)	Captive	Sexual, excretory, ovulation	Bullock & Gurevich, 1979; Kuznetsov, 1974; Nachtigall, 1986; Sokolov & Volkova, 1973; Wells, 1984
Variety of delphinids	Taste (gustatory)	Wild	Courtship	Herzing, 2000; St. Aubin et al., 2013
Guiana dolphin, *Sotalia guianensis*	Electrical	Captive	Experimental	Czech-Damal et al., 2012
Variety of delphinids	Magnetic	Wild	Orientation	Bauer et al., 1985; Roitblat, 1990
Atlantic spotted dolphin, *Stenella frontalis* Common dolphin, *Delphinus delphis*	Touching with sound	Captive and wild	Courtship and mating	Herzing, 2004; Kolchin & Bel'kovich, 1973

Table 6.1 (continued)

Species	Sensory Modalities Special Issues	Captive or Wild	Behavioral Context	References
Bottlenose dolphin, *Tursiops truncatus* Atlantic spotted dolphin, *Stenella frontalis*	Learning	Wild	Development, teaching	Bender et al., 2009; Fripp et al., 2005; Herzing, 2005; Janik, 1999
Variety of species	Information content	Wild	Foraging, referential labeling, Interspecies communication	Boisseau, 2005; Herzing, 1996, 2015; Johnson et al., 2006; King & Janik, 2013; May-Collado, 2010; Papale et al., 2013
Bottlenose dolphin, *Tursiops truncatus*	Cross-modal acoustic and vision	Captive and wild	Cognitive, sensory, communication signals	Harley et al., 2003; Herzing et al., 2012; Pack & Herman, 1995; Reiss & McCowan, 1993; Xitco & Roitblat, 1996; Xitco et al., 2001
Variety of species	Mimicry and synchrony	Captive and wild	Physical and vocal	Brownlee & Norris, 1994; Connor et al., 1992, 2006; Connor & Smolker, 1996; Cusick & Herzing, 2014; DeRuiter et al., 2012; dos Santos et al., 1995; Herzing, 1996, 2015; Herzing & Johnson, 1997; Janik et al., 2011; Lilly, 1965; Pryor & Kang-Shallenberger, 1991; Reiss & McCowan, 1993; Ridgway et al., 2012, 2014; Sakai et al., 2009; Sayigh et al., 2012; Whitehead, 1996; Wood, 1953
Variety of species	Technology-assisted discoveries	Captive and wild	Orientation, velocity of travel, phantom echoes, degraded images, vocalizer identification	Amundin et al., 2008; Au & Moore, 1988; Aubauer & Au, 1998; Aubauer et al., 2000; Azzuli et al., 1995; Ball & Buck, 2005; Blomqvist & Amundin, 2004b; Herman et al., 1990; Herzing, 1996; Lammers et al., 2003, 2006; Madsen et al., 2005; Sayigh et al., 2012; Starkhammar et al., 2009; Thomas et al., 2002; Tyack, 1985; Tyack et al., 2006; Tyack & Recchia, 1991

1990 to Present: Nonvocal Communication Signals
Researchers have also tried to correlate chemical signal use in the wild from behavioral observations, although measurements are difficult to come by. Herzing (2000) described the open-mouth behavior of solitary Atlantic spotted dolphins in the water column, suggesting that they were either (1) tasting the water and following a chemical trail or (2) passively listening through their lower jaw as a directional receiver or (3) doing both. Herzing also described open-mouth postures during courtship pursuit by males to females. Whether the dolphin is "tasting" or sensing chemicals in the water, or whether this signal functions as a visual cue, or both, is unknown, but it exemplifies the difficulties inherent in examining multimodal communication signals in the wild. Detecting chemicals and hormone use by dolphins provides researchers a great challenge in a medium such as water that rapidly disperses chemicals. Recent measures of cortisol levels and hematological factors in wild and captive dolphins (e.g., St. Aubin et al., 2013), collected while following stressed animals during chases and entanglements, has provided some new information on physiological changes relative to behavior. And although Roitblat (1990) commented on the possibility of dolphins following magnetic lines, as had been postulated during stranding events, little work has been done on the magnetic sense in dolphins. The use of magnetic cues had been discussed in earlier work (Bauer, Fuller, Perry, Dunn, & Zoeger, 1985), and some relevant anatomical structures were described (see Hanke & Erdsack, chap. 3, this vol.). Only recently has a new sense, electroreception, been verified in the Guiana dolphin (Czech-Damal et al., 2012), but this sense remains largely untested in most species.

Mimicry and Synchrony as Mechanisms in Communication
Early studies suggested that bottlenose dolphins had the ability to mimic many sound types (Lilly, 1965), which opened the door to potential multimodal uses of mimicry. Spontaneous mimicry of vocal signals during human interactive contexts and cognitive experiments was also noted during the training of captive beluga whales (Ridgway, Carder, Jeffries, & Todd, 2012) and during the use of a keyboard (Reiss & McCowan, 1993). Such displays suggested that dolphins had complex cognitive flexibility and were applying it in unique situations of cross-species communication. And as researchers took to the field, the importance of synchrony in real-life communication was revealed. Such synchrony was observed during collaborative activities, including spinner dolphin behavior while dispersing from bays (Brownlee & Norris, 1994), during surfacing behavior of Indian Ocean bottlenose dolphins (Connor, Smolker, & Bejder, 2006; Sakai, Morisaka, Kogi, Hishii, & Kohshima, 2009), and

during herding behavior of females (Connor & Smolker, 1996; Connor, Smolker, & Richards, 1992). Other collaborative contexts of vocal synchrony include the chorusing of bottlenose dolphins in Scotland (Janik et al., 2011), the rhythmic braying of bottlenose dolphins (dos Santos et al., 1995; Herzing, 2015), and the diving of sperm whales (Whitehead, 1996).

Dolphins also display synchrony during defensive activities, as observed in male pantropical spotted dolphins while being herded into tuna nets (Pryor & Kang-Shallenberger, 1991). Dolphins also used synchrony in aggressive contexts, including interspecific interactions between Atlantic spotted dolphins and bottlenose dolphins (Cusick & Herzing, 2014; Herzing, 1996; Herzing & Johnson, 1997), and during intraspecific aggression in bottlenose dolphins (Herzing, 2015).

Observations concerning mimicry, which can be seen as a kind of "delayed synchronization," have been reported, including spontaneous mimicry in the wild of humans, of whistle playbacks, and of human mechanical signals like sonar (DeRuiter et al., 2012; Sayigh, Quick, Hastie, & Tyack, 2012). Together these observations suggest that mimicry and synchrony are important communication signal mechanisms for many odontocetes.

New Studies and Technologies for Multimodal Research

Exciting new work continues to emerge with new species in new geographical locations (e.g., Boisseau, 2005; May-Collado, 2010; Papale, Azzolin, Cascão, & Gannier, 2013) that report additional elements in the animals' vocal repertoires and sophisticated analyses of sound use (King & Janik, 2013). For all of this, the multimodal study of dolphin communication still presents many challenges. In the 1990s, researchers were starting to understand how dolphins used their overlapping senses to negotiate their aquatic world. Researchers began investigating the process of visual and sound overlap, including eavesdropping on the echolocation signals of others (Xitco & Roitblat, 1996), the cross-modal abilities of dolphins (Harley, Putman, & Roitblat, 2003; Pack & Herman, 1995), and the visual recognition of shapes and the perception of complex echoes (Au & Moore, 1988; Azzuli, Manzini, & Buracchi, 1995). The potential for synesthesia as documented in humans across multiple senses (Marks, 1978) is a possibility in dolphins; the occurrence of other modality combinations has not been tested (touch/sound, vision/touch, chemical/vision, etc.). While assessing the body postures of others implies the use of the visual sense, other senses have rarely been correlated with specific body postures or sounds.

Some studies have demonstrated that degraded or modified visual signals are still understood by dolphins. This includes the acoustic phantom echo work—essentially playback of a reflected echo off an object (Aubauer & Au, 1998)—and the use of degraded video images (Herman, Morrel-Sammuels, & Pack, 1990). Secondary sensory systems or use of degraded or modified communication signals may vary with the environment and conditions and should be looked at within the context of signal discrimination.

Scientists also began to think about the myriad of ways dolphins could share information, including the transmission and type of social learning (Herzing, 2005), potential teaching mechanisms (Bender, Herzing, & Bjorklund, 2009), implications of vocal learning (Janik, 1999), exposure to information from a larger community (Fripp et al., 2005), and potential "enculturation" issues, as identified by primate researchers (Bjorklund, Yunger, Bering, & Ragan, 2002). Placing these possible transmission mechanisms in the context of multimodal signal use may provide insights into the process of communication.

The emergence of new technologies has also been critical in detecting and correlating communication signals. Single hydrophones or multiple hydrophone systems or arrays have been tested over the decades (e.g., Clark, 1980; Sayigh, Tyack, & Wells, 1993; Watkins & Schevill, 1974). Other technology has been developed for high-frequency acquisition with directional or multiple hydrophone units (Lammers & Au, 1996; Lammers et al., 2003; Schotten, Lammers, Sexton, & Au, 2005). In some cases, these systems have successfully been deployed to measure spatial interactions of vocalizing dolphins, through hydrophone triangulation, to determine the distance between animals during signal exchanges (Aubauer, Lammers, & Au, 2000; Lammers, Schotten, & Au, 2006).

To explore dolphin-to-dolphin exchange in captivity, Tyack (1985) and Blomqvist and Amundin (2004b) developed acoustic tags, worn by dolphins, which indicate the sender or receiver of an acoustic signal. Other high-resolution synchronized video and sound systems in captivity have also been attempted but have had difficulty localizing the vocalizer to a productive degree (Ball & Buck, 2005; Thomas, Fristrup, & Tyack, 2002; Tyack & Recchia, 1991). In the wild, underwater video recordings synchronized with sound were first used by Herzing in 1985 and continue to be used extensively to correlate underwater sound and behavior of Atlantic spotted dolphins in the Bahamas (Herzing, 1996, 2000, 2004). Although some field sites have been conducive to close underwater work, the correlation and specific behavioral function of sounds, as well as other communicative modalities, within dolphin society still elude us for most species. Identifying the vocalizer still remains one of the greatest challenges to the study of dolphin communication signals today.

The Future of Multimodal Research

The Use of Technology

Although the use of data tags (DTAGs) for deep-diving odontocetes was originally limited to spatial orientation and oceanographic data recordings, tags with hydrophones for general behavioral correlations to sound have recently been developed (Madsen, Johnson, de Soto, Zimmer, & Tyack, 2005; Tyack, Johnson, Soto, Sturlese, & Madsen, 2006). These have been used, for example, to study the coordinated use of sound during pilot whale exchanges, demonstrating the potential for recording vocal exchanges between tagged individuals in the wild (Sayigh et al., 2012), and synchronous diving (Aoki, Sakai, Miller, Visser, & Sato, 2013). Such technology offers unique access and provides a glimpse of unstudied open-ocean cetacean vocal exchange on an individual level.

Technology created for other studies, such as of cognition or echolocation, might be repurposed for studying communication signals across modalities. For example, interface systems created for sound studies (Amundin et al., 2008; Starkhammar et al., 2009) provide the potential for tracking individually vocalizing dolphins. Other potential interfaces illuminating dolphin communication include underwater keyboards (Reiss & McCowan, 1993; Xitco, Gory, & Kuczaj, 2001), especially ones that provide for acoustic exchanges (Herzing, Delfour, & Pack, 2012).

With the ongoing development of hardware including underwater equipment, DTags, and underwater robots (AUVs, ROVs), it is possible that we will gain insight into worlds that have eluded us in the past. Sophisticated software programs (McCowan & Reiss, 1995c), neural networks (e.g., Deecke, Ford, & Spong, 1999; Oswald et al., 2003), and pattern discovery algorithms (e.g., Doyle, McCowan, Johnston, & Hanser, 2011; Kershenbaum, Sayigh, & Janik, 2013; Kohlsdorf, Mason, Herzing, & Starner, 2013; Magnusson, 1996) offer us ways to expedite the processing of data and identify units of information to categorize signals (e.g., Marler, 1982) and correlate them with behavior. Sequential and contextual analysis should also be a priority for the biologically or ecologically significant interpretation of behavior using data acquired by any of these methods.

New techniques, including software or observational frameworks, are needed to determine natural categories of sound (Herzing, 2006; van Hooff, 1982). The need to illuminate how the organism itself categorizes a signal should also be investigated (e.g., Harley, 2008), as it has been for other species (e.g., vervet monkeys: Cheney & Seyfarth, 1982; Placer & Slobodchikoff, 2000). Aspects of sound, including prosodic, rhythmic, and graded signal elements, are neglected areas of study and remain an issue in animal communication studies (Smith, 1965). Furthermore, the obvious need for modalities outside sound and physical postures, including chemical, tactile, magnetic, and other senses should be measured creatively and systematically integrated into the exploration of multimodal communication research in dolphins.

Multimodal behavioral analysis has been successful in human communication analysis and interpretation (e.g., Goodwin, 2000; see Johnson, chap. 9, this vol.) and provides a potential framework for many species. The importance of correlating behavioral contexts with signal measurements should not be underestimated. Methods similar to those Collins and colleagues (2011) used to study mother-pup vocalizations of Weddell seals (*Leptonychotes weddellii*) offer a good example of what could be done with delphinid signals. Whether universal rules of information design and expression apply to dolphins, as reported for other mammals and birds (Morton, 1977), is yet undetermined. Understanding the mechanisms and processes of coordinating signals using multiple modalities may be critical in the understanding and interpretation of communication information and may require new frameworks (e.g., Di Paolo, 2000; Di Paolo, Rohde, & Iizuka, 2008; Nolfi, 2005).

The Cognitive Overlap with Communication

The overlap and interplay of communication and cognition have also been noted through the decades (Johnson, 1993; Norris & Evans, 1988). The perceptual world of dolphins suggests that their senses evolved to suit their environment, and their overlap and function can be complicated (Jerison, 1986). As described by Mercado and DeLong (2010), moving from sensory abilities to complex communication is a natural progression. The importance of contextual saliency and vocal behavior (Hooper, Reiss, & Carter, 2006; Reiss, McCowan, & Marino, 1997), the discussion of the culture of cetaceans relative to their communicative abilities (Rendell & Whitehead, 2001), and the cognitive implications of synchrony (Fellner, Bauer, & Harley, 2006) have not escaped the eye of cetacean researchers. Much can be learned from carefully reviewing the cognitive complexity issues in other social mammals to gain perspective (Johnson, 2010). Two new frameworks that could further inform the cognitive-communicative overlap include Dynamic Systems Theory (King & Shanker, 2003) and Distributed Cognition (Herzing & Johnson, 2006; Hutchins & Johnson, 2009; Johnson, 2001; Johnson, chap. 9, this vol.). These approaches could complement or be integrated with the foregoing behavioral, sequential, and pattern analyses as emerging frameworks for future research.

References

Amundin, M., Starkhammar, J., Evander, M., Almqvist, M., Lindström, K., & Persson, H. W. (2008). An echolocation visualization and interface system for dolphin research. *Journal of the Acoustical Society of America, 123*, 1188.

Aoki, K., Sakai, M., Miller, P. J., Visser, F., & Sato, K. (2013). Body contact and synchronous diving in long-finned pilot whales. *Behavioural Processes, 99*, 12–20.

Au, W. W. L. (1993). *The sonar of dolphins*. New York: Springer.

Au, W. W. L., Floyd, R. W., & Haun, J. E. (1978). Propagation of Atlantic bottlenose dolphin echolocation signals. *Journal of the Acoustical Society of America, 64*, 411–412.

Au, W. W. L., & Moore, P. W. (1988). The perception of complex echoes by an echolocating dolphin. In P. E. Nachtigall & P. W. B. Moore (Eds.), *Animal sonar: Processes and performance* (pp. 295–299). New York: Plenum Press.

Aubauer, R., & Au, W. W. (1998). Phantom echo generation: A new technique for investigating dolphin echolocation. *Journal of the Acoustical Society of America, 104*, 1165.

Aubauer, R., Lammers, M. O., & Au, W. W. (2000). One-hydrophone method of estimating distance and depth of phonating dolphins in shallow water. *Journal of the Acoustical Society of America, 107*, 2744.

Azzuli, B., Manzini, A., & Buracchi, G. (1995). Acoustic recognition by a dolphin of shapes. In R. A. Kastelein, J. A. Thomas, & P. E. Nachtigall (Eds.), *Sensory systems of aquatic mammals* (pp. 137–156). Woerden, the Netherlands: De Spil.

Ball, K. R., & Buck, J. R. (2005). A beamforming video recorder for integrated observations of dolphin behavior and vocalizations. *Journal of the Acoustical Society of America, 117*, 1005.

Bateson, G. (1965). Porpoise community research: Final report. National Institutes of Health Contract No. N 60530-C-1098.

Bauer, G. B., Fuller, M., Perry, A., Dunn, J. R., & Zoeger, J. (1985). Magnetoreception and biomineralization of magnetite in cetaceans. In J. L. Kirschvink, D. S. Jones, & B. J. MacFadden (Eds.), *Magnetite biomineralization and magnetoreception in organisms* (pp. 489–507). New York: Plenum Press.

Bebus, S. E., & Herzing, D. L. (2015). Mother-offspring signature whistle similarity and patterns of association in Atlantic spotted dolphins (*Stenella frontalis*). *Animal Behavior and Cognition, 2*(1), 71–87.

Bender, C. E., Herzing, D. L., & Bjorklund, D. F. (2009). Evidence of teaching in Atlantic spotted dolphins (*Stenella frontalis*) by mother dolphins foraging in the presence of their calves. *Animal Cognition, 12*(1), 43–53.

Bjorklund, D. F., Yunger, J. L., Bering, J. M., & Ragan, P. (2002). The generalization of deferred imitation in enculturated chimpanzees (*Pan troglodytes*). *Animal Cognition, 5*(1), 49–58.

Blomqvist, C., & Amundin, M. (2004a). High-frequency burst-pulse sounds in agonistic/aggressive interactions in bottlenose dolphins, *Tursiops truncatus*. In J. A. Thomas, C. F. Moss, & M. Vater (Eds.), *Echolocation in bats and dolphins* (pp. 425–431). Chicago: University of Chicago Press.

Blomqvist, C., & Amundin, M. (2004b). An acoustic tag for recording directional pulsed ultrasounds aimed at free-swimming bottlenose dolphins (*Tursiops truncatus*) by conspecifics. *Aquatic Mammals, 30*(3), 345–356.

Blomqvist, C., Mello, I., & Amundin, M. (2005). An acoustic play-fight signal in bottlenose dolphins (*Tursiops truncatus*) in human care. *Aquatic Mammals, 31*(2), 187–194.

Boisseau, O. (2005). Quantifying the acoustic repertoire of a population: The vocalizations of free-ranging bottlenose dolphins in Fiordland, New Zealand. *Journal of the Acoustical Society of America, 117*(4), 2318.

Brown, D. H., Caldwell, D. K., & Caldwell, M. C. (1966). Observations on the behavior of wild and captive false killer whales with notes on associated behavior of other genera of captive delphinids. *Los Angeles City Museum Contributions in Science, 96*, 1–32.

Brown, D. H., & Norris, K. S. (1956). Observations of captive and wild cetaceans. *Journal of Mammalogy, 37*, 311–326.

Brownlee, S. M., & Norris, K. S. (1994). The acoustic domain. In K. S. Norris, B. Würsig, R. S. Wells, & M. Würsig (Eds.), *The Hawaiian spinner dolphin* (pp. 161–185). Berkeley: University of California Press.

Bullock, T. H., & Gurevich, V. S. (1979). Soviet literature on the nervous system and psychobiology of cetacea. *International Review of Neurobiology, 21*, 47–127.

Burris, J. (2004). *Signature whistle stability in wild female Atlantic spotted dolphins, Stenella frontalis.* Master's thesis, Florida Atlantic University.

Busnel, R. G., & Dziedzic, A. (1966). Acoustic signals of the pilot whale *Globicephala melaena* and of the porpoises *Delphinus delphis* and *Phocoena phocoena*. In K. S. Norris (Ed.), *Whales, dolphins, and porpoises* (pp. 607–648). Berkeley: University of California Press.

Caldwell, D. K., & Caldwell, M. C. (1966). Observations on the distribution, coloration, behavior and audible sound production of the spotted dolphin, *Stenella plagiodon. Contributions in Science, 104*, 1–28.

Caldwell, D. K., & Caldwell, M. C. (1971). Underwater pulsed sounds produced by captive spotted dolphins, *Stenella plagiodon. Cetology, 1*, 1–7.

Caldwell, D. K., & Caldwell, M. C. (1972). Senses and communication. In S. Ridgway (Ed.), *Mammals of the sea* (pp. 466–502). Springfield, IL: Thomas.

Caldwell, M. C., & Caldwell, D. K. (1965). Individualized whistle contours in bottlenose dolphins (*Tursiops truncatus*). *Nature, 207,* 434–435.

Caldwell, M. C., & Caldwell, D. K. (1967). Intraspecific transfer of information via the pulsed sound in captive odontocete cetaceans. In R. G. Busnel (Ed.), *Animal sonar systems, biology, and bionics* (pp. 879–936). Jouy-en-Josas: Laborative de Physiologie Acoustique France.

Caldwell, M. C., Caldwell, D. K., & Evans, W. E. (1966). Sounds and behavior of captive Amazon freshwater dolphins, *Inia geoffrensis. Los Angeles County of California Museum Contributions in Science, 108,* 1–24.

Caldwell, M. C., Caldwell, D. K., & Miller, J. F. (1973). Statistical evidence for individual signature whistles in the spotted dolphin, *Stenella plagiodon. Cetology, 16,* 1–21.

Caldwell, M. C., Caldwell, D. K., & Tyack, P. L. (1990). Review of the signature whistle hypothesis for the Atlantic bottlenose dolphin. In S. Leatherwood & R. R. Reeves (Eds.), *The bottlenose dolphin* (pp. 199–233). San Diego: Academic Press.

Caldwell, M. C., Haugen, R. M., & Caldwell, D. K. (1962). High-energy sound associated with fright in the dolphin. *Science, 138,* 907–908.

Cheney, D. L., & Seyfarth, R. M. (1982). How vervet monkeys perceive their grunts: Field playback experiments. *Animal Behaviour, 30,* 739–751.

Clark, C. W. (1980). A real-time direction finding device for determining the bearing to the underwater sounds of the southern right whales, *Eubalaena australis. Journal of the Acoustical Society of America, 68,* 508–511.

Collins, K. T., McGreevy, P. D., Wheatley, K. E., & Harcourt, R. G. (2011). The influence of behavioural context on Weddell seal (*Leptonychotes weddellii*) airborne mother-pup vocalization. *Behavioural Processes, 87*(3), 286–290.

Connor, R. C., & Smolker, R. A. (1996). "Pop" goes the dolphin: A vocalization male bottlenose dolphins produce during consortships. *Behaviour, 133,* 643–662.

Connor, R. C., Smolker, R., & Bejder, L. (2006). Synchrony, social behaviour, and alliance affiliation in Indian Ocean bottlenose dolphins (*Tursiops aduncus*). *Animal Behaviour, 72*(6), 1371–1378.

Connor, R. C., Smolker, R. A., & Richards, A. F. (1992). Two levels of alliance formation among male bottlenose dolphins (*Tursiops* sp.). *Proceedings of the National Academy of Sciences of the United States of America, 89,* 987–990.

Corkeron, P. J., & van Parijs, S. M. (2001). Vocalizations of eastern Australian Risso's dolphins, *Grampus griseus. Canadian Journal of Zoology, 79*(1), 160–164.

Cusick, J. A., & Herzing, D. L. (2014). The dynamic of aggression: How individual and group factors affect the long-term interspecific aggression between two sympatric species of dolphin. *Ethology, 120*(3), 287–303.

Czech-Damal, N. U., Liebschner, A., Miersch, L., Klauer, G., Hanke, F. D., Marshall, C., et al. (2012). Electroreception in the Guiana dolphin (*Sotalia guianensis*). *Proceedings of the Royal Society*, *279*(1729), 663–668.

Dawson, S. M. (1991). Clicks and communication: The behavioural and social contexts of Hector's dolphin vocalizations. *Ethology*, *84*, 265–276.

Deecke, V. C., Ford, J. K. B., & Spong, P. (1999). Quantifying complex patterns of bioacoustics variation: Use of a neural network to compare killer whale (*Orcinus orca*) dialects. *Journal of the Acoustical Society of America*, *105*(4), 2499–2507.

DeRuiter, S. L., Boyd, I. L., Claridge, D. E., Clark, C. W., Gagnon, C., Southall, B. L., et al. (2012). Delphinid whistle production and call matching during playback of simulated military sonar. *Marine Mammal Science*, *29*(2), E46–E59.

Di Paolo, E. A. (2000). Behavioral coordination, structural congruence and entrainment in a simulation of acoustically coupled agents. *Adaptive Behavior*, *8*(1), 27–48.

Di Paolo, E. A., Rohde, M., & Iizuka, H. (2008). Sensitivity to social contingency or stability of interaction? Modelling the dynamics of perceptual crossing. *New Ideas in Psychology*, *26*(2), 278–294.

dos Santos, M. E., Caporin, G., Moreira, H. O., Ferreira, A. J., & Coelho, J. L. B. (1990). Acoustic behavior in a local population of bottlenose dolphins. In J. A. Thomas & R. A. Kastelein (Eds.), *Sensory abilities of cetaceans* (pp. 585–598). New York: Plenum Press.

dos Santos, M. E., Ferreira, A. J., & Harzen, S. (1995). Rhythmic sound sequences emitted by aroused bottlenose dolphins in the Sado estuary, Portugal. In J. A. Thomas, R. A. Kastelein, & P. E. Nachtigall (Eds.), *Sensory abilities of cetaceans* (pp. 325–334). De Woerden: Spil Publisher.

Doyle, L. R., McCowan, B., Johnston, S., & Hanser, S. F. (2011). Information theory, animal communication, and the search for extraterrestrial intelligence. *Acta Astronautica*, *68*(3), 406–417.

Dreher, J. J., & Evans, W. E. (1964). Cetacean communication. In W. N. Tavolga (Ed.), *Marine bioacoustics* (pp. 373–399). Oxford: Pergamon Press.

Driscoll, A. D. (1995). *Categorizing the whistles and choruses of Hawaiian spinner dolphins*. Doctoral dissertation, University of California, Santa Cruz.

Essapian, F. S. (1953). The birth and growth of a porpoise. *Natural History*, *58*, 385–392.

Estes, R. D., & Goddard, J. (1967). Prey selection and hunting behavior of the African wild dog. *Journal of Wildlife Management*, *31*, 52–70.

Evans, W. E. (1966). Vocalizations among marine mammals. In W. N. Tavolga (Ed.), *Marine bioacoustics* (Vol. 2, pp. 159–185). Oxford: Pergamon Press.

Favaro, L., Gnome, G., & Pessani, D. (2013). Postnatal development of echolocation abilities in a bottlenose dolphin (*Tursiops truncatus*): Temporal organization. *Zoo Biology*, *32*(1), 210–215.

Fellner, W., Bauer, G. B., & Harley, H. E. (2006). Cognitive implications of synchrony in dolphins: A review. *Aquatic Mammals, 32*(4), 511–516.

Ford, J. (1991). Vocal traditions among resident killer whales (*Orcinus orca*) in coastal waters of British Columbia. *Canadian Journal of Zoology, 69*(6), 1454–1483.

Ford, J. K., & Fisher, H. D. (1978). Underwater acoustic signals of the narwhal (*Monodon monoceros*). *Canadian Journal of Zoology, 56*(4), 552–560.

Fripp, D. (2006). Bubblestream whistles are not representative of a bottlenose dolphin's vocal repertoire. *Marine Mammal Science, 21*(1), 29–44.

Fripp, D., Owen, C., Quintana-Rizzo, E., Shapiro, A., Buckstaff, K., Jankowski, K., et al. (2005). Bottlenose dolphin (*Tursiops truncatus*) calves appear to model their signature whistles on the signature whistles of community members. *Animal Cognition, 8,* 17–26.

Gish, S. L. (1979). *Quantitative description of two-way acoustic communication between captive Atlantic bottlenose dolphins (Tursiops truncatus Montagu).* Doctoral dissertation, University of California, Santa Cruz.

Gnome, G., Moriconi, T., & Gambini, G. (2005). Sleep behavior: Activity and sleep in dolphins. *Nature, 435*(7046), 1177.

Goodwin, C. (2000). Action and embodiment within situated human interaction. *Journal of Pragmatics, 32*(10), 1489–1522.

Grahman, M., & Noonan, M. (2010). Call types and acoustic features associated with aggressive chase in killer whale (*Orcinus orca*). *Aquatic Mammals, 36*(1), 9–18.

Harley, H. E. (2008). Whistle discrimination and categorization by the Atlantic bottlenose dolphin (*Tursiops truncatus*): A review of the signature whistle framework and a perceptual test. *Behavioural Processes, 77,* 243–268.

Harley, H. E., Putman, E. A., & Roitblat, H. L. (2003). Bottlenose dolphins perceive object features through echolocation. *Nature, 424*(6949), 667–669.

Henderson, E. E., Hildebrand, J. A., Smith, M. H., & Falcone, E. A. (2012). The behavioral context of common dolphin (*Delphinus* sp.) vocalizations. *Marine Mammal Science, 28*(3), 439–460.

Herman, L. M. (Ed.). (1980). *Cetacean behavior: Mechanisms and functions.* New York: John Wiley & Sons.

Herman, L. M., Morrel-Sammuels, P., & Pack, A. A. (1990). Bottlenosed dolphin and human recognition of veridical and degraded video displays of an artificial gestural language. *Journal of Experimental Psychology, 119,* 215–230.

Herman, L. M., & Tavolga, W. N. (1980). Communication systems of cetaceans. In L. M. Herman (Ed.), *Cetacean behavior: Mechanisms and functions* (pp. 149–197). New York: John Wiley & Sons.

Herzing, D. L. (1988). *A quantitative description and behavioral associations of a burst-pulsed sound, the squawk, in captive bottlenose dolphins, Tursiops truncatus.* Master's thesis, San Francisco State University.

Herzing, D. L. (1996). Vocalizations and associated underwater behavior of free-ranging Atlantic spotted dolphins, *Stenella frontalis*, and bottlenose dolphins, *Tursiops truncatus*. *Aquatic Mammals*, *22*, 61–80.

Herzing, D. L. (2000). Acoustics and social behavior of wild dolphins: Implications for a sound society. In W. W. L. Au, A. N. Popper, & R. R. Fay (Eds.), *Springer handbook of auditory research: Vol. 12, Hearing by whales and dolphins* (pp. 225–272). New York: Springer.

Herzing, D. L. (2004). Social and nonsocial uses of echolocation in free-ranging *Stenella frontalis* and *Tursiops truncatus*. In J. A. Thomas, C. F. Moss, & M. Vater (Eds.), *Echolocation in bats and dolphins* (pp. 404–410). Chicago: University of Chicago Press.

Herzing, D. L. (2005). Transmission mechanisms of social learning in dolphins: Underwater observations of free-ranging dolphins in the Bahamas. In *Autour de l'ethologie et de la cognition animale* (pp. 185–193). Lyon, France: Presses Universitaires de Lyon.

Herzing, D. L. (2006). The currency of cognition: Assessing tools, techniques, and media for complex behavioral analysis. *Aquatic Mammals*, *32*(4), 544–553.

Herzing, D. L. (2015). Synchronous and rhythmic vocalizations and correlated underwater behavior of free-ranging Atlantic spotted dolphins (*Stenella frontalis*) and bottlenose dolphins (*Tursiops truncatus*) in the Bahamas. *Animal Behavior and Cognition*, *2*(1), 14–29.

Herzing, D. L., Delfour, F., & Pack, A. A. (2012). Responses of human-habituated wild Atlantic spotted dolphins to play behaviors using a two-way interface. *International Journal of Comparative Psychology*, *25*, 137–165.

Herzing, D. L., & Johnson, C. J. (1997). Interspecific interactions between Atlantic spotted dolphins (*Stenella frontalis*) and bottlenose dolphins (*Tursiops truncatus*) in the Bahamas, 1985–1995. *Aquatic Mammals*, *23*, 85–99.

Herzing, D. L., & Johnson, C. M. (2006). Conclusions and possibilities of new frameworks and techniques for research on marine mammal cognition. *Aquatic Mammals*, *32*(4), 554.

Hofmann, B., Scheer, M., & Behr, I. P. (2004). Underwater behaviors of short-finned pilot whales (*Globicephala macrorhynchus*) off Tenerife. *Mammalia*, *68*(2–3), 221–224.

Hooper, S., Reiss, D., & Carter, M. (2006). Importance of contextual saliency on vocal imitation by bottlenose dolphins. *International Journal of Comparative Psychology*, *19*, 116–128.

Horback, K. M., Friedman, W. R., & Johnson, C. M. (2010). The occurrences and context of s-posture display by captive belugas (*Delphinapterus leucas*). *International Journal of Comparative Psychology*, *23*, 689–700.

Hult, R. W. (1982). Another function of echolocation for bottlenose dolphins. *Cetology*, *47*, 1–7.

Hutchins, E., & Johnson, C. M. (2009). Modeling the emergence of language as an embodied collective cognitive activity. *Topics in Cognitive Science, 1*(3), 523–546.

Janik, V. M. (1999). Origins and implications of vocal learning in bottlenose dolphins. In H. O. Box & K. R. Gibson (Eds.), *Mammalian social learning: Comparative and ecological perspectives* (Vol. 72, pp. 308–324). Cambridge: Cambridge University Press.

Janik, V. M. (2000). Food-related bray calls in wild bottlenose dolphins (*Tursiops truncatus*). *Proceedings of the Royal Society, 267*(1446), 923–927.

Janik, V. M., Simard, P., Sayigh, L. S., Mann, D., & Frankel, A. (2011). Chorussing in delphinids. *Journal of the Acoustical Society of America, 130*(4), 2322.

Janik, V. M., & Slater, P. J. B. (1998). Context-specific use suggests that bottlenose dolphin signature whistles are cohesion calls. *Animal Behaviour, 56*(4), 829–838.

Jerison, H. J. (1986). The perceptual worlds of dolphins. In R. J. Schusterman, J. A. Thomas, & F. G. Wood (Eds.), *Dolphin cognition and behavior: A comparative approach* (pp. 141–166). Hillsdale, NJ: Erlbaum.

Johnson, C. M. (1993). Animal communication by way of coordinated cognitive systems. In P. P. G. Bateson, P. H. Klopfer, & N. S. Thompson (Eds.), *Perspectives in ethology* (pp. 187–205). New York: Plenum Press.

Johnson, C. M. (2001). Distributed primate cognition: A review. *Animal Cognition, 3*(4), 167–183.

Johnson, C. M. (2010). Observing cognitive complexity in primates and cetaceans. *International Journal of Comparative Psychology, 23*, 587–624.

Johnson, C. M., & Moewe, K. (1999). Pectoral fin preference during contact in Commerson's dolphins (*Cephalorhynchus commersoni*). *Aquatic Mammals, 25*, 73–78.

Johnson, M., Madsen, P. T., Zimmer, W. M. X., Aguilar de Soto, N., & Tyack, P. L. (2006). Foraging Blainville's beaked whales (*Mesoplodon densirostris*) produce distinct click types matched to different phases of echolocation. *Journal of Experimental Biology, 209*, 5038–5050.

Kershenbaum, A., Sayigh, L. S., & Janik, V. M. (2013). The encoding of individual identity in dolphin signature whistles: How much information is needed? *PLoS ONE, 8*(10), e77671.

King, B. J., & Shanker, S. G. (2003). How can we know the dancer from the dance? The dynamic nature of African great ape social communication. *Anthropological Theory, 3*(1), 5–26.

King, S. L., & Janik, V. M. (2013). Bottlenose dolphins can use learned vocal labels to address each other. *Proceedings of the National Academy of Sciences of the United States of America, 110*(32), 13216–13221.

King, S. L., Sayigh, L. S., Wells, R. S., Fellner, W., & Janik, V. M. (2013). Vocal copying of individually distinctive signature whistles in bottlenose dolphins. *Proceedings of the Royal Society B: Biological Sciences, 280*(1757).

Kohlsdorf, D., Mason, C., Herzing, D., & Starner, T. (2014). Probabilistic extraction and discovery of fundamental units in dolphin whistles. In *2014 IEEE International Conference on Acoustics, Speech and Signal Processing (ICASSP)* (pp. 8242–8246). Los Alamitos, CA: IEEE Publications.

Kolchin, S. P., & Bel'kovich, V. M. (1973). Tactile sensitivity in *Delphinus delphis*. *Zoologicheskiy zhurnal, 52*, 620–622.

Kuznetsov, V. B. (1974). A method of studying chemoreception in the Black Sea bottlenose dolphin (*Tursiops truncatus*). In *Morphology, physiology, and acoustics of marine mammals* (pp. 27–45). Moscow: Nauka Publishing House.

Lammers, M. O., & Au, W. W. (1996). Broadband recording of social acoustic signals of the Hawaiian spinner and spotted dolphins. *Journal of the Acoustical Society of America, 100*, 2609.

Lammers, M. O., Au, W. W., & Herzing, D. L. (2003). The broadband social acoustic signaling behavior of spinner and spotted dolphins. *Journal of the Acoustical Society of America, 114*, 1629.

Lammers, M. O., Schotten, M., & Au, W. W. L. (2006). The spatial context of free-ranging Hawaiian spinner dolphins (*Stenella longirostris*) producing acoustic signals. *Journal of the Acoustical Society of America, 119*(2), 1244–1250.

Lilly, J. C. (1965). Vocal mimicry in *Tursiops*: Ability to match numbers and durations of human vocal bursts. *Science, 147*, 300–301.

Lilly, J. C., & Miller, A. M. (1961). Vocal exchanges between dolphins. *Science, 134*, 1873–1876.

López, B. D. (2011). Whistle characteristics in free-ranging bottlenose dolphins (*Tursiops truncatus*) in the Mediterranean Sea: Influence of behavior. *Mammalian Biology: Zeitschrift für Säugetierkunde, 76*(2), 180–189.

Madsen, P. T., Jensen, F. H., Carder, D., & Ridgway, S. (2012). Dolphin whistles: A functional misnomer revealed by heliox breathing. *Biology Letters, 8*(2), 211–213.

Madsen, P. T., Johnson, M., de Soto, N. A., Zimmer, W. M. X., & Tyack, P. (2005). Biosonar performance of foraging beaked whales (*Mesoplodon densirostris*). *Journal of Experimental Biology, 208*(2), 181–194.

Magnusson, M. S. (1996). Hidden real-time patterns in intra- and inter-individual behavior: Description and detection. *European Journal of Psychological Assessment, 12*, 112–123.

Marks, L. E. (1978). *The unity of the senses: Interrelations among the modalities.* New York: Academic Press.

Marler, P. R. (1982). Avian and primate communication: The problem of natural categories. *Neuroscience and Biobehavioral Reviews, 6*, 87–92.

Marten, K., Herzing, D. L., Poole, M., & Newman-Allman, K. (2001). The acoustic predation hypothesis: Linking underwater observations and recordings during odontocete predation and observing the effects of loud impulsive sounds on fish. *Aquatic Mammals, 27*(1), 56–66.

Marten, K., Norris, K. S., Moore, P. W. B., & Englund, K. A. (1988). Loud impulse sounds in odontocete predation and social behavior. In P. E. Nachtigall & P. W. B. Moore (Eds.), *Animal sonar: Processes and performance* (pp. 567–579). New York: Plenum Press.

Marten, K., Shariff, K., & Psarakos, S. (1996). Ring bubbles of dolphins. *Scientific American, 275*, 82–87.

May-Collado, L. J. (2010). Changes in whistle structure of two dolphin species during interspecific associations. *Ethology, 116*(11), 1065–1074.

McBride, A. F., & Hebb, D. O. (1948). Behavior of the captive bottlenose dolphin, *Tursiops truncatus*. *Journal of Comparative and Physiological Psychology, 41*, 111–123.

McCowan, B., Marino, L., Vance, E., Walke, L., & Reiss, D. (2000). Bubble ring play of bottlenose dolphins (*Tursiops truncatus*): Implications for cognition. *Journal of Comparative Psychology, 114*(1), 98–106.

McCowan, B., & Reiss, D. (1995a). Maternal aggressive contact vocalizations in captive bottlenose dolphins (*Tursiops truncatus*): Wide-band, low-frequency signals during mother/aunt–infant interactions. *Zoo Biology, 14*, 293–309.

McCowan, B., & Reiss, D. (1995b). Whistle contour development in captive-born infant bottlenose dolphins (*Tursiops truncatus*): Role of learning. *Zoo Biology, 109*(3), 242–260.

McCowan, B., & Reiss, D. (1995c). Quantitative comparison of whistle repertoires from captive adult bottlenose dolphins (Delphinidae, *Tursiops truncatus*): A re-evaluation of the signature whistle hypothesis. *Ethology, 100*, 194–209.

Mercado, E., & DeLong, C. M. (2010). Dolphin cognition: Representation and processes in memory and perception. *International Journal of Comparative Psychology, 23*, 344–378.

Morton, E. S. (1977). On the occurrence and significance of motivation: Structural rules in some bird and mammal sounds. *American Naturalist, 111*, 855–869.

Mountcastle, V. B. (1974). Sensory receptors and neural coding: Introduction to sensory processes. In V. B. Mountcastle (Ed.), *Medical physiology* (Vol. 1, pp. 285–306). St. Louis: Mosby.

Murray, S. O., Mercado, E., & Roitblat, H. L. (1998). Characterizing the graded structure of false killer whale (*Pseudorca crassidens*) vocalizations. *Journal of the Acoustical Society of America, 104*, 1679.

Nachtigall, P. E. (1986). Vision, audition, and chemoreception in dolphins and other marine mammals. In R. S. Schusterman, J. A. Thomas, & F. G. Wood (Eds.), *Dolphin cognition and behavior: A comparative approach* (pp. 79–113). Hillsdale, NJ: Erlbaum.

Nolfi, S. (2005). Emergence of communication in embodied agents: Co-adapting communicative and non-communicative behaviours. *Connection Science, 17*(3–4), 231–248.

Norris, K. S. (1966). *Whales, dolphins, and porpoises*. Berkeley: University of California Press.

Norris, K. S., & Dohl, T. (1980). Structure and function of cetacean schools. In L. M. Herman (Ed.), *Cetacean behavior: Mechanisms and functions* (pp. 230–244). New York: John Wiley & Sons.

Norris, K. S., & Evans, E. C. (1988). On the evolution of acoustic communication systems in vertebrates, Part II: Cognitive aspects. In P. E. Nachtigall & P. W. B. Moore (Eds.), *Animal sonar: Processes and performance* (pp. 671–682). New York: Plenum Press.

Norris, K. S., & Mohl, B. (1981). Can odontocetes stun prey with sound? *American Naturalist, 122,* 85–104.

Norris, K. S., Würsig, B., Wells, R. S., & Würsig, M. (1994). *The Hawaiian spinner dolphin.* Berkeley: University of California Press.

Norris, K. S., Prescott, J. H., Asa-Dorian, P. V., & Perkins, P. (1961). An experimental demonstration of echolocation behavior in the porpoise, *Tursiops truncatus,* Montagu. *Biological Bulletin, 120,* 163–176.

Oehen, S. (1996). *The acoustic behavior of the bottlenose dolphins Tursiops truncatus in the Northern Adriatic Sea.* Master's thesis, Zoo Museum, University of Zurich.

Ostman, J. (1994). *Social organization and social behavior of Hawaiian spinner dolphins (Stenella longirostris).* Doctoral dissertation, University of California, Santa Cruz.

Oswald, J. N., Barlow, J., & Norris, T. F. (2003). Acoustic identification of nine delphinid species in the eastern tropical Pacific Ocean. *Marine Mammal Science, 19*(1), 20–037.

Overstrom, N. A. (1983). Association between burst-pulse sounds and aggressive behavior in captive Atlantic bottlenose dolphins (*Tursiops truncatus*). *Zoo Biology, 2,* 93–103.

Pack, A. A., & Herman, L. M. (1995). Sensory integration in the bottlenose dolphin: Immediate recognition of complex shapes across the sense of echolocation and vision. *Journal of the Acoustical Society of America, 98,* 722–733.

Palmer, E., & Weddell, G. (1964). The relationship between structure, innervation, and function of the skin of the bottlenose dolphin, *Tursiops truncatus. Proceedings of the Zoological Society of London, 143,* 553–568.

Papale, E., Azzolin, M., Cascao, I., Gannier, A., Lammers, M. O., Martin, V. M., ... & Giacoma, C. (2014). Macro- and micro-geographic variation of short-beaked common dolphin's whistles in the Mediterranean Sea and Atlantic Ocean. *Ethology, Ecology, and Evolution, 26*(4), 392–404.

Paulos, R. D., Dudzinski, K. M., & Kuczaj, S. A., II. (2008). The role of touch in select social interactions of Atlantic spotted dolphin (*Stenella frontalis*) and Indo-Pacific bottlenose dolphin (*Tursiops aduncus*). *Journal of Ethology, 26*(1), 153–164.

Placer, J., & Slobodchikoff, C. N. (2000). A fuzzy-neural system for identification of species-specific alarm calls of Gunnison's prairie dogs. *Behavioural Processes, 52*(1), 1–9.

Pryor, K. W. (1990). Non-acoustic communication in small cetaceans: Glance, touch, position, gesture, and bubbles. In J. A. Thomas & R. A. Kastelien (Eds.), *Sensory abilities of cetaceans* (pp. 537–544). New York: Plenum Press.

Pryor, K., & Kang, I. (1980). *Social behavior and school structure in pelagic porpoises (Stenella attenuata and S. longirostris) during purse seining for tuna.* Adm. Rpt. LJ-80-11c. La Jolla, CA: National Marine Fisheries Service.

Pryor, K., & Kang-Shallenberger, I. (1991). Social structure in spotted dolphins (*Stenella attenuata*) in the tuna purse seine fishery in the eastern tropical Pacific. In K. Pryor & K. S. Norris (Eds.), *Dolphin societies: Discoveries and puzzles* (pp. 161–196). Berkeley: University of California Press.

Puente, A. E., & Dewsbury, D. A. (1976). Courtship and copulatory behavior of bottlenosed dolphins, *Tursiops truncatus*. *Cetology, 21*, 1–9.

Quick, N. J., & Janik, V. M. (2012). Bottlenose dolphins exchange signature whistles when meeting at sea. *Proceedings of the Royal Society B: Biological Sciences, 279*(1738), 2539–2545.

Rehn, N., Teichert, S., & Thomsen, F. (2007). Structural and temporal emission patterns of variable pulsed calls in free-ranging killer whales (*Orcinus orca*). *Behaviour, 144*, 307–329.

Reiss, D. (1988). Observations on the development of echolocation in young bottlenose dolphins. In P. E. Nachtigall & P. W. B. Moore (Eds.), *Animal sonar systems* (pp. 121–127). Helsingør, Denmark: Plenum Press.

Reiss, D., & McCowan, B. (1993). Spontaneous vocal mimicry and production by bottlenose dolphins (*Tursiops truncatus*): Evidence for vocal learning. *Journal of Comparative Psychology, 107*(3), 301–312.

Reiss, D., McCowan, B., & Marino, L. (1997). Communicative and other cognitive characteristics of bottlenose dolphins. *Trends in Cognitive Sciences, 1*(4), 140–145.

Rendell, L., & Whitehead, H. (2001). Cetacean culture: Still afloat after the first naval engagement of the culture wars. *Behavioral and Brain Sciences, 24*(2), 360–382.

Ridgway, S., Carder, D., Jeffries, M., & Todd, M. (2012). Spontaneous human speech mimicry by a cetacean. *Current Biology, 22*(20), R860–R861.

Ridgway, S. H., Moore, P. W., Carder, D. A., & Romano, T. A. (2014). Forward shift of feeding buzz components of dolphins and belugas during associative learning reveals a likely connection to reward expectation, pleasure, and brain dopamine activation. *Journal of Experimental Biology, 217*(16), 2910–2919.

Robinson, K. P. (2014). Agonistic intraspecific behavior in free-ranging bottlenose dolphins: Calf-directed aggression and infanticidal tendencies by adult males. *Marine Mammal Science, 30*(1), 381–388.

Roitblat, H. L. (1990). Concluding comments on other sensory abilities. In J. A. Thomas & R. A. Kastelein (Eds.), *Sensory abilities of cetaceans* (pp. 699–702). New York: Plenum Press.

Roitblat, H. L., Au, W. W. L., Nachtigall, P. E., Shizumura, R., & Moons, G. (1995). Sonar recognition of targets embedded in sediment. *Neural Networks, 8*, 1263–1273.

Rossbach, K. A., & Herzing, D. L. (1997). Underwater observations of benthic-feeding bottlenose dolphins (*Tursiops truncatus*) near Grand Bahama Island, Bahamas. *Marine Mammal Science, 13*, 498–504.

Saayman, G. S., & Tayler, C. K. (1979). The socioecology of humpback dolphins (*Sousa* sp.). In H. E. Winn & B. L. Olla (Eds.), *Behavior of marine animals: Current perspectives in research: Vol. 3. Cetaceans* (pp. 165–226). New York: Plenum.

Saayman, G. S., Tayler, C. K., & Bower, D. (1973). Diurnal activity cycles in captive and free ranging Indian Ocean bottlenose dolphins, *Tursiops aduncus* Ehrenburg. *Behaviour, 44*, 212–233.

Sakai, M., Hishii, T., Takeda, S., & Kohshima, S. (2006). Flipper rubbing behaviors in wild bottlenose dolphins (*Tursiops aduncus*). *Marine Mammal Science, 22*, 966–978.

Sakai, M., Morisaka, T., Kogi, K., Hishii, T., & Kohshima, S. (2009). Fine-scale analysis of synchronous breathing in wild Indo-Pacific bottlenose dolphins (*Tursiops aduncus*). *Behavioural Processes, 83*, 48–53.

Sayigh, L. S., Esch, H. C., Wells, R. S., & Janik, V. M. (2007). Facts about signature whistles of bottlenose dolphins, *Tursiops truncatus. Animal Behaviour, 74*(6), 1631–1642.

Sayigh, L., Quick, N., Hastie, G., & Tyack, P. (2012). Repeated call types in short-finned pilot whales, *Globicephala macrorhynchus. Marine Mammal Science, 29*(2), 312–324.

Sayigh, L. S., Tyack, P. L., & Wells, R. S. (1993). Recording underwater sounds of free-ranging dolphins while underway in a small boat. *Marine Mammal Science, 9*, 209–213.

Sayigh, L. S., Tyack, P. L., Wells, R. S., & Scott, M. D. (1990). Signature whistles of free-ranging bottlenose dolphins *Tursiops truncatus*: Stability and mother-offspring comparisons. *Behavioral Ecology and Sociobiology, 26*, 247–260.

Scheer, M., Hofmann, B., & Behr, I. P. (2004). Ethogram of selected behaviors initiated by free-ranging short-finned pilot whales (*Globicephala macrorhynchus*) and directed to human swimmers during open water encounters. *Anthrozoos, 17*(3), 244–258.

Schotten, M., Lammers, M. O., Sexton, K., & Au, W. W. L. (2005). Application of a diver-operated 4-channel acoustic/video recording device to study wild dolphin echolocation and communication. *Journal of the Acoustical Society of America, 117*(4), 2552.

Schultz, K. W., Cato, D. H., Corkeron, P. J., & Bryden, M. M. (1995). Low frequency narrow-band sounds produced by bottlenose dolphins. *Marine Mammal Science, 11*, 503–509.

Schusterman, R. J., Thomas, J. A., & Wood, F. G. (1986). *Dolphin cognition and behavior: A comparative approach*. Hillsdale, NJ: Erlbaum.

Simila, T., & Ugarte, F. (1993). Surface and underwater observations of cooperatively feeding killer whales in northern Norway. *Canadian Journal of Zoology, 71*, 1494–1499.

Sjare, B. L., & Smith, T. G. (1986). The vocal repertoire of white whales, *Delphinapterus leucas*, summering in Cunningham Inlet, Northwest Territories. *Canadian Journal of Zoology, 64*, 407–415.

Slooten, E. (1994). Behavior of Hector's dolphin: Classifying behavior by sequence analysis. *Journal of Mammalogy, 75*, 956–964.

Smith, W. J. (1965). Message, meaning, and context in ethology. *American Naturalist, 99*(108), 405–409.

Smolker, R. A., Mann, J., & Smuts, B. B. (1993). Use of signature whistles during separations and reunions by wild bottlenose dolphin mothers and infants. *Behavioral Ecology and Sociobiology, 33*, 393–402.

Smolker, R., & Pepper, J. W. (1999). Whistle convergence among allied male bottlenose dolphins (Delphinidae, *Tursiops* sp.). *Ethology, 105*(7), 595–617.

Smuts, B. B. (1985). *Sex and friendship in baboons.* Piscataway, NJ: Transaction Publishers.

Sokolov, V. E., & Volkova, O. V. (1973). Structure of the dolphin's tongue. In K. K. Chapskii & V. E. Sokolov (Eds.), *Morphology and ecology of marine mammals* (pp. 119–127). New York: Wiley.

St. Aubin, D. J., Forney, K. A., Chivers, S. J., Scott, M. D., Danil, K., Romano, T. A., et al. (2013). Hematological, serum, and plasma chemical constituents in pantropical spotted dolphins (*Stenella attenuata*) following chase, encirclement, and tagging. *Marine Mammal Science, 29*(1), 14–35.

Starkhammar, J., Amundin, M., Nilsson, J., Jansson, T., Kuczaj, S. A., Almqvist, M., et al. (2009). Forty-seven-channel burst-mode recording hydrophone system enabling measurements of the dynamic echolocation behavior of free-swimming dolphins. *Journal of the Acoustical Society of America, 126*, 959.

Tamaki, N., Morisaka, T., & Taki, M. (2006). Does body contact contribute towards repairing relationships? The association between flipper-rubbing and aggressive behavior in captive bottlenose dolphins. *Behavioural Processes, 73*, 209–215.

Taruski, A. G. (1979). The whistle repertoire of the North Atlantic pilot whale (*Globicephala melaena*) and its relationship to behavior and environment. In H. E. Winn & B. L. Olla (Eds.), *Behavior of marine animals: Current perspectives in research* (pp. 345–368). New York: Plenum Press.

Tavolga, M. C. (1966). Behavior of the bottlenose dolphin, *Tursiops truncatus*: Social interactions in a captive colony. In K. S. Norris (Ed.), *Whales, dolphins, and porpoises* (pp. 718–730). Berkeley: University of California Press.

Tavolga, M. C., & Essapian, F. S. (1957). The behavior of the bottlenose dolphin, *Tursiops truncatus*: Mating, pregnancy and parturition, mother-infant behavior. *Zoologica, 42*, 11–31.

Tavolga, W. N. (1983). Theoretical principles for the study of communication in cetaceans. *Mammalia, 47*, 3–26.

Thomas, R. E., Fristrup, K. M., & Tyack, P. L. (2002). Linking the sounds of dolphins to their locations and behavior using video and multichannel acoustic recordings. *Journal of the Acoustical Society of America, 112,* 1692.

Thomsen, F., Franck, D., & Ford, J. K. (2002). On the communicative significance of whistles in wild killer whales (*Orcinus orca*). *Naturwissenschaften, 89,* 404–407.

Tizzi, R., & Accorsi, P. A. (2010). Non-invasive multidisciplinary approach to the study of reproduction and calf development in bottlenose dolphin (*Tursiops truncatus*): The Rimini Delfinario experience. *International Journal of Comparative Psychology, 23,* 734–776.

Tyack, P. (1985). An optical telemetry device to identify which dolphin produces a sound. *Journal of the Acoustical Society of America, 78,* 1892–1895.

Tyack, P. (1993). Animal language research needs a broader comparative and evolutionary framework. In H. L. Roitblat, L. M. Herman, & P. E. Nachtigall (Eds.), *Language and communication: Comparative perspectives* (pp. 1115–1138). Hillsdale, NJ: Erlbaum.

Tyack, P. L., Johnson, M., Soto, N. A., Sturlese, A., & Madsen, P. T. (2006). Extreme diving of beaked whales. *Journal of Experimental Biology, 209*(21), 4238–4253.

Tyack, P. L., & Recchia, C. A. (1991). A datalogger to identify vocalizing dolphins. *Journal of the Acoustical Society of America, 90,* 1668.

van Hooff, J. A. R. A. M. (1982). Categories and sequences of behavior: Methods of description and analysis. In K. R. Scherer & P. Ekman (Eds.), *Handbook of methods in nonverbal behavior research* (pp. 362–439). New York: Cambridge University Press.

van Parijs, S. M., & Corkeron, P. J. (2001). Vocalizations and behaviour of Pacific humpback dolphins *Sousa chinensis*. *Ethology, 107*(8), 701–716.

Wang, D., Würsig, B., & Evans, W. (1995). Comparisons of whistles among seven odontocete species. In R. A. Kastelein, J. A. Thomas, & P. E. Nachtigall (Eds.), *Sensory systems of aquatic mammals* (pp. 299–324). Woerden, the Netherlands: De Spil.

Watkins, W. A., & Schevill, W. E. (1974). Listening to Hawaiian spinner porpoise, *Stenella cf. longirostris*, with a three-dimensional hydrophone array. *Journal of Mammalogy, 55,* 319–328.

Watkins, W. A., & Schevill, W. E. (1977). Sperm whale codas. *Journal of the Acoustical Society of America, 62,* 1485–1490.

Watwood, S. L., Tyack, P. L., & Wells, R. S. (2004). Whistle sharing in paired male bottlenose dolphins, *Tursiops truncatus*. *Behavioral Ecology and Sociobiology, 55*(6), 531–543.

Wells, R. S. (1984). Reproductive behavior and hormonal correlates in Hawaiian spinner dolphins, *Stenella longirostris*. In W. F. Perrin, R. L. Brownell Jr., & D. P. DeMasters (Eds.), *Reproduction of whales, dolphins, and porpoises* (pp. 465–472). Report of the International Whaling Commission, Special Issue No. 6. Cambridge: International Whaling Commission.

Whitehead, H. (1996). Babysitting, dive synchrony, and indications of alloparental care in sperm whales. *Behavioral Ecology and Sociobiology, 38*(4), 237–244.

Wood, F. G., Jr. (1953). Underwater sound production and concurrent behavior of captive porpoise, *Tursiops truncatus* and *Stenella plagiodon*. *Bulletin of Marine Science of the Gulf and Caribbean, 3*, 120–133.

Wood, F. G., Caldwell, D. K., & Caldwell, M. C. (1970). Behavioral interactions between porpoises and sharks. In G. Pilleri (Ed.), *Investigations on cetacea* (Vol. 2, pp. 264–277). University of Berne, Switzerland: Brain Anatomy Institute.

Würsig, B., Kieckhefer, T. R., & Jefferson, T. A. (1990). Visual displays for communication in cetaceans. In J. A. Thomas & R. A. Kastelein (Eds.), *Sensory abilities of cetaceans* (pp. 545–559). New York: Plenum Press.

Xiao, J., & Wang, D. (2005). Construction of ethogram of the captive Yangtze finless porpoises, *Neophocaena phocaenoides asiaeorientalis*. *Acta Hydrobiologica Sinica, 29*(3), 253.

Xitco, M. J., Gory, J. D., & Kuczaj, S. A. (2001). Spontaneous pointing by bottlenose dolphins (*Tursiops truncatus*). *Animal Cognition, 4*(2), 115–123.

Xitco, M. J., & Roitblat, H. L. (1996). Object recognition through eavesdropping: Passive echolocation in bottlenose dolphins. *Animal Learning and Behavior, 24*, 355–365.

Yablokov, A. V., Bel'kovich, V. M., & Borisov, V. I. (1972). *Whales and dolphins*. Jerusalem: Israel Program for Scientific Translations.

III Dolphin Cognition

7 Experimental Studies of Dolphin Cognitive Abilities

Adam A. Pack

Experimental investigations of dolphin cognition come under the broad umbrella of cognitive psychology, the study of perception, attention, memory, mental representations, problem solving, concept formation, knowledge, language skills, reasoning, and decision making (Goldstein, 2011). Before 1970, little information in these areas was available on dolphins. However, over the past forty years, researchers have made considerable progress. Early interest in the mind of the dolphin was inspired by observations of the large dolphin brain, which, like the brains of humans and anthropoid species, has a considerable degree of cortical fissuring, and an impressive ratio of brain-to-body mass, resulting in an encephalization quotient that is below humans but above apes (chap. 1, this vol.). More recently, researchers have noted the enlarged neocortical surface area, as well as the relatively high synaptic density and high numbers of neurons and synapses in the dolphin's neocortex (reviewed in Huggenberger, 2008). However, as Herman (1980) pointed out, it is ultimately an animal's behavior, and not simply the structure or architecture of its neurological hardware, that is the true measure of its intellect. In the late 1960s, Herman and colleagues set out to describe what Herman (1980) called the dolphin's "cognitive characteristics," meaning the breadth of its cognitive abilities, specializations, and limitations. This approach, which continues today, strives to describe the types of intelligence a dolphin possesses, as well as the evolutionary pressures driving the selection of abilities that may be characterized as intelligent (Connor, 2007; Herman, 2006). In this chapter, I summarize key experimental findings in various domains in dolphin cognition. Because the majority of these studies have focused on bottlenose dolphins (*Tursiops* sp.), unless otherwise noted, the term "dolphin" refers to that species. Also, unless otherwise stated, reports of performance accuracy were significantly above expectations based on responding by chance.

Sensory Perception

Some of the earliest inquiries into dolphin cognition focused on investigations of sensory perception, including the types of information dolphins can perceive through each sense, and the detection abilities and fine distinctions that can be achieved (reviewed in chaps. 1 and 3, this vol.). Here I focus on more recent studies examining cross-modal perception.

In the 1990s, Pack and colleagues began a series of studies examining object perception across the dolphin's senses of echolocation and vision, to determine how dolphins perceive and mentally represent objects inspected through echolocation (see Pack, Herman, & Hoffmann-Kuhnt, 2004). They developed a cross-modal task in which a shape inspected through echolocation alone could be matched to an identical shape inspected through vision alone and vice versa. Objects exposed to vision alone were held in air, a medium in which dolphin echolocation is ineffective. Objects exposed to echolocation alone were presented within visually opaque boxes through which dolphins could echolocate. Two cross-modal matching-to-sample (MTS) procedures were developed. In echolocation-to-vision (E-V) matching, a dolphin first inspected echoically a sample object inside the box and then chose the match visually from two comparisons held in air. In vision-to-echolocation (V-E) matching, visual inspection of the sample preceded finding a match through echolocation. Pack and Herman (1995) hypothesized that if the dolphin matched an object accurately across the senses of echolocation and vision on the first trial with a variety of complex shapes that had never been exposed to both senses simultaneously (thus eliminating matching through associative learning), then this would support its formation of a shape percept through echolocation. First-trial matching of complexly shaped objects was found cross-modally in both V-E and E-V, including novel objects that had never been exposed previously to echolocation or vision. Overall the dolphin was correct on 19 of 24 first trials in E-V and 21 of 26 first trials in V-E (Pack et al., 2004). In follow-up work using objects with overlapping features and a response paddle to indicate that none of the comparisons matched the sample, Pack, Herman, Hoffmann-Kuhnt, and Branstetter (2002) showed that objects were perceived through echolocation holistically.

In a separate study, Harley, Putman, and Roitblat (2003) also used a cross-modal task to examine dolphin echolocation perception, but with a different methodology from Pack and colleagues. With some sets of objects, the dolphin was rewarded for selecting a comparison that physically matched the sample. With other sets, however, a choice of a consistent nonmatching comparison was rewarded even though a physical match was present among the comparisons. Harley et al. (2003) reasoned that if the dolphin is directly perceiving shape characteristics through echolocation, then the former task should be easier for the dolphin than the latter because only in the former task is shape consistent from sample to rewarded comparison. If, however, the dolphin is simply learning to associate sounds (echoes) with the sights of objects, then the tasks should be equivalently challenging. Despite the differences in reward contingency, the dolphin tended to select the comparison object that physically matched the sample. Taken together, these studies support the theory that dolphins directly perceive object shape through echolocation. They also indicate a substantial degree of sensory integration in the dolphin such that mental representations of shape derived through either echolocation or vision are readily accessible to the alternate sense.

Attention

In the daily lives of most species, individuals are bombarded with a myriad of stimuli. Judging which ones to focus on and which to ignore occurs early in cognitive processing and is the work of selective attention (Goldstein, 2011). Studies measuring selective attention often use vigilance tasks in which a subject must maintain close attention to detect target stimuli that occur unpredictably among successively presented distracters.

Ridgway et al. (2006) used an auditory vigilance task to test the sustained attention abilities of two dolphins. Two 70 kHz sounds differing only in duration were used, a long 1.5-second target sound and a short 0.5-second distracter sound. The distracter sound was projected every 30 seconds. At random intervals between 4 and 24 minutes, the target sound was substituted for a distracter. The dolphin received a reward when it pressed a paddle within 21 seconds of hearing the target sound, and thus was required to maintain attention, as the target's appearance within the series of distracters was unpredictable. One dolphin was tested for two 120-hour sessions and the other for three 120-hour sessions. Mean performance levels in correct target detections across all sessions were over 94 percent for both dolphins. Importantly, attention was sustained over the 120 hours with no significant decrement in target detection over the course of the final testing session for each dolphin, including comparisons of day versus night detection. These data not only provide strong evidence for the ability for sustained attention over several days but also indicate that dolphin brains are able to remain alert and attentive during the entire diurnal cycle.

Memory

Memory is essential for cognitive functioning and depends on the formation of internal representations of experiences and ideas. Early studies of dolphin memory (see Herman, 1980) focused on characterizing working memory, a mental storage system that, in humans, is of limited capacity and durability, with separate memory buffers that maintain modality-specific information (Baddeley, Eysenck, & Anderson, 2009).

Working Memory Durability for Sounds and Objects

Tests of dolphin working memory durability for sounds heard passively and objects inspected visually used a delayed MTS procedure. A sample stimulus was presented briefly, and after a delay interval, several comparison stimuli appeared. The dolphin's task was to indicate which comparison matched the sample. Memory durability for samples was determined from performance accuracy with increasing delay. In studies of auditory MTS, Herman and Gordon (1974) found that matching accuracy remained at or near ceiling levels across hundreds of pairs of novel sounds for nearly all delays tested up to 120 seconds (the longest delay tested). In follow-up studies (see Herman, 1980), working memory for sounds was shown to be subject to both proactive interference (i.e., memories of sounds from earlier trials interfered with memory for the current sample) and retroactive interference (i.e., irrelevant sounds occurring during the delay interval interfered with memory of the previous sample), and memory span for serially presented lists of sounds appeared to be between four and five sounds.

In studies of dolphin visual MTS, Herman, Hovancik, Gory, and Bradshaw (1989) reported that matching accuracy for objects remained at or near ceiling levels through delays of 30 seconds and then gradually declined with greater delays, remaining above 70 percent correct with delays through 80 seconds (the longest delay tested). Collectively, these findings indicate that dolphin working memory is well developed and can accommodate information arriving in different modalities. Durability for both sounds and objects appears more similar to visual working memory in nonhuman primates than in pigeons (Herman & Gordon, 1974).

Working Memory for Behaviors Performed by Others and Self

The ability to remember actions and their consequences is important for social and individual learning. Herman (2002) reported on tests of a dolphin's working memory for behaviors that it witnessed another dolphin perform over delays up to 80 seconds. Two dolphins were positioned on opposite sides of a large, vertically oriented board, each facing its own trainer. The board allowed the dolphins to view each other, but not each other's trainer. Each dolphin could serve as a model or as a witness. In each trial, the model's trainer produced a sign associated with one of seven different motor behaviors. After the model successfully completed the requested action and returned to the board, a delay ensued, and on its termination the trainer of the witness dolphin either requested through a "mimic" sign that the dolphin imitate the behavior performed earlier by the model, or produced a gesture requesting that the dolphin perform one of a variety of other behaviors. Both dolphins successfully performed the delayed imitation task (Herman, 2002). The more accurate of the two was 85 percent correct or better with delays up to 25 seconds. Performance accuracy dropped to approximately 60 percent correct at the 80 second delay (see hereafter for additional findings on imitation).

Mercado, Murray, Uyeyama, Pack, and Herman (1998) tested working memory for a dolphin's own behavior. Two dolphins were trained to maintain a representation of the behavior they had just performed and to either carry out the same behavior (in response to a "repeat" gesture) or carry out another behavior, different from the one just performed (in response to a specific gesture associated with that behavior). Both dolphins learned the repeat instruction and applied it to trainer-signed familiar and novel behaviors, and combinations of behaviors, as well as to self-selected behaviors in response to a "create" instruction (see the section on creativity later in this chapter). Mercado et al. (1998) reported that the two dolphins correctly carried out the "repeat" instruction to 32 familiar behaviors on 114 of 127 trials (90 percent), and 62 of 109 trials (57 percent). The dolphin with the superior overall performance correctly repeated 28 of the 32 behaviors on at least three of four test trials and was errorless in repeating 24 behaviors. Also, she correctly repeated three out of four self-selected behaviors and two out of four novel behaviors. Her ability to repeat self-selected behaviors, as well as to perform without error on 16 trials involving successive repeat gestures, indicated that she was recalling her own past actions rather than the specific instructions of the trainer. In further tests, Mercado, Uyeyama, Pack, and Herman (1999) showed that this dolphin spontaneously generalized its repeat concept to specified actions on objects.

Management of Working Memory

Within working memory, the "central executive" plays a key managerial role (Baddeley et al., 2009). Two important functions are monitoring stored material and creating updates as new information becomes available. Herman (2002) recounted a study in which the "repeat" instruction described earlier was contrasted with an "any" sign to create lengthy strings of instructions that tested these executive functions in the dolphin. The "any" gesture instructed the dolphin to perform any of five known behaviors, with the condition that the one chosen should not have just been performed. The dolphin was tested on trials consisting of sequences of "repeat" and "any," which required that it maintain a mental representation of the behavior it had last performed, regularly update that representation with each new behavior performed, and use that representation to choose a subsequent behavior depending on the gesture received from the trainer. Herman (2002) reported the dolphin's performance accuracy on eight types of four-element sequences, including those in which the dolphin was required to either repeat or not repeat a self-selected behavior. The dolphin was correct on 279 of 318 (87.7 percent) four-element trials ending in "any" and 306 of 320 trials (95.6 percent) ending in "repeat." These data show high proficiency in maintaining, monitoring, and continually updating representations.

Long-Term Memory

The dolphin's ability to perform behaviors and complex cognitive tasks after delays of weeks or even months indicates that its long-term memory can store various perceptual and motor skills, procedures, and factually based information. However, until recently, researchers had conducted no formal experiments on dolphin long-term memory.

Bruck (2013) examined dolphin long-term memory for previous social affiliates using naturally learned whistles, termed "signature whistles," that are unique to an individual dolphin and are sometimes produced when companions are separated (King & Janik, 2013). Forty-three dolphins that were housed together with other dolphins for varying periods of time and then separated from them were tested for their recognition of familiar whistles from their previous associates. A given subject was first habituated to unfamiliar whistles played through an underwater speaker until it ignored them for thirty seconds. After habituation, the subject was then tested with either another unfamiliar signature whistle or a familiar signature whistle from a previous associate. Response (e.g., turning the head toward the speaker or approaching the speaker within a few seconds of the playback) or lack of response was recorded. The results showed greater responding to familiar signature whistles than to unfamiliar whistles, for periods of separation between the subject and its previous associate ranging from 0 to 5 years to 15 or more years (Bruck, 2013). There was no effect of duration of separation on response level up to 20 years, the maximum separation time tested.

Learning Generalized Rules and Abstract Concepts

A fundamental question in studies of problem-solving abilities is whether a subject solves a class of problems by learning stimulus-specific regularities for each individual problem, or whether it perceives, from a limited set of exemplars the equivalencies among the problems and abstracts, a generalized rule or concept that it then applies to novel problems within that class. The latter skill is more efficient and effective than the former and has been referred to as "learning how to learn" (or "learning set formation") or second-order relational learning (Herman, 1980).

Win-Stay, Lose-Shift Rule

Herman and Arbeit (1973) demonstrated learning set formation in the dolphin by showing, in a series of auditory two-choice discrimination problems, that it could form a "win-stay, lose-shift rule." Pairs of different sounds were projected sequentially underwater, each through a separate speaker. Only one sound was correct and rewarded. The location of the rewarded sound and whether it was presented first or second varied randomly across trials. The most efficient strategy for solving this class of problem is that for each pair, if in the first trial the sound selected is rewarded, the subject should continue choosing that sound for the remaining trials (i.e., if win, then stay). However, if in the first trial the sound selected is not rewarded, the subject should switch choices to the alternate sound in all remaining trials (i.e., if lose, then shift). Thus, across different pairs of sounds, performance on Trial 2 is a key measure of efficiency in rule learning. The dolphin was 90 percent correct or better on Trial 2 across numerous pairs and was errorless on 23 of the final 25. It was also highly accurate when several reversals occurred within the same pairs of sounds (Herman, 1980).

Same–Different Concept

Concept formation has often been tested in animals using judgments of physical similarity in MTS tasks. A fundamental question is whether a subject is able to abstract a matching concept that can be applied to novel problems or must rely on learning the contingencies of each separate problem. Typically, first-trial performance on tests with completely novel materials provides the most convincing evidence of concept formation, inasmuch as in first trials there has been no opportunity for associative learning. If the subject is able to abstract a matching rule, two secondary questions are (a) how rapidly does this abstraction occur and after how many exemplars, and (b) how broadly does the concept extend?

Herman and Gordon (1974) tested for acquisition of a matching concept using auditory stimuli heard passively (described earlier). Training of auditory MTS was carried out with 21 unique pairs of sounds. Testing was then conducted with an additional 346 unique pairs. The dolphin's first-trial performance accuracy remained near ceiling levels throughout testing. It was correct on the first trial of 96 of the initial 121 novel pairs, 42 of the following 50 novel pairs, and 173 of the final 175 novel pairs. Clearly the dolphin had abstracted a matching concept from its initial training that it readily applied across hundreds of pairs of novel sounds. The same dolphin then extended the identity-matching concept with high accuracy across procedures to a serial-probe recognition task in which it was required to report whether a "probe" sound had been present in, or absent from, a previously heard list of sounds (see Herman, 1980).

In terms of MTS concept formation based on active echolocation, the cross-modal tests described earlier provide positive evidence of first-trial transfer across numerous pairs of objects. Also, preceding these tests, the dolphin performed accurately on 25 of 32 first trials with 16 pairs of novel objects in pure echolocation MTS (Pack & Herman, 1995).

Herman et al. (1989) tested a dolphin's ability to acquire a visual matching concept. After MTS training and testing of different combinations of familiar and novel objects, the dolphin matched accurately on 31 of 36 novel combinations of novel objects, including 100 percent choice accuracy on the final 12 combinations. Considering each of the first trials in which a novel object appeared as the matching comparison and as the nonmatching comparison, across the final 8 novel objects, the dolphin was correct on all 16 first trials. The same dolphin that was tested in these procedures then successfully extended its matching concept from three-dimensional objects to two-dimensional figures (Herman, Pack, & Morrel-Samuels, 1993a), as well as to a symbolic same/different task in which it indicated whether two objects presented successively were either "same" or "different" (Herman, Pack, & Wood, 1994). Other studies have shown that dolphins are capable of first-trial transfer in making symbolic same/different judgments using *simultaneously* presented novel visual stimuli (Herman et al., 1994; Mercado, Killebrew, Pack, Macha, & Herman, 2000), a task considered conceptually more complex than the successive same/different task (Premack, 1983).

How rapidly can a dolphin abstract a matching rule? Herman, Pack, and Morrel-Samuels (1993) reported that in a visual MTS task, one dolphin demonstrated first-trial transfer to 6 novel objects after only 72 trials of prior experience with 6 familiar objects. It was nearly errorless on these initial trials, matching correctly on 63 (88 percent) of them, including 19 of the first 24. In contrast, although pigeons may eventually perform significantly above chance in visual MTS with novel stimuli, this often requires thousands of trials with trial-unique stimuli (Wright, Cook, Rivera, Sands, & Delius, 1988).

Concepts of "Less" and "More"

Jaakkola, Fellner, Erb, Rodrigues, and Guarino (2005) studied the dolphin's understanding of relative numerosity on an ordinal scale. Different numbers of small, medium, and large dots were arranged into arrays on two backgrounds controlling for nonnumeric feature differences. Two dolphins were first trained to select the array that contained fewer ("less") dots using several exemplars (e.g., 1:6, 1:3, and 3:7 for one dolphin; 1:8, 3:7, 2:4, and 4:7 for the other). After training, each dolphin was tested with every numerical contrast using the numerical quantities 1 through 8. Both dolphins demonstrated spontaneous transfer of the "less" concept to novel contrasts overall (83 percent and 82 percent correct responses respectively), as well as on their first exposures (76 percent and 88 percent respectively).

Evidence demonstrating the dolphin's flexibility in judging relative numerosity was reported by Yaman, Kilian, von Fersen, and Güntürkün (2012). After a dolphin was trained with limited exemplars to respond to whichever of two arrays contained the greater number of items, it transferred the "more" concept to novel numerical contrasts. The reward contingencies were then reversed, requiring the dolphin to respond to the array containing the least number of items. After training with one numerical contrast, the dolphin readily applied the new "choose less" rule at performance levels of 85 percent or greater to other contrasts, including some it had responded to previously by choosing "more."

Concept of Behavioral "Creativity"

Pryor, Haag, and O'Reilly (1969) first studied dolphin innovation by showing that two rough-toothed dolphins (*Steno bredanensis*) could learn to "create" novel behaviors (i.e., those not previously performed) across training sessions. The dolphins learned that in each new session, novel behaviors, not those repeated from a previous session, would be reinforced. The dolphin's eventual success in learning this rule provided evidence of its long-term memory of previously reinforced behaviors, as well as its ability to create novel behaviors. More recently, Herman (2002) reported that a concept of "innovate" could also operate in dolphin working memory and could be formally put under the control of a "create" gesture. For example, through training, a dolphin learned that within sessions, the appropriate response to an experimenter's "create" gesture was to perform a self-selected behavior that was different from the behavior just performed. Herman (2002) reported that in response to 144 trials in which the "create" gesture was given (across multiple sessions), the dolphin performed 72 unique behaviors, 38 of which were novel.

Communication and Language

Human language is a unique form of communication that has at its core an understanding of semantics (arbitrary symbols that represent and refer to things, including those displaced in space or time) and syntax (grammatical rules that govern how symbols are arranged in different orders to create different meanings) (Herman, 2009). Several species beyond humans have been shown to possess rudimentary forms of either semantics (e.g., unique alarm calls associated with different predators) or syntax (e.g., regularities in the ordering of nonreferential vocal signals) in their natural communication systems (see Snowden, 1993).

As noted earlier, dolphins naturally produce signature whistles. An individual dolphin's signature whistle contains a frequency-modulated contour that is unique to that individual (i.e., it contains nonvoice identity information), and is developed through a combination of vocal learning and individualized modification. Signature whistles may be produced both in the wild and in the laboratory when a dolphin is visually separated from its close social companions, and sometimes occur during reunions and may also be imitated by close companions (Janik & Sayigh, 2013; King, Sayigh, Wells, Fellner, & Janik, 2013). Harley (2008) showed that dolphins can maintain a mental inventory of several different individual signature whistles and their learned conditional associates, and they maintain these associations when confronted with novel exemplars from the same whistlers that retain contour information. Dolphins appear to use signature whistles referentially to address each other (King & Janik, 2013). However, thus far there is no evidence in the wild for the referential use of nonsignature whistles or nonwhistle sounds, or for the use of syntax (Herman, 2009). This prompts two questions: (a) can learned symbols, beyond signature whistles, be understood referentially; and (b) are dolphins capable of syntactic competence?

To answer these questions, Herman, Richards, and Wolz (1984) taught two dolphins artificial symbolic language systems that contained both semantic and syntactic components. Different abstract symbols (sounds projected underwater for one dolphin, and human gestures for the other) were associated with different objects, agents, actions, modifiers, and relationships. Within each system, a set of syntactic rules governed the creation of "sentences" (either imperative sentences exclusively for the dolphin educated in the acoustic system, or imperative and interrogative sentences for the dolphin educated in the visual system). For both dolphins, the rules governing sentences that required an action to be taken directly to an object were the same: (Modifier) + Object + Action.[1] For example, the sequence "Bottom Hoop Under" instructs the dolphin to locate the hoop on the bottom of the habitat (avoiding all the other objects, including the hoop at the surface and any other object on the bottom) and swim under it. The rules governing more complex sequences requiring that a relationship be created between objects were different for each dolphin. In the acoustic system, the sound sequence Direct Object + Relationship + Indirect Object directly mirrored the order in which the actions were to be carried out. Thus to request that the surfboard be transported to the Frisbee required the sound sequence "Surfboard Fetch Frisbee." In the visual system, an inverse grammar was used for creating relational instructions of the form Indirect Object + Direct Object + Relationship. Thus the gestural sequence to carry out the same relational instruction as described earlier would be "Frisbee Surfboard Fetch." As with the simpler direct object sequences, modifiers could be placed just before either or both object symbols. A critical feature of both grammars that embodies the functionality of syntactic comprehension is that the same symbols can be used in different orders to create different meaning. Thus to request that the dolphin carry out the opposite sequence of actions specified earlier would require "Frisbee Fetch Surfboard" in the acoustic system and "Surfboard Frisbee Fetch" in the visual system. Strong evidence for syntactic comprehension would require the dolphin to perform these and other reversible sentences on their first exposure.

Both dolphins were highly accurate in responding to imperative sentences, including those that were reversible. Herman (1986) reported that across 191 novel sentences of the types described earlier, the dolphin educated in the acoustic system was entirely correct (i.e., all parts of the sentence responded to correctly) on 67.5 percent. Similarly, across 214 novel sentences, the dolphin educated in the visual system was entirely correct on 65.4 percent. Both dolphins' high performance accuracy relative to chance (4 percent or less) makes a compelling case for their spontaneous comprehension of novel sentences and for their taking into account both the semantic and syntactic components of these sentences.

In a follow-up study, Herman and Forestell (1985) used the dolphin trained to respond to human gestures to investigate its ability to respond to a trainer's interrogatives about the presence or absence of named objects in the dolphin's habitat. In response to a trainer producing the gesture for a named object followed by a "question" gesture, the dolphin pressed either a paddle to its right (to indicate "presence") or a paddle to its left (to indicate "absence"). After learning the contingencies of the paddles using a small set of exemplars, the dolphin spontaneously responded accurately when novel queries were made about named objects. In ten first trials, five reporting on objects present and five reporting on absent objects, the dolphin was correct on all but one (Herman & Forestell, 1985). These findings indicate that object symbols and their referents were represented mentally such that when the symbol was produced, it evoked the mental representation—a "search image"—of the referenced item. Herman, Pack, and Morrel-Samuels (1993) summarized these and other uses and semantic functions of object symbols taught by Herman et al. (1984) and from this provided a compelling argument that these symbols were understood as having a referring function.

Two other studies of dolphin communication used underwater keyboards to examine if dolphins would spontaneously use keys in communicative exchanges. Both keyboards were equipped with physical symbols representing items (e.g., objects, events, locations). Each symbol, when activated, also resulted in an associated sound being played underwater. The two dolphins in a study by Xitco, Gory, and Kuczaj (2001) appeared to understand the items associated with the symbols. When symbols were activated (by a human companion), a dolphin swam ahead of the human to the appropriate item and oriented to it while gazing back at the human. Dolphins also spontaneously activated the keys. However, the study provided no details on the validity of the dolphin's productions as demonstrating referential competence. In Reiss and McCowan (1993), two dolphins activated keys on a keyboard. They also spontaneously produced both imitations of symbol-associated sounds and facsimiles of sounds in conjunction with their associated objects or actions. However, data on symbol production were insufficient to consider the communication referential.

Social Knowledge and Awareness

Social knowledge is possession of information about another and may include an understanding of another's actions, as well as the networks of individuals and their strengths of association in a society (Connor & Mann, 2006). Social awareness may include the understanding of another's attention state and knowledge, as well as how these may be manipulated.

Social Imitation

Early experiments of social knowledge in dolphins focused on social learning through dolphin-to-dolphin motor imitation, the procedures for which were described earlier (Herman, 2002). In that study, after both dolphins were trained in the imitation task using a set of familiar behaviors, they were tested for imitation of an additional twelve familiar behaviors (not used during training) and three novel behaviors. The novel behaviors were different for each dolphin. Herman (2002) reported that one dolphin imitated three familiar behaviors in the model's first demonstration and within the first six demonstrations imitated another four behaviors. This dolphin also imitated two of three novel behaviors after several demonstrations. The second dolphin imitated two familiar behaviors in the model's first demonstration and within the first five demonstrations imitated another four behaviors. This dolphin also imitated one of three novel behaviors after several demonstrations. The successful performances of both dolphins, relative to chance, indicate that they were operating based on a concept of imitation (see also Bauer & Johnson, 1994; Jaakkola, Guarino, & Rodriguez, 2010). Similar performance levels to those reported by Herman (2002), including imitation of some novel behaviors on first trials, has also recently been shown with captive killer whales (*Orcinus orca*) (Abramson, Hernandez-Lloreda, Call, & Colmenares, 2012).

As noted earlier, the degree to which a concept may be extended is an indication of its sophistication and complexity. Herman (2002) summarized additional studies demonstrating the breadth of the dolphin's imitative capabilities. These include the ability to imitate arbitrary synthetic sounds both within and across octaves, and the ability to imitate the motor actions performed by a human, either "live" or displayed on a television monitor.

Social Awareness

Social awareness in animals has often been examined through studies of the management of joint attention between an informant and receiver through the comprehension and production of pointing and gazing toward an object, event, or place of interest (Pack & Herman, 2006). Studies investigating the management of joint attention in dolphins have shown that they respond accurately on first trials to dynamic brief pointing cues by humans toward distal objects, including objects placed behind the dolphin (Herman et al., 1999). They also show first-trial success in response to dynamic gazing cues, static gazing cues, and static pointing cues all performed by humans toward objects (Pack & Herman, 2004). Dolphins can also form mental representations of objects pointed to or gazed at by a human. For example, Herman et al. (1999) substituted brief human points (both direct points and cross-body points) for object symbols within relational sequences of an artificial language (see earlier section on communication and language) that requested the dolphin to transport the second indicated object to the first indicated object with three potential objects present in the dolphin's habitat. The dolphin's above-chance performance levels when carrying out such sequences indicate that it formed a mental representation of the destination object while processing the remainder of the sequence. In another study, Pack and Herman (2007) showed that a dolphin could accurately find an identical match for an object initially gazed at or pointed toward by a human based on a mental representation of that object. Regarding the production of pointing, after exposure to human divers using a keyboard to communicate about objects and their locations, each of two dolphins spontaneously began pointing (with its rostrum and aligned body) at an object of interest while monitoring (through gaze) an approaching diver from whom they needed assistance to obtain the object (Xitco et al., 2001). The dolphin's pointing behavior appeared both referential and conditional on the attentiveness of a human (Xitco, Gory, & Kuczaj, 2004).

Self-Knowledge and Awareness

The findings presented earlier on the dolphin's ability to successfully recall its past actions (Herman, 2002) are consistent with a theory that dolphins are aware of their own actions (Herman, 2012). Additional evidence supporting a broader theory of self-knowledge and awareness is presented in the following sections.

Using a Mirror for Self-Inspection

The ability to recognize oneself in a mirror (i.e., mirror self-recognition, or MSR) appears to be rare in animals, and among noncetaceans has been demonstrated only in humans and apes (reviewed in Herman, 2012), as well as in elephants (Plotnick, de Waal, & Reiss, 2006) and magpies (Prior, Schwarz, & Güntürkün, 2008).

Reiss and Marino (2001) investigated MSR in two dolphins by using markers during a feed to temporarily draw shapes on the dolphin's bodies in locations that they could not see directly. After the feed, the animals' behaviors were observed when mirrors and other reflective surfaces were available or unavailable in their habitat. The observed dolphin behaviors on "mark" trials were compared with those that occurred during "sham-mark" trials (i.e., trials identical to mark trials except that the marker contained clear water) and on "no-mark" trials. Overall, the greatest amount of time a dolphin spent performing self-directed behaviors occurred during the marked condition. In particular, the researchers noted if the dolphin swam to a reflective surface and adopted a posture or performed a behavior in front of it that allowed the dolphin to see the area of the body that had been marked or sham marked. Sham markings that occurred after a dolphin had experienced being marked were characterized by self-directed behaviors that were terminated when it apparently became clear (to the dolphin) that indeed no mark had been made. In the second part of the study, two dolphins were tested separately in pools with nonreflective walls and exposed to conditions in which a mirror was either available for use or unavailable. Reiss and Marino (2001) pointed to several lines of evidence that they considered consistent with an interpretation of self-recognition. First, there was an absence of social behaviors (i.e., behaviors typically performed toward other dolphins) when a dolphin was at the mirror. Second, a dolphin spent more time at the mirror location when it was marked and also when the mirror was available (compared to either being present but covered or absent). Finally, upon being released from station at the end of a feed, a dolphin was quicker to commence mirror use after being marked or sham marked than when neither marking procedure occurred. Based on a consideration of the collective findings, Reiss and Marino (2001) asserted that their dolphins had demonstrated MSR. However,

Harley (2013) has recently questioned whether the data are robust enough to warrant this interpretation.

Understanding the Use of One's Body Parts

Herman, Matus, Herman, Ivancic, and Pack (2001) investigated the dolphin's awareness of its body parts by associating different human gestures with nine different body parts and then testing the dolphin's comprehension of different symbolic sequences that requested the dolphin to use these body parts in different ways, including ways that were novel. Experiments included tests of novel symbolic sequences in which a named object was followed by a named body part and the requested action "touch" or "toss." The dolphin could also be requested to simply display or shake a named body part. Overall, the dolphin performed at near-ceiling levels and was able to immediately transfer its use of named body parts successfully to novel objects and to objects positioned in novel ways. The dolphin's consistently high levels of performance demonstrated that it understood the body part symbols as representing its own body parts. Importantly, its ability to spontaneously modify its use of the same body part in novel ways to correctly execute instructions when objects were positioned in unusual locations suggests that it was consciously aware of its body parts and not simply responding to instructions by using body parts in rote fashion (Herman, 2012).

Reasoning and Decision Making

Reasoning involves drawing conclusions, based on an assessment of preliminary information, that extend beyond the particulars of that information. Decision making often relies on deductive and inductive reasoning (Goldstein, 2011). Each of the studies described earlier required a dolphin to make some type of decision. This final section provides additional evidence of decision making, including decisions that appear to require some degree of reasoning. Herman (2006) has described some of these results as evidence for rational behavior.

Decisions in Response to Novel Instructions during Language Tasks

As described earlier, Herman et al. (1984) educated one of their dolphins in a visual language system. Herman, Uyeyama, and Pack (2008) pointed out that this dolphin's ability to spontaneously understand novel long grammatical frames (e.g., Modifier + Indirect Object + Modifier + Direct Object + Relationship) after having only been taught shorter grammatical frames (e.g., Modifier + Direct Object + Action and Indirect Object + Direct Object + Relationship) provided evidence that the dolphin was able to infer (presumably through some type of reasoning process) a higher-order relation that resulted from the conjoining of first-order relations. As noted earlier, in a separate study, the same dolphin was also able to spontaneously respond accurately to sentences when a human pointing gesture toward one or more objects was substituted for object symbols (Herman et al., 1999). In such instances, the dolphin's response, to immediately carry out these sequences as it would have with the object symbols, revealed a decision based on logical inference and likely inductive reasoning.

Decisions When Faced with Language Anomalies

Herman, Kuczaj, and Holder (1993) tested the dolphin educated in the visual language system on its responses to both syntactic and semantic violations. Syntactic violations involved extended strings of symbols that violated the familiar syntactic structure of the language. In such cases, the dolphin spontaneously extracted subsets of elements that made up a "legitimate" grammatical rule. Semantic violations involved proper grammatical strings that violated a semantic rule, for example, a sequence instructing the dolphin to transport a nontransportable object. In such instances, the dolphin rejected the instruction, either offering no response at all or sometimes substituting a transportable object for the specified nontransportable. The dolphin's responses to violations of both syntactic structure and semantic rules provide evidence of decision making, reflecting its understanding of the functional relationship of different grammatical frames to each other, of elemental positions within these frames, and of different classes of objects represented by different elements.

Decisions to "Escape" When Faced with Uncertainty

Smith et al. (1995) compared human and dolphin performance in an uncertainty task involving the classification of tones as "high-pitched" or "low-pitched." The rules were as follows: For a 2,100 Hz tone, respond by selecting the response signifying "high." For any tone below 2,100 Hz (the researchers used 1,200 Hz to 2,099 Hz), respond by selecting the response signifying "low." The researchers also provided the dolphin (and the humans) with a third option as an "uncertain" response that they could learn to use to "escape" from the current trial to an easier one (which, in theory, they would use if the discrimination became too difficult). The study was designed so that correct judgments of high or low resulted in a reward, incorrect judgments were associated with nonreward and a brief time-out interval in which no trials could be initiated, and overuse of the "uncertain response" option (e.g., as a default) was discouraged through its association with a much more costly delay than incorrect judgments. For the dolphin, twelve consecutive sessions of trials were conducted. The results indicated that presses of the "high" paddle occurred most often to the 2,100 Hz tone, presses to the low paddle occurred most often to lower-frequency tones, and the discrimination crossover was at 2,086 Hz. The dolphin used the "uncertain response" sparingly and most often when lower-frequency "probe" trials were near or at the crossover point (i.e., within the frequency window in which discriminations between the high tone and lower tones apparently became difficult). Smith et al. (1995) found that the occurrence of specific dolphin behaviors (e.g., slow approaches to the paddles, wavering between them, or aggressive open-mouth displays at them) was correlated with these more difficult discriminations. The appearance of such behaviors primarily during these trials, and not during easier trials, is consistent with interpretations of dolphin uncertainty and possibly frustration. The dolphin's performance in the task was very similar to what was observed with human subjects who described their use of the escape response, by introspecting on their mental state, in terms of uncertainty and doubt. Thus the dolphin's performance, in combination with its behavior, suggests that the dolphin may also have been aware of its own knowledge state.

Decisions Related to Task Efficiency

Kuczaj and Walker (2006) reported on a task in which a minimum weight placed into a container was required to mechanically release a fish. Two dolphins were shown how to collect four weights, one at a time, and drop each into the container until the fish was released. The dolphins mastered this task. Then the distance from the weights to the container was increased by an amount that would substantially increase the round-trip travel time and distance covered by the dolphins if they continued to pick up and drop one weight at a time. When presented with this new configuration, the dolphins spontaneously began transporting more than one weight on each trip to the container. Herman (2006) reported similar observations of dolphins spontaneously fetching multiple objects simultaneously after being trained to fetch a single object. The dolphins' decision to carry multiple objects and thus decrease the number of round trips both promoted efficiency and demonstrated the ability to adapt means to ends, apparently through reasoning.

Decisions Requiring Inferential Reasoning about Hidden Objects

In the wild, it is not uncommon for prey, predators, and conspecifics being tracked by a dolphin to sometimes move out of its direct visual perception. Such events, if they also render the object hidden to echolocation, naturally lead to questions of whether the dolphin understands that the visibly displaced entity continues to exist, and can infer its location (Mitchell & Hoban, 2010). A further complication may occur if the barrier behind which the entity "disappeared" itself disappears behind a larger occluder and then reemerges without the entity. In such cases, when it is made apparent that the entity is no longer hidden behind the barrier, for the dolphin to accurately track the entity, it must infer that the entity was invisibly displaced behind the occluder. Although these types of cognitive problems have been used to examine "object permanence" and mental representation in human infants, they are also relevant to any species that hunts or lives in social groups.

Jaakkola, Guarino, Rodriguez, Erb, and Trone (2010) examined whether dolphins demonstrate an understanding of visible and invisible displacement using an object containment procedure and three different types of hiding: (a) single visible displacement, in which a toy was placed in a single bucket; (b) double visible displacement, in which a toy was placed in one bucket, then removed and visibly placed in a second bucket, where it remained; and (c) invisible displacement, in which a toy was first placed in an opaque plastic cylinder, the cylinder was then placed in one of the buckets where the toy was invisibly (to the dolphin) moved from the container to the bucket, the empty container was removed from the bucket, and the dolphin was shown that it was empty. Jaakkola, Guarino, Rodriguez, Erb, and Trone (2010) found correct bucket choice for three of the six dolphins in the single visible displacement task, for one dolphin in the double visible displacement task, but for no dolphins in the invisible displacement task.

Success by dolphins in understanding invisible displacement has, however, recently been demonstrated in an innovative task involving hiding targets by *occlusion* rather than by containment. Johnson, Sullivan, Buck, Trexel, and Scarpuzzi (2015) created dynamic video display events that dolphins could view through an underwater window and respond to at their own discretion. For these events, a projection of a target shape traveled around a screen and at times was either momentarily fully occluded behind a barrier (visible displacement) or disappeared fully behind two larger shapes, which themselves momentarily disappeared behind one of two occluders, then emerged and split apart to reveal no target shape behind. The appropriate inference in the latter case would be that the target shape was displaced invisibly behind the occluder, while the larger shapes were behind it. Johnson et al. (2015) videotaped the dolphins' head movements during these events and had coders blind to the location of the hidden target judge these movements. They then compared the results to the actual trajectory and location of the target that had been either visibly or invisibly displaced. In the first trial of visible displacement, the head movements of 9 of 10 dolphins were consistent with following the correct trajectory of where the hidden object would emerge. Likewise, in the first trial of invisible displacement, the head movements of 8 of 10 dolphins were consistent with the location behind which the target shape was invisibly displaced. These performances indicate that dolphins can understand invisible, as well as visible, displacement.

Conclusions and Future Directions

I began this review by noting the lack of scientific studies of dolphin cognition before 1970. Since that time, a wealth of information has been generated in various domains, helping to illuminate dolphin mental abilities and processes and relate them as adaptations to challenges that individual dolphins face from both their physical and their social environments. In summary, dolphins have well-developed perceptual abilities. Although they may be considered specialists in the acoustic domain, their ability to perform a variety of conceptual tasks through vision as well as passive hearing and active echolocation suggests that dolphins may be best characterized as cognitive generalists. Dolphin abilities for sustained attention are impressive and well suited for diurnal and nocturnal activities in an open marine environment in which continuous vigilance is beneficial. Dolphin working memory is flexible and able to handle information both within and across modalities. Working-memory processes and capacity are similar to those seen in nonhuman primates (Herman, 1980). Monitoring, updating, and manipulating mental representations in working memory are tasks typically associated with the central executive and seem well developed in the dolphin. Long-term memory span for social affiliates, broad imitative capabilities, and social and self-awareness skills are all well suited for life within a complex social matrix involving cooperative and competitive ventures and relationships that may be transient or endure for decades. Dolphins also have the capacity to solve a variety of problems rapidly, and decisions often appear to involve some form of reasoning. Their problem-solving strategies suggest that when it comes to judging relationships such as similarity and difference, the dolphin may be more inclined toward solutions based on conceptual learning than on learning rote rules of association. Indeed, experimental studies suggest that dolphins have a strong generalized natural aptitude toward understanding a variety of relationships conceptually, including those involving mental representations.

As an example, earlier I described the dolphin's competence in syntactic processing within a language paradigm. Sentence processing relies heavily on monitoring and flexibly manipulating mental representations, as well as understanding complex and hierarchically organized relationships. This is exactly the type of cognitive ability that may be beneficial to an individual living in a fission–fusion society with large social networks and varying patterns of groupings, including first-order and second-order alliances (Connor & Mann, 2006; Cusick & Herzing, 2014; Elliser & Herzing, 2014). As Connor (2007) has pointed out, these two factors (social network size and varying patterns of groupings) present dolphins with continuous uncertainty in their knowledge of relationships, including third-party relationships. If a dolphin's knowledge of its social relationships is ultimately reproductively beneficial, then the cognitive traits that promote the acquisition, mental modeling, maintenance, manipulation, tracking, and updating of this knowledge, especially inasmuch as they generate behavior that mediates those relationships and reduces their uncertainty, should be evolutionarily favored. The following are four areas of experimental inquiry into dolphin cognition that may promote further understanding in this fascinating area.

• *Episodic long-term memory.* The studies in which a dolphin responded accurately to instructions to repeat or not repeat its own behaviors, including those that were self-selected, provide evidence of the dolphin's working memory of its past actions. Whether such memory extends to long-term episodic memory—temporal episodes including information on what, when, and where—is as yet unknown (Tulving & Craik, 2000). Such episodic memory could certainly be beneficial in the dolphin's decisions about its long-term relationships and the relationships of others.

• *Inferential reasoning about social knowledge.* The largely positive findings of dolphins in cognitive tasks involving the formation and use of secondary representations (Perner, 1991), as well as more recent studies showing inferential reasoning about invisibly hidden objects (Johnson et al., 2015), would appear to indicate that dolphins possess many of the requisite skills for inferential reasoning about another's mental state. Yet tests by Tschudin (2006) using a nonverbal "false-belief task" (Call & Tomasello, 1999) provided only equivocal support for the dolphin's understanding of the belief state of another (for a critique, see Pack & Herman, 2006). Researchers should devote more effort to studies of this type, as they would seem highly relevant to a dolphin's ability to track third-party relationships.

• *Cooperative problem solving.* To date, studies of problem solving in dolphins have focused on the individual. Yet, laboratory studies indicate that dolphins are well adapted to closely coordinate their behaviors including those involving innovation (Herman, 2002), and evidence from the wild suggests that synchrony between males (Conner & Mann, 2006) can be instrumental in their ability to sequester a female (i.e., obtain a goal). Future laboratory studies should examine problem solving from a broader, social perspective in which the achievement of goals can be maximized by dolphins coordinating their efforts. If dolphins are able to work together to solve problems, it would be interesting to examine the cognitive basis for such coordination (e.g., communication, motor imitation, goal emulation).

• *Extensions of studies of language abilities.* The work of Herman and colleagues demonstrating the dolphin's ability to carry out novel sentences within artificial languages, taking into account both their semantic and syntactic components, is arguably one of the most important achievements in studies of dolphin communication and cognition (Herman et al., 1984; Herman & Forestell, 1985; Herman, 1986; Herman et al. 1993ab). That work focused on receptive language competencies by examining responses to a wide variety of imperative sentences and a more limited number of interrogatives. The dolphins' sensitivity to other sentence forms such as the declarative have yet to be examined as well as its ability to use a language system referentially to communicate not only with humans (cf. Reiss & McCowan, 1993; Xitco et al., 2001) but also with conspecifics. Continuation of dolphin language studies into these and other new realms will not only enhance our understanding of the dolphins' linguistic competency but also the cognitive mechanisms and processes supporting language acquisition itself.

Acknowledgments

Many of the studies I have summarized here were carried out in Hawaii at the Kewalo Basin Marine Mammal Laboratory, founded and directed by Louis M. Herman, PhD. It was through his scientific acumen, innovative ideas, and unwavering devotion and support that numerous groundbreaking discoveries on dolphin cognition were made, and thousands of students, including this author, were educated. This chapter is dedicated to him with "aloha" and gratitude.

Note

1. Using this notation, the + marks the space between sequenced symbols, and the parentheses indicate that the modifier is optional. Modifiers in the acoustic system were "bottom and surface" and in the visual system were the dolphin's "left" and "right."

References

Au, W. W. L. (1993). *The sonar of dolphins.* New York: Springer-Verlag.

Abramson, J. Z., Hernandez-Lloreda, V., Call, J., & Colmenares, F. (2012). Experimental evidence for action imitation in killer whales (*Orcinus orca*). *Animal Cognition, 16,* 11–22.

Baddeley, A. D., Eysenck, M., & Anderson, M. C. (2009). *Memory.* New York: Psychology Press.

Bauer, G. B., & Johnson, C. M. (1994). Trained motor imitation by bottlenose dolphins (*Tursiops truncatus*). *Perceptual and Motor Skills, 79,* 1307–1315.

Bruck, J. N. (2013). Decades-long social memory in bottlenose dolphins. *Proceedings of the Royal Society B: Biological Sciences, 280,* 1–6.

Call, J., & Tomasello, M. (1999). A nonverbal false belief task: The performance of children and great apes. *Child Development, 70,* 381–395.

Connor, R. C. (2007). Dolphin social intelligence: Complex alliance relationships in bottlenose dolphins and a consideration of selective environments for extreme brain size evolution in mammals. *Philosophical Transactions of the Royal Society of London B: Biological Sciences, 362,* 587–602.

Connor, R., & Mann, J. (2006). Social cognition in the wild: Machiavellian dolphins? In S. Hurley & M. Nudds (Eds.), *Rational animals?* (pp. 329–367). Oxford: Oxford University Press.

Cusick, J. A., & Herzing, D. L. (2014). The dynamic of aggression: How individual and group factors affect the long-term interspecific aggression between two sympatric species of dolphin. *Ethology, 120,* 287–303.

Elliser, C. R., & Herzing, D. L. (2014). Long-term social structure of a resident community of Atlantic spotted dolphins, *Stenella frontalis,* in the Bahamas, 1991–2002. *Marine Mammal Science, 30,* 308–328.

Goldstein, E. B. (2011). *Cognitive psychology.* Belmont, CA: Wadsworth.

Harley, H. E. (2008). Whistle discrimination and categorization by the Atlantic bottlenose dolphin (*Tursiops truncatus*): A review of the signature whistle framework and a perceptual test. *Behavioural Processes, 77,* 243–268.

Harley, H. E. (2013). Consciousness in dolphins? A review of recent evidence. *Journal of Comparative Physiology A: Neuroethology, Sensory, Neural, and Behavioral Physiology, 199,* 565–582.

Harley, H. E., Putman, E. A., & Roitblat, H. L. (2003). Bottlenose dolphins perceive object features through echolocation. *Nature, 424,* 667–669.

Herman, L. M. (1980). Cognitive characteristics of dolphins. In L. M. Herman (Ed.), *Cetacean behavior: Mechanisms and functions* (pp. 363–429). New York: Wiley Interscience.

Herman, L. M. (1986). Cognition and language competencies of bottlenosed dolphins. In R. J. Schusterman, J. Thomas, & F. G. Wood (Eds.), *Dolphin cognition and behavior: A comparative approach* (pp. 221–251). Hillsdale, NJ: Erlbaum.

Herman, L. M. (2002). Vocal, social, and self-imitation by bottlenosed dolphins. In C. Nehaniv & K. Dautenhahn (Eds.), *Imitation in animals and artifacts* (pp. 63–108). Cambridge, MA: MIT Press.

Herman, L. M. (2006). Intelligence and rational behaviour in the bottlenosed dolphin. In S. Hurley & M. Nudds (Eds.), *Rational animals?* (pp. 439–467). Oxford: Oxford University Press.

Herman, L. M. (2009). Language learning and cognitive skills. In W. F. Perrin, B. Würsig, & J. G. M. Thewissen (Eds.), *Encyclopedia of marine mammals* (2nd ed., pp. 665–672). New York: Academic Press.

Herman, L. M. (2012). Body and self in dolphins. *Consciousness and Cognition, 21*, 526–545.

Herman, L. M., Abichandani, S. L., Elhajj, A. N., Herman, E. Y. K., Sanchez, J. L., & Pack, A. A. (1999). Dolphins (*Tursiops truncatus*) comprehend the referential character of the human pointing gesture. *Journal of Comparative Psychology, 113*, 1–18.

Herman, L. M., & Arbeit, W. R. (1973). Stimulus control and auditory discrimination learning sets in the bottlenosed dolphin. *Journal of the Experimental Analysis of Behavior, 19*, 379–394.

Herman, L. M., & Forestell, P. H. (1985). Reporting presence or absence of named objects by a language-trained dolphin. *Neuroscience and Biobehavioral Reviews, 9*, 667–691.

Herman, L. M., & Gordon, J. A. (1974). Auditory delayed matching in the bottlenosed dolphin. *Journal of the Experimental Analysis of Behavior, 21*, 19–26.

Herman, L. M., Hovancik, J. R., Gory, J. D., & Bradshaw, G. L. (1989). Generalization of visual matching by a bottlenosed dolphin (*Tursiops truncatus*): Evidence for invariance of cognitive performance with visual or auditory materials. *Journal of Experimental Psychology: Animal Behavior Processes, 15*, 124–136.

Herman, L. M., Kuczaj, S. A., II, & Holder, M. D. (1993). Responses to anomalous gestural sequences by a language-trained dolphin: Evidence for processing of semantic and syntactic information. *Journal of Experimental Psychology: General, 122*, 184–194.

Herman, L. M., Matus, D., Herman, E. Y. K., Ivancic, M., & Pack, A. A. (2001). The bottlenosed dolphin's (*Tursiops truncatus*) understanding of gestures as symbolic representations of body parts. *Animal Learning and Behavior, 29*, 250–264.

Herman, L. M., Pack, A. A., & Morrel-Samuels, P. (1993). Conceptual and representational abilities in bottlenosed dolphins. In H. R. Roitblat, L. M. Herman, & P. Nachtigall (Eds.), *Language and communication: Comparative perspectives* (pp. 403–442). Hillsdale, NJ: Erlbaum.

Herman, L. M., Pack, A. A., & Wood, A. M. (1994). Bottlenosed dolphins can generalize rules and develop abstract concepts. *Marine Mammal Science, 10*, 70–86.

Herman, L. M., Richards, D. G., & Wolz, J. P. (1984). Comprehension of sentences by bottlenosed dolphins. *Cognition, 16*, 129–219.

Herman, L. M., Uyeyama, R. K., & Pack, A. A. (2008). Bottlenose dolphins understand relationships between concepts. *Behavioral and Brain Sciences, 31*, 139–140.

Huggenberger, S. (2008). The size and complexity of dolphin brains—A paradox? *Journal of the Marine Biological Association of the United Kingdom, 88*, 1103–1108.

Jaakkola, K., Fellner, W., Erb, L., Rodrigues, M., & Guarino, E. (2005). Understanding of the concept of numerically "less" by bottlenose dolphins (*Tursiops truncatus*). *Journal of Comparative Psychology, 119*, 296–303.

Jaakkola, K., Guarino, E., & Rodriguez, M. (2010). Blindfolded imitation in a bottlenose dolphin (*Tursiops truncatus*). *International Journal of Comparative Psychology, 23*, 671–688.

Jaakkola, K., Guarino, E., Rodriguez, M., Erb, L., & Trone, M. (2010). What do dolphins (*Tursiops truncatus*) understand about hidden objects? *Animal Cognition, 13*, 103–120.

Janik, V. M., & Sayigh, L. S. (2013). Communication in bottlenose dolphins: Fifty years of signature whistle research. *Journal of Comparative Physiology A: Neuroethology, Sensory, Neural, and Behavioral Physiology, 199*, 479–489.

Johnson, C. M., Sullivan, J., Buck, C. L., Trexel, J., & Scarpuzzi, M. (2015). Object permanence and invisible displacement with dynamic visual occlusion in bottlenose dolphins (*Tursiops* spp.). *Animal Cognition, 18*, 179–193.

King, S. L., & Janik, V. (2013). Bottlenose dolphins can use learned vocal labels to address each other. *Proceedings of the National Academy of Sciences of the United States of America, 32*, 13216–13221.

King, S. L., Sayigh, L. S., Wells, R. S., Fellner, W., & Janik, V. M. (2013). Vocal copying of individually distinctive signature whistles in bottlenose dolphins. *Proceedings of the Royal Society B: Biological Sciences, 280*, 1–9.

Kuczaj, S. A., II, & Walker, R. T. (2006). How do dolphins solve problems? In E. A. Wasserman & T. R. Zentall (Eds.), *Comparative cognition: Experimental explorations of animal intelligence* (pp. 580–600). New York: Oxford University Press.

Mercado, E. M., III, Killebrew, D. A., Pack, A. A., Macha, I. V. B., & Herman, L. M. (2000). Generalization of same–different classification abilities in bottlenosed dolphins. *Behavioural Processes, 50*, 79–94.

Mercado, E. M., III, Murray, S. O., Uyeyama, R. K., Pack, A. A., & Herman, L. M. (1998). Memory for recent actions in the bottlenosed dolphin (*Tursiops truncatus*): Repetition of arbitrary behaviors using an abstract rule. *Animal Learning and Behavior, 26*, 210–218.

Mercado, E. M., III, Uyeyama, R. K., Pack, A. A., & Herman, L. M. (1999). Memory for action events in the bottlenosed dolphin. *Animal Cognition, 2*, 17–25.

Mitchell, R. W., & Hoban, E. (2010). Does echolocation make understanding object permanence unnecessary? Failure to find object permanence understanding in dolphins and beluga whales. In F. L. Dolins & R. W. Mitchell (Eds.), *Spatial cognition, spatial perception: Mapping the self and space* (pp. 259–280). Cambridge: Cambridge University Press.

Pack, A. A., & Herman, L. M. (1995). Sensory integration in the bottlenosed dolphin: Immediate recognition of complex shapes across the senses of echolocation and vision. *Journal of the Acoustical Society of America, 98*, 722–733.

Pack, A. A., & Herman, L. M. (2004). Bottlenosed dolphins (*Tursiops truncatus*) comprehend the referent of both static and dynamic human gazing and pointing in an object-choice task. *Journal of Comparative Psychology, 118*, 160–171.

Pack, A. A., & Herman, L. M. (2006). Dolphin social cognition: Our current understanding. *Aquatic Mammals, 32*, 443–460.

Pack, A. A., & Herman, L. M. (2007). The dolphin's (*Tursiops truncatus*) understanding of human gazing and pointing: Knowing *what* and *where*. *Journal of Comparative Psychology, 121*, 34–45.

Pack, A. A., Herman, L. M., & Hoffmann-Kuhnt, M. (2004). Dolphin echolocation shape perception: From sound to object. In J. Thomas, C. Moss, & M. Vater (Eds.), *Advances in the study of echolocation in bats and dolphins* (pp. 288–298). Chicago: University of Chicago Press.

Pack, A. A., Herman, L. M., Hoffmann-Kuhnt, M., & Branstetter, B. (2002). The object behind the echo: Dolphins (*Tursiops truncatus*) perceive object shape globally through echolocation. *Behavioural Processes, 58*, 1–26.

Perner, J. (1991). *Understanding the representational mind*. Cambridge, MA: MIT Press.

Plotnick, J. M., de Waal, F. B. M., & Reiss, D. (2006). Self-recognition in an Asian elephant. *Proceedings of the National Academy of Sciences of the United States of America, 103*, 17043–17057.

Premack, D. (1983). The codes of man and beast. *Behavioral and Brain Sciences, 6*, 125–167.

Prior, H., Schwarz, A., & Güntürkün, O. (2008). Mirror-induced behavior in the magpie (*Pica pica*): Evidence of self-recognition. *PLoS Biology, 6*(8), e202. doi:10.1371/journal.pbio.0060202.

Pryor, K., Haag, R., & O'Reilly, J. (1969). The creative porpoise: Training for novel behavior. *Journal of the Experimental Analysis of Behavior, 12*, 653–661.

Reiss, D., & Marino, L. (2001). Mirror self-recognition in the bottlenose dolphin: A case for cognitive convergence. *Proceedings of the National Academy of Sciences of the United States of America, 89*, 5937–5942.

Reiss, D., & McCowan, B. (1993). Spontaneous vocal mimicry and production by bottlenosed dolphins (*Tursiops truncatus*): Evidence for vocal learning. *Journal of Comparative Psychology, 107*, 301–312.

Ridgway, S., Carder, D., Finneran, J., Keogh, M., Kamolnick, T., Todd, M., et al. (2006). Dolphin continuous auditory vigilance for five days. *Journal of Experimental Biology, 209*, 3621–3628.

Smith, J. D., Schull, J., Jared, S., McGee, K., Egnor, R., & Erb, L. (1995). The uncertainty response in the bottlenosed dolphin (*Tursiops truncatus*). *Journal of Experimental Psychology: General, 124*, 391–408.

Snowden, C. T. (1993). Linguistic phenomena in the natural communication of animals. In H. R. Roitblat, L. M. Herman, & P. Nachtigall (Eds.), *Language and communication: Comparative perspectives* (pp. 175–194). Hillside, NJ: Erlbaum.

Tschudin, A. (2006). Belief attribution tasks with dolphins: What social minds can reveal about animal rationality. In S. Hurley & M. Nudds (Eds.), *Rational Animals?* (pp. 413–436). Oxford: Oxford University Press.

Tulving, E., & Craik, F. I. M. (2000). *The Oxford handbook of memory*. New York: Oxford University Press.

Wright, A. A., Cook, R. G., Rivera, J. J., Sands, S. F., & Delius, J. D. (1988). Concept learning by pigeons: Matching-to-sample with trial-unique video picture stimuli. *Animal Learning and Behavior, 16*, 436–444.

Xitco, M. J., Jr., Gory, J. D., & Kuczaj, S. A., II. (2001). Spontaneous pointing by bottlenose dolphins (*Tursiops truncatus*). *Animal Cognition, 4*, 115–123.

Xitco, M. J., Jr., Gory, J. D., & Kuczaj, S. A., II. (2004). Dolphin pointing is linked to the attentional behavior of a receiver. *Animal Cognition, 7*, 231–238.

Yaman, S., Kilian, A., von Fersen, L., & Güntürkün, O. (2012). Evidence for a numerosity category that is based on abstract qualities of "few" vs. "many" in the bottlenose dolphin (*Tursiops truncatus*). *Frontiers in Psychology, 3*, 473.

8 How Do Dolphin Calves Make Sense of Their World?

Stan A. Kuczaj II and Kelley A. Winship

In this chapter, we consider the processes involved in the acquisition and use of information about oneself, others, and the environment. The fetal dolphin, cocooned in its mother's womb, likely learns something about sound before birth and perhaps even learns to associate particular sounds with its mother, but the cognitive demands on the fetus are minimal. Once it is born, the young dolphin's life changes dramatically. It must now deal with a much more complex world than the one it inhabited before birth. How does the calf make sense of this new world? Does it rely on instincts, as is the case in a variety of other species from arthropods to mammals (Thorpe, 1963)? Or does the calf require experience to learn the behaviors and strategies necessary for survival?

The extent to which innate predispositions and experience affect ontogeny has long been debated. James (1890) believed that experience was necessary for human infants to successfully interpret the vast amounts of information that bombarded their senses. Consequently James characterized the world of infants as a "blooming, buzzing confusion." Although developmental psychologists have determined that human infants are much more capable of understanding their world than James suggested (Gopnik, Meltzoff, & Kuhl, 1999), comparable work with the young of other species is lacking. One of the mysteries concerning dolphin ontogeny concerns the manner in which dolphin calves make sense of their world. Are they born into the "blooming, buzzing confusion" that James believed characterized the experience of human infants? Or, like human infants, are they born with certain predispositions that either are present at birth or enable them to quickly acquire the cognitive skills that are necessary for survival?

Although our knowledge of calf behavior has advanced in the past fifty years, there is still much that we do not know. Herman and Tavolga (1980) lamented the lack of knowledge concerning dolphin calf sensory systems around thirty-five years ago, and unfortunately the situation is not much improved today. We know little about the ontogeny of dolphin cognitive and sensory systems and must speculate about calf capabilities in these areas based on what we know about calf behavior rather than on systematic investigations of these systems per se. For example, calves begin to produce the sorts of clicks that are used in echolocation in the first month or so of life (Favaro, Gnone, & Pessani, 2013). A calf's early efforts to produce click trains appear to be triggered by the mother and so begin shortly after the mother initiates her click train, suggesting that mothers cue their calves' early attempts to produce clicks. However, we know nothing about the calf's use of clicks or when the calf becomes capable of gleaning information about its immediate environment from the echoes that result from its clicks.

Acquiring Basic Skills

Most dolphin calves enter the world tail first (Caldwell & Caldwell, 1972), and their first task after birth is to reach the surface and take a breath of air, an accomplishment that most dolphin calves achieve without assistance (McBride & Kritzler, 1951; Wells, 1991). In addition to this instinct to swim toward the surface and breathe, dolphin calves appear to be programmed to follow rapidly moving objects (Connor, Wells, Mann, & Read, 2000; Mann & Smuts, 1998). This behavior is reminiscent of the imprinting behavior observed by Lorenz (1979) in his research on goslings, and has likely evolved to facilitate bonding between a dolphin calf and its mother, as well as to enhance the mother's ability to nurture and protect the calf. However, this predisposition also enables other dolphins to lure the calf away from its mother (Mann & Smuts, 1998; Tavolga & Essapian, 1957). As a result, mothers must diligently ensure that their young calves do not wander off with other dolphins. Young dolphins are also easy prey for sharks (Fearnbach, Durban, Parsons, & Claridge, 2012; Mann & Barnett, 1999) and may also fall victim to male dolphins that engage in infanticide to facilitate mating (Kaplan, Lentell, & Lange, 2009; Robinson, 2014). Close proximity enables a mother to better protect her calf and perhaps teach it about these and other dangers. Early in life, the burden for proximity maintenance rests on the mother, which she accomplishes by either herding or following calves (Hill, Greer, Solangi, & Kuczaj, 2007). A calf must learn to distinguish its mother from other dolphins, as well as to develop a preference for her, a process that may begin in utero as the calf becomes familiar with the mother's vocalizations. As a calf develops, it assumes more of the responsibility for remaining close to the mother or returning to her if they become separated (Chirighin, 1987; Connor et al., 2000; Hill et al., 2007; Mann & Smuts, 1999; Reid, Mann, Weiner, & Hecker, 1995).

Dolphin mothers rarely stop swimming for the calf to nurse, so the calf must both locate and secure a moving target to obtain its mother's milk. Dolphin calves begin nursing shortly after birth (Wells, 1991; Mann & Smuts, 1998; Mann, Connor, Barre, & Heithaus, 2000), and these early nursing bouts may be cued by the mother's presentation of her mammary slits to the calf (Tavolga, 1966). As they mature, calves learn to initiate nursing behaviors by bumping the mother's mammary area (Cockcroft & Ross, 1990; Reid et al., 1995; Sakai et al., 2013; Asper, Young, & Walsh, 1988) and may also learn to acoustically signal the mother that the calf would like to nurse (Morisaka, Shinohara, & Taki, 2005b).

The strenuous activity of swimming through water is both a challenge and a danger to calves. A calf that exhausts itself could drown or become an easy meal for a predator. Infant dolphins do not swim as well as adults and do not reach adultlike levels until at least one year of age (Noren, Biedenback, & Edwards, 2006). However, calves quickly come to ride in the echelon position, where the calf is positioned to the side of the mother near the dorsal fin (e.g., Tavolga & Essapian, 1957; Mann & Smuts, 1999; Saayman & Tayler, 1979; Tardin, Espécie, Lodi, & Simão, 2013). This position places calves in their mother's slipstream and so reduces the swimming effort they must exert (Noren et al., 2006). It is unclear if calves instinctually adopt this position, are maneuvered to do so by their mothers, or learn via trial and error the energetic benefits of this position.

Calves do not spend all their time with the mother in the echelon position but also swim in other positions, including the infant position, where a calf swims underneath its mother, slightly behind the mother's pectoral fins (Tavolga & Essapian, 1957). As calves age, they typically spend less time in the echelon position and more time in the infant position (Gubbins, McCowan, Lynn, Hooper, & Reiss, 1999). This developmental pattern likely reflects the protective benefits of the infant position, as well as the calf's improved swimming abilities (Cockcroft & Ross, 1990; Reid et al., 1995). The infant position seems to serve as a secure base from which calves can observe their world, as well as a safe haven to return to after a calf has left to explore; but some calves seem to prefer other swimming positions with their mothers as a secure base.

The preferred swimming positions and the frequency of mother-calf pair swimming per se throughout the first year of life reflect both the mother's maternal style (Hill et al., 2007) and the calf's temperament. Dolphins have personalities (Highfill & Kuczaj, 2007, 2010; Kuczaj, Highfill, & Byerly, 2012), and these personalities begin to emerge shortly after birth. Although the ontogeny of personality in dolphins has not systematically been investigated, differences in temperament are evident relatively early. Bold calves are more likely to wander away from their mothers to explore and play, while cautious calves are more likely to stay with their mothers (Kuczaj, Yeater, & Highfill, 2012). But overly protective mothers will prevent a bold calf from exploring as much as it may wish to, and a nurturing but less restrictive mother may encourage even her cautious calves to interact with others and to explore their environment. Unfortunately we know little about how these early temperamental differences are reflected in later personality. Human infant temperament predicts later personality (Caspi et al., 2003), and it seems likely that the same would be true for dolphins. But we simply do not know at this point.

Regardless of personality type and maternal style, during the first week of a calf's life, it does not stray far from its mother's side, nor are other dolphins typically allowed to approach the calf (Tavolga & Essapian, 1957). Mann and Smuts (1998) suggested that it is during this week that dolphin calves become imprinted on their mother, and after this time the mother is more accepting of allomaternal care from females that had previously been denied proximity to the calf. Dolphin calves become more independent as they age, typically spending more time away from their mother as they become more physically mature and more curious about their world (Cockcroft & Ross, 1990; Gubbins et al., 1999; Tavolga & Essapian, 1957).

Learning about Others

Dolphin calves quickly learn to synchronize aspects of their behavior with that of their mother and even other dolphins, perhaps because synchrony in swimming provides energetic benefits for calves (Noren et al., 2006). Synchronicity among human mothers and infants reflects the nature of the relationship between mother and child and facilitates normal development of the infant (Feldman, 2007). In dolphins, synchronous swimming and breathing may increase the bond between calf and mother. Such behaviors may also facilitate positive interactions between the calf and other dolphins. Dolphin calves engage in synchronous behaviors shortly after birth, suggesting that calves may be predisposed to engage in synchronous activities (Fellner, Bauer, & Harley, 2006; Sakai et al., 2013). The overwhelming majority of calf behavior is synchronous during the first month of life and then gradually decreases as calves mature (Fellner, Bauer, Stamper, Losch, & Dahood, 2013). Observations of dolphins near Mikura Island in Japan found synchronous breathing in mother-calf and escort-calf pairs and demonstrated that adult dolphins are likely to engage in identical behaviors before and after synchronous breathing, consistent with the notion that synchronous breathing is affiliative in nature for adult dolphins as well as younger animals (Sakai, Morisaka, Kogi, Hishii, & Koshima, 2010). In a study of wild Atlantic spotted dolphins (*Stenella frontalis*), Miles and Herzing (2003) reported that mother-and-calf synchronous traveling and breathing increased throughout the first three years of life but decreased during the fourth year of the calf's life as the calf began to spend more time alone.

Bottlenose dolphins live in fission–fusion societies, a form of social organization in which small groups fuse into larger groups and large groups split into smaller groups, these changes in group size and composition being the modus operandi in dolphin society. Animals in this sort of social environment must recognize a large number of individuals and understand and remember the nature and quality of the social interactions that occur when certain animals are present. This places considerable cognitive burdens on calves. They must somehow sort out the myriad array of individuals they encounter and the nature of the social interactions they either take part in or witness. Younger calves tend to interact with more dolphins than do older calves, but older calves tend to spend more time in groups than do younger calves (Gibson & Mann, 2008a). Younger calves seem to be less discriminating than older calves in associating with other dolphins, with the older calves forming stronger bonds with a smaller group of individuals.

Gender also influences the number of dolphins with which a calf associates. Female calves that were more independent (i.e., spent less time with their mothers) had fewer associates than did independent male calves (Gibson & Mann, 2008b). Moreover, the more independent a male calf was, the greater the number of associates. This pattern is strikingly similar to that found for human children, where females form a small group of close friendships, and males have more friends but not necessarily close ones (Maccoby, 1990). In dolphins, this pattern changes in at least some populations, where males form pair bonds that last a lifetime (Connor et al., 2000).

Sex differences are also evident in the behavior of calves during separations from their mothers. Male calves are more likely to seek social interactions with other dolphins than are female calves (Gibson & Mann, 2008b). This difference may reflect later sex differences in the social interactions that characterize male and female dolphins (e.g., Connor et al., 2000). Males in Shark Bay, Australia, will eventually form cooperative alliances, and their early efforts to interact with others may help them to explore potential companions for later alliances. Females typically remain in same-sex groups or live more solitary lives than do males, which may lessen the need for exploration in early life.

The number of associates the mother has also appears to influence the number of associates that calves will experience. However, this may reflect the locations in which the mothers live rather than maternal style (Gibson & Mann, 2008b). Mothers with more associates may inhabit areas where more dolphins are available for the calf to interact with. The relative roles of habitat and maternal "friendliness" on the number of calf associates warrant further investigation.

Atlantic spotted dolphins are more likely to associate with kin than with other members of their social groups (Welsh & Herzing, 2008). Such preferences are common in the animal kingdom, perhaps because it is evolutionarily advantageous to help one's kin survive. Associating with kin makes it easier for an individual to aid kin, helping to ensure the survival of the genetic line. Perhaps this is why biases toward sex and age segregation observed in the general population are not always found in kin dyads (Welsh & Herzing, 2008).

Regardless of their gender, calves must learn to navigate their social world. They must learn to deal with both affiliative and agonistic behaviors directed toward them, and learn when and whom to direct such behaviors toward. Both affiliative and agonistic interactions often involve tactile exchanges. Aggressive interactions can include biting, raking, ramming with the rostrum, and slapping with the fluke (Herzing, 1996; Miles & Herzing, 2003; Samuels & Gifford, 1997; Scott, Mann, Watson-Capps, Sargeant, & Connor, 2005; Weaver, 2003) but may also occur without physical contact. For example, a dolphin might chase another dolphin but cease the pursuit once the other dolphin has moved away. Or it might produce a jaw clap as a warning that it will attack if the situation does not change. Calves must learn the signals that warn of an impending attack, the level and type of response that is appropriate if one is attacked, and the distinction between a play fight and an actual fight.

Although we are currently unable to specify exactly how calves learn about the signals and behaviors used in agonistic interactions, it is clear that calves engage in many tactile exchanges, most of which are affiliative in nature. Initially, the mother is the primary source of affiliative tactile interactions, but calves quickly engage in tactile interactions with other dolphins, especially other calves. Mothers will discipline their calves using tactile behavior such as making contact to the flank of a misbehaving calf with their rostrum, and even pinning their calf to the floor in more serious cases (Herzing, 1996). As calves mature, they become increasingly likely to touch and be touched by peers in the same age class (Dudzinski, 1998). Tactile contact facilitates feelings of security in many species (Dunbar, 2010; Frank, 1957) and in some cases can increase the likelihood of risk-taking behavior in humans (Levav & Argo, 2010), and we suspect that touch is one of the more significant aspects of dolphin social life, beginning immediately after the calf is born.

The most studied aspect of tactile interactions has been that in which one dolphin touches the pectoral fin of another dolphin. Pectoral fin contact in dolphins is an affiliative behavior that has been used to assess social relationships (Sakai, Hishii, Takeda,

& Koshima, 2006), the basic idea being that touch may facilitate the formation and maintenance of dolphin social bonds (Dudzinski, Gregg, Ribic, & Kuczaj, 2009). Some of these pectoral interactions involve rubbing, where one dolphin rubs the pectoral fin of another with its own. Spotted dolphin subadults and calves were more likely to be the ones doing the rubbing, and calves initiated this type of contact more frequently with other calves than with the other four age classes, excluding mother-calf pairs (Dudzinski et al., 2009). Observations in Shark Bay found that calves often initiated rubbing bouts with their mother, focusing on the head (Mann & Smuts, 1999).

Pectoral fin rubbing appears to play a role in reconciliation between individuals after aggressive interactions and may also serve to reduce conflicts in juvenile-adult female relationships (Tamaki, Morisaka, & Taki, 2006). We have observed dolphin calves initiating pectoral rubs with their mothers following maternal discipline. In these cases, calves may use pectoral rubbing as a means to reassure themselves that the bond with their mothers has not been irreparably damaged. Or it may signal their acceptance and understanding of the mother's disciplinary actions. We have also observed calves initiating pectoral rubs with their mothers immediately after reunions in which the mother has called the calf, perhaps as an attempt by the calves to reduce the possibility of maternal discipline for having wandered too far away. However, we know little about the ontogenetic progression of calf pectoral rubbing, and longitudinal studies are needed to sort out all these possibilities.

Learning to Communicate

One of the things that calves must do is learn the significance of all communicative signals used by dolphins. The dolphin acoustic repertoire is diverse and includes barks, clicks, chirps, jaw pops, squawks, yelps, and whistles (Caldwell & Caldwell, 1972; Evans, 1967). Dolphins also use other modes of communication, including visual signals such as postures and aerial displays, and nonvocal sounds such as those produced by tail slaps and hitting the surface after leaping out of the water.

Although dolphins communicate in a myriad of ways, their use of whistles has been the most studied. Moreover, a particular class of whistles, signature whistles, has been studied much more extensively than any other sort of whistle (Caldwell & Caldwell, 1979; Tyack & Sayigh, 1997; see Lammers & Oswald, chap. 5, this vol.). Tyack (1997) suggested that dolphins do not use individual voice characteristics to distinguish and identify individuals and so instead rely on the acoustic differences between signature whistles for individual identification. How, though, do young dolphins acquire and apprehend the significance of such whistles?

Dolphins are one of the few mammalian species capable of vocal learning (Janik & Slater, 1997). Dolphin calves spontaneously mimic sounds produced by their mothers and other dolphins and may even imitate computer-generated whistles (McCowan & Reiss, 1993; Tyack & Sayigh, 1997). As one might expect, given calves' interest in sounds and their propensity for vocal mimicry, the sounds that calves learn are influenced by the sounds they hear, with exposure to sounds likely beginning before birth. Pregnant dolphins whistle more often during pregnancy, with a dramatic increase in whistle rates occurring in the few days before birth (Mello & Amundin, 2005). We know that human infants learn to recognize their mother's voice and even particular passages that mothers have read out loud during pregnancy (DeCasper & Fifer, 1980), and it seems likely that dolphin mothers are providing acoustic models for their unborn calves, information that may make it easier for a calf to recognize its mother's signature and other whistles.

Dolphin calves begin producing sounds shortly after birth, including whistles and burst pulse sounds (Caldwell & Caldwell, 1979; McBride & Kritzler, 1951; Morisaka, Shinohara, & Taki, 2005a). Both whistle squawks and click trains have been recorded in the first few days of life (Favaro, Gnone, & Pessani, 2013; Reiss, 1988). All these sounds are less complex than those the calf will produce as it matures (Morisaka et al., 2005a). Given their capacity for vocal learning, it is possible that calves play with sounds during the first years of life as a means to facilitate the acquisition of mature whistles and other acoustic signals (Kuczaj & Makecha, 2008), just as human children play with sounds as they acquire their first language (Kuczaj, 1983; Weir, 1962). Atlantic spotted dolphin calves have been recorded making vocalizations in instances of excitement, such calls sometimes eliciting calming tactile responses from conspecifics (Herzing, 1996).

Young calves typically develop a signature whistle by eighteen months (Tyack & Sayigh, 1997). However, as is the case for most developmental norms, there is considerable variation in the age of acquisition. For some calves, signature whistles appear in the first weeks of life, but signature whistles are not apparent in other calves' repertoires until the second year of life.

Both a calf's relationship with its mother and its acoustic environment influence the whistle that eventually becomes the calf's signature (Tyack & Sayigh, 1997), and both influence the rate with which that whistle is acquired. Calves with strong attachments to their mothers seem more likely to develop a signature whistle that resembles that of their mother, and may also acquire that whistle at a relatively early age. But the acoustic environment also influences both what whistle a calf acquires and when it acquires this whistle. Calves with rich and complex acoustic environments seem more likely to develop signature whistles that differ from their mothers. Given this pattern, calves with strong bonds to the mother and few whistle models from which to choose should be most likely to develop signature whistles that resemble that of their mother. But weaker mother-calf bonds and more whistle types should result in a calf acquiring a whistle that is distinct from its mother's.

The richness of a calf's acoustic environment is strongly correlated with the complexity of its whistle repertoire (Tyack & Sayigh, 1997). Exposure to a rich and complex acoustic environment may provide calves with more raw material to interpret, organize, and play with, which could in turn result in more complex whistle repertoires (Kuczaj, 2014). However, this richness of environment may also hinder calves' ability to identify individual calls. Dolphins live in groups, and oftentimes multiple dolphins are vocalizing at the same time. The acoustic interference and auditory masking inherent in such situations can be challenging for mature individuals (Bee & Micheyl, 2008), and the manner in which dolphin calves resolve these problems is unknown.

Kuczaj (2014) noted that it would be interesting to learn how dolphin personality and maternal style interact with the acoustic environment to shape a dolphin's whistle repertoire. Are calves more likely to develop signature whistles that resemble their mother's signature whistles if the mother is overly protective and restrictive? Are shy dolphin calves less likely to imitate sounds they hear than are bold calves? The interaction between maternal style and calf personality has received little study in general, and the way in which these two factors influence a calf's acquisition of its signature whistle and its overall acoustic repertoire deserves additional study.

It is important to remember that dolphin calves are selective vocal learners, and the sounds they decide to mimic and play with are determined by a variety of environmental, personality, and social factors (Kuczaj, Yeater, & Highfill, 2012; Kuczaj, 2014). For example, dolphin calves tend to develop signature whistles that are similar to those of other dolphins in their community. Fripp et al. (2005) found that, on average, the signature whistles of six other dolphins in the group were similar to those of a given calf. However, these six dolphins were most likely to be the ones that rarely interacted with the calf. Why do calves select model whistles from individuals with whom they rarely interact? Perhaps they are predisposed to acquire signature whistles that are rare in their environment to avoid confusion among conspecifics. If so, calves must somehow keep track of the whistles they have heard and the relative frequency of interactions between other dolphins and the calf and its mother. Or it may simply be the case that calves are predisposed to attend to novelty, the result being that infrequent models are more interesting and so more likely to be mimicked than are more frequent models (Kuczaj, 2014; Kuczaj & Walker, 2012).

Kuczaj (2014) suggested that dolphin calves' preference for novel whistles may have evolved to ensure that signature whistles are sufficiently variable within a population. Although calves acquire signature whistles in the first year or two of life, it is not the case that every calf develops its own truly unique whistle. Instead the foundation for a calf's signature whistle comes from its mother and other members of its community. This suggests that there are constraints on the number and types of possible signature whistles. Determining these constraints will be necessary for us to completely understand why individual calves select the whistles that they do.

Adult dolphins engage in vocal matching, a phenomenon in which an animal mimics the sound produced by another. Vocal matching has significant social implications, such as identifying oneself or one's group, attracting another's attention, helping to maintain contact, and enhancing cooperative behavior (Sewall, 2012). So in addition to learning how to produce whistles (including signature whistles) and other sounds, calves must learn how to use these sounds to communicate. Little is known about this process, but it seems likely that mothers and peers play intimate roles in dolphin calves' acquisition of a communication system (Kuczaj, 2014). For example, calves may learn via observation of their mothers how to adapt their signals to noisy environments, including those produced by human activities.

Learning from Others

Although the exact processes are yet to be determined, it is clear that calves learn about their communication system via experience with others. But there is much more to being a dolphin than communication, and social learning is an important aspect of the transformation from a newborn calf to a mature adult.

For example, as dolphin calves age, they nurse less and rely more on fish as a source of nutrition. The transition from nursing to a solid-food diet is not always an easy one, particularly for wild dolphins that must forage for their own food. Although nursing seems at least partly instinctual, capturing prey requires considerable amounts of experience. To be successful hunters, dolphin calves must learn to recognize, hunt, and capture prey. Dolphin foraging is quite flexible, with different groups using different strategies, as well as focusing on different sorts of prey (Weiss, 2006; Mann & Sargeant, 2003). Dolphins in Shark Bay have at least thirteen different foraging strategies (Mann & Sargeant, 2003), and a mother's preferred foraging strategy is the best predictor of an offspring's preferred method (Sargeant & Mann, 2009). Dolphins in Sarasota, Florida, also use different foraging strategies, but individual dolphins appear to specialize in one strategy, which dictates where they exert most of their foraging effort (Weiss, 2006).

Given that any individual dolphin calf should be able to learn any one of these many foraging techniques, the significance of a calf's learning environment is apparent. Dolphin calves spend the first several years of life with their mothers, and opportunities to observe and learn successful foraging methods likely contribute to this relatively long relationship (Weiss, 2006; Sargeant & Mann, 2009).

Sponge feeding, a foraging tactic found in the Shark Bay population, involves a dolphin using a sponge on its rostrum as it forages in the substrate (Krützen et al., 2005). Sponging is most often transmitted from mother to daughter (Krützen et al., 2005), but some male calves also acquire this strategy (Mann, Stanton, Patterson, Bienenstock, & Singh, 2012). In this population, differences in foraging strategies are correlated with social segregation, with spongers primarily interacting with other spongers (Mann et al., 2012), suggesting that foraging differences among dolphins may have long-lasting effects on the social lives of calves.

Although the available evidence strongly suggests that sponging behavior is transmitted vertically from mother to offspring, the precise mechanisms involved in this cultural transmission are not known. Mothers certainly provide models of sponging, and these observational opportunities should be especially salient for calves with close bonds to their mothers (see Kuczaj, Yeater, & Highfill, 2012). But is a calf's attention drawn to the sponge per se, to the activity of wearing the sponge, or to the use of the sponge for foraging? Does the mother do more than simply provide a model? Does she teach her calf how to forage with a sponge? We lack the detailed observations that we need to answer these questions. The discovery that foraging behaviors tend to occur within a matrilineal line is just the beginning: exactly how is this information passed from generation to generation?

Parental teaching does occur in nonhuman animals (for a review, see Hoppitt et al., 2008), and before we discuss this issue, it is worth reiterating Caro and Hauser's (1992) criteria for teaching. For Caro and Hauser, teaching can be said to occur only if (1) the teacher modifies its behavior only in the presence of a naive observer, (2) there is some cost for the teacher, and (3) there is no immediate benefit for the teacher. The teacher provides an example for the learner and may actively encourage or discourage aspects of the learner's behavior. Teaching results in the learner acquiring knowledge or behaviors more quickly than it would via trial-and-error learning, and may even result in acquisitions that would not have been learned at all without the intervention of the teacher.

Teaching, then, is more than observational learning. Teaching requires that the teacher actively participate in the transfer of a skill, rather than simply provide a model for the learner to emulate. Caro and Hauser also distinguished "opportunity teaching" and "coaching." Opportunity teaching occurs when offspring are provided opportunities to practice a particular behavior, whereas coaching involves the teacher actively reinforcing or punishing behavior (Caro & Hauser, 1992). Opportunity teaching is much more common than coaching, and it is sometimes difficult to distinguish opportunity teaching from observational learning that does not involve teaching. For example, mother chimpanzees provide models of termite-fishing techniques to their offspring (Lonsdorf, 2006). However, the chimpanzee mothers in this study did not teach their young proper termite-fishing techniques; the infants could only watch. Being curious, they sometimes took the mother's tool, a behavior that the mother typically did not appreciate. So even though the mother provided opportunities for her offspring to observe termite fishing, she did not engage in behaviors that constitute teaching.

Dolphin calves are curious about the prey caught by adults, including individuals other than their mothers. This curiosity results in opportunities to gather information about appropriate prey items (Mann, Sargeant, & Minor, 2007) and may enable young dolphins to create object categories that distinguish edible and nonedible items. However, as was the case with the mother chimpanzees discussed earlier, calf inspections of captured prey seem to involve observational learning more so than teaching. In Brazil, an adult female and her offspring work together with local fishermen to feed off the fishermen's catch, suggesting that this cooperative relationship is somehow transferred from mother to offspring (Boran & Heimlich, 1999). Once again, however, there is no evidence that teaching is involved in sharing foraging strategies across generations. Is cultural transmission in dolphins limited to observational learning? Or is teaching involved in at least some cases?

Killer whales forage on a variety of marine species, including fish, sharks, birds, and mammals. At least some of the techniques used to hunt these animals may be taught to young orcas. In the Antarctic, groups of killer whales sometimes submerge and swim rapidly under ice floes to create waves that wash seals and penguins into the water (Visser et al., 2008). Killer whale calves also attempt to wash animals off ice floes, and adults occasionally place a captured and still living seal back on top of the ice, creating an opportunity for the young to practice this behavior, apparently teaching youngsters this technique (Visser et al., 2008). Maniscalco, Matkin, Maldini, Calkins, and Atkinson (2007) provide another example of possible teaching, reporting that adult killer whales released a captured stellar sea lion pup approximately 500 meters from the shoreline, allowing a calf to observe and practice hunting behavior.

Some killer whales use a hunting technique in which the animals strand themselves temporarily on the shore to capture prey (López & López, 1985). Juveniles were observed practicing this behavior while being monitored by an adult male that later demonstrated the proper technique, after which the juveniles tried again (López & López, 1985). In other instances, both a juvenile and an adult would strand at the same time, and if the juvenile was unsuccessful in its prey capture attempt, the adult would fling its captured prey to the juvenile, which would either push the prey with its head or catch it in its mouth. Guinet (1991) documented an instance of a mother pushing her juvenile offspring onto the beach, then stranding herself to push the calf back into the water, apparently teaching her offspring the proper methods of stranding and returning to the ocean and even creating a wave to wash the calf back into the ocean when it became stuck.

Although several populations of *Tursiops* sp. (Silber & Fertl, 1995; Hoese, 1971; Duffy-Echevarria, Connor, & St. Aubin, 2008) also specialize in a beach-stranding foraging technique, evidence for teaching among these animals is rare. Hoese (1971) observed pairs working together in Georgia that differed in body length and speculated that calves intentionally practiced the beaching technique. However, Duffy-Echevarria, Connor, and St. Aubin (2008) noted only one instance in which a juvenile was involved in a strand-feeding event in South Carolina. In Shark Bay, Australia, dolphins hydro-plane to reach fish in very shallow waters, a strategy that is limited to a small number of adults and their calves. However, we have no evidence that the adults engage in any form of teaching to facilitate the acquisition of such behavior by their calves (Sargeant, Mann, Berggren, & Krutzen, 2005).

However, evidence also points to teaching in Atlantic spotted dolphins (*Stenella frontalis*) (Bender, Herzing, & Bjorklund, 2009). Mothers chased their prey significantly longer in the presence of their calves and made significantly more referential body movements toward the prey when their calves were nearby than at any other time, suggesting that the mothers were actively teaching their calves appropriate foraging skills. In such cases, the burden of learning is shared by mother and calf rather than relying only on the calves' ability to observe.

Dolphin calves must learn other behaviors in addition to those associated with foraging. Similar to foraging, observation plays an important role in the acquisition of many behaviors, but direct teaching is relatively rare. Kuczaj and Yeater (2006) reported the following observation of apparent teaching by a bottlenose dolphin in a bow-riding context:

The adult dolphin performed a barrel roll and then turned its head toward the juvenile. The adult repeated this behavior several times, after which the juvenile attempted a barrel roll, but lost its position on the pressure wave while doing so. The adult immediately left, but both animals returned within a few minutes. The adult then performed two barrel rolls, looking toward the juvenile after each roll was completed. The juvenile again attempted a roll, and once again fell off the wave. The adult followed, and both animals quickly returned to the wave. The adult produced one roll following their return, after which the juvenile attempted a roll, once again losing its position in the wave. The adult did not follow the juvenile this time, and the juvenile soon returned. At this point, the juvenile completed a roll and managed to stay on the wave, after which it looked toward the adult. It then produced several successive rolls in a row. (p. 417)

A more typical example of observational learning comes from dusky dolphins (*Lagenorhynchus obscurus*). These animals are known for their acrobatic leaps, and although calves may learn via trial and error to perform these leaps, learning the context in which to perform each type of leap appears to be facilitated by observing others (Weir, Deutsch, & Pearson, 2010). However, little if any teaching has been reported.

Observational learning is common in both captive and wild dolphins (for a review, see Yeater & Kuczaj, 2010), but individual animals differ in both what behaviors and which animals they will imitate. Dolphins are one of the few nonhuman species capable of both vocal and behavioral imitation, and social learning appears to be an important aspect of dolphin social life and dolphin behavioral development. In addition to vocal learning, dolphins discover behaviors for foraging, play, and social interactions by observing other members of their social group.

But dolphins neither indiscriminately observe nor mindlessly imitate other dolphins. To the contrary, dolphin calves are quite selective in their choices of whom to observe and imitate (Kuczaj, Yeater, & Highfill, 2012). Although calves are most likely to learn foraging behaviors from their mothers, they are more likely to watch and imitate the play behaviors of other calves (Kuczaj et al., 2006). But not all calves are equally likely to be good models. There is a general tendency for calves to watch and learn from calves that are older than they are, given that the older calves produce behaviors that the observing calf does not know (Chirighin, 1987; Kuczaj et al., 2006). Dolphin calves produce more complex play behaviors at an earlier age if there are older calves in their social groups (Kuczaj et al., 2006). Observing older calves at play provides younger calves with model behaviors to attempt to reproduce and also seems to motivate the younger animals to attempt behaviors that they might otherwise not attempt until weeks or months later. Boran and Heimlich (1999) reported that juvenile dolphins demonstrated techniques for playing with (chasing and harassing) stingrays, with young animals attempting to duplicate the modeled play behavior. But differences in age are only part of the story. In fact, personality may be more important than age (Kuczaj, Highfill, & Byerly, 2012).

Dolphins exhibit individual personalities from an early age, and the more curious and bold animals are the ones most likely to be observed and mimicked. For example, when a novel object was encountered, the bolder calves and juveniles were the first to examine the object, with the more cautious calves hovering a short distance behind the bolder animals, looking over the bolder dolphins' "shoulders," and keeping the bold calf between themselves and the novel object (Kuczaj, Yeater, & Highfill, 2012). The bold animals began to more actively explore and even manipulate the novel object, and only after the cautious animals had witnessed the bold animals' successful manipulations of the object did they approach the object and attempt to replicate the model's behaviors. These results led Kuczaj and colleagues to suggest that the selectivity of dolphin social learning is influenced by behavioral context, novelty of behavior, the significance of the model, and personality.

The relationship between personality and learning has received little attention, most likely because there have been no longitudinal studies of learning or personality in calves. Some mothers may be more likely to engage in teaching behaviors than others, and some calves may benefit more from teaching than others. Exploring the effects of individual differences on development is essential for increasing our understanding of the factors that influence ontogeny.

In addition to aiding the acquisition of necessary survival skills, imitation may also influence social interactions. Humans who feel excluded from a group tend to mimic others more often than do humans who are members of the group (Lakin, Chartrand, & Arkin, 2008), even selectively mimicking others who are part of the in-group rather than other excluded individuals. Capuchin monkeys react more positively to humans who imitate them (Paukner et al., 2009), suggesting that nonhuman primates can recognize imitation though they rarely engage in imitation themselves (Call & Carpenter, 2009; Paukner, Suomi, Visalberghi, & Ferrari, 2009). Given the significance of synchronous behavior for dolphins and dolphin calves, determining the ontogenetic relationship of synchrony and imitation in dolphins is an important step in understanding the social significance of imitation for dolphins. Do dolphins mimic others as a means to facilitate positive social interactions? Does being mimicked result in a dolphin becoming more affiliative and prosocial toward conspecifics in general? Such seems to be the case for humans (Van Baaren, Holland, Kawakami, & Van Knippenberg, 2004). If imitation promotes positive social exchanges, one might expect imitation to be relatively common in dolphins and an important skill for dolphin calves to acquire.

The Significance of Play

The significance of play for development cannot be overstated. If young members of a species are not able to engage in sufficient amounts of play behavior, abnormal social and cognitive development may occur (Pellis & Pellis, 2009), with absence of play in otherwise healthy animals often reflecting poor animal welfare (Held & Špinka, 2011; Kuczaj & Horback, 2013). Dolphins are somewhat unique among animals in that they continue to play throughout their life span, although young dolphins are much more likely to play than are adults. Dolphin play includes the modification of behaviors, manipulation of inanimate objects, interaction with other species, and social play with one another, all of which have been observed in wild and captive populations (for a review, see Paulos, Trone, & Kuczaj, 2010).

Bekoff (2001) suggested that play facilitates social development. Social play helps young animals learn more about other members of their group (Thompson, 1998) and may even influence the social relationships of young animals (Bekoff & Byers, 1981), affecting relationships that may last throughout the animals' lives. In addition, social play may facilitate the development of an animal's social behavioral repertoire, assisting in the acquisition of appropriate and successful behaviors used in affiliation, aggression, communication, foraging, and mating (Coelho & Bramblett, 1982; Kuczaj et al., 2006; West, 1974).

Social play bouts often involve role reversals in which a dominant animal and a sub-ordinate animal switch dominance relations, a phenomenon never observed outside play. Such reversals are possible because social play often involves some sort of play signal to express the playfulness of the ensuing event, which frees the play partners from the normal (nonplay) rules that govern social interaction in their species. For play signals to be effective, the signaler must use them honestly, and the listener must be able to recognize and correctly interpret the intent of the signaler and be able to adapt its behavior to playful contexts (Bekoff & Allen, 1998).

Many species use play signals to ensure that the playful intent of their behavior is recognized (Bateson, 1955, 1972; Kuczaj & Horback, 2013; Palagi, 2008; Pellis & Pellis, 2011). In captive dolphins, an acoustic play-fight signal, consisting of a burst pulse followed by a frequency-modulated whistle, was recorded during affiliative, but not aggressive, interactions (Blomqvist, Mello, & Amundin, 2005). Similar signals likely exist in wild populations, though such information has yet to be reported. Adult rough-toothed dolphins sometimes encourage the participation of young animals in play activities (Kuczaj & Highfill, 2005), and it is possible that some sort of play signal is used to communicate the adult's message. But more research is sorely needed in this area. When do dolphins use a play signal? If it exists, is it innate or learned?

Dolphin play is creative and becomes more complex with increasing age (Kuczaj, Makecha, Trone, Paulos, & Ramos, 2006; Kuczaj & Makecha, 2008; Kuczaj & Walker, 2012; McBride & Hebb, 1948; Tavolga, 1966). Piaget (1952) suggested that moderately discrepant events, events that are both familiar and novel, are essential aspects of cognitive development for human children. Such events provide a familiar basis for interpreting novel information but also require the organism experiencing the event to learn something new if it is to understand the event. Play provides a context in which dolphins create their own moderately discrepant events and so enhance their cognitive development by providing stimulating environments. In turn, this may facilitate the ontogeny of flexible problem-solving skills, an important set of abilities for a species that inhabits a wide range of habitats and preys on a diverse range of species. Given that adult dolphins also play, dolphin play may help to maintain cognitive functioning throughout the life span. If this is true, play may have evolved to enhance the ability to adapt to novel situations (Kuczaj & Walker, 2012; Spinka, Newberry, & Bekoff, 2001).

Although the presence of other calves is not necessary for an individual's play to become more complex, the opportunity to interact with and observe other calves seems to drive the increasing complexity of an individual's play behaviors (Kuczaj et al., 2006). This may occur because dolphin calves strive to outdo one another (although there are undoubtedly individual differences in this regard), perhaps as they try to establish their social rank within the group.

Future Directions: Unanswered Questions and Research Needs

One of the recurring themes of this chapter has been how little we know about dolphin calf development. Research to date provides tantalizing glimpses of possible aspects of dolphin calf cognition and behavior, but more detailed investigations of dolphin ontogeny are sorely needed.

As we have seen, both instinct and experience are necessary to produce the behavioral similarities and behavioral diversity that characterize dolphin populations around the world. Future work should focus on identifying the set of instincts and predispositions that young dolphins bring to the task of making sense of their world and the manner in which these interact with experience to produce the competencies of an adult dolphin. Doing so should provide answers about the ontogeny of sensory systems, communication systems, and social understanding.

Some questions that need to be answered:

1. How and when do calves learn the meanings and use of the signals that make up their communicative system?

2. How and when does competency with echolocation emerge?

3. How and when do calves recognize their mothers? Other dolphins?

4. Are there additional gender differences in development beyond those that have already been identified?

5. How do calves distinguish friend from foe?

6. How do individual differences in synchrony affect later development and behavior in dolphins? Parent-infant synchrony in humans facilitates the parent-infant bond, emotional self-regulation, symbol use, and capacity for empathy (Feldman, 2007). Is the same true for dolphins?

7. What is the relationship of personality development, cognitive development, and social development? Chimpanzees form friendships based on homophily in personality (Massen & Koski, 2013), perhaps because friendships based on personality similarities make it easier to predict friends' behaviors. Are dolphin calves more likely to choose friends that have personalities like theirs? If so, why?

Answers to these questions will require a variety of methodological approaches. Longitudinal studies of individual animals will provide the sorts of detailed information that is necessary both to describe and to explain development, and can be supplemented with cross-sectional studies that compare different animals at different ages. Observational studies can provide a wealth of descriptive data, but more controlled approaches may be necessary to determine sensory and cognitive capabilities. The innovative research techniques that have been used to study human infant cognition take advantage of behaviors that infants produce spontaneously when they are interested in some aspect of their world, and such techniques provide excellent models for potential studies of calf cognition. Although it is possible to conduct research on dolphin calf cognition in the wild insofar as it relates to social behavior, play, and communication, advances in understanding the ontogeny of dolphin sensory systems and the manner in which young dolphins represent and make sense of their world require more controlled settings. Such research will be successful only to the extent that we are able to design situations that intrigue these curious animals.

References

Asper, E. D., Young, W. G., & Walsh, T. (1988). Observations on the birth and development of a captive-born killer whale, *Orcinus orca. International Zoo Yearbook, 27*, 295–304.

Bateson, G. (1955). A theory of play and fantasy. *Psychiatric Research Reports, 2*(39), 39–51.

Bateson, G. (1972). *Steps to an ecology of mind.* New York: Ballantine Books.

Bee, M. A., & Micheyl, C. (2008). The cocktail party problem: What is it? How can it be solved? And why should animal behaviorists study it? *Journal of Comparative Psychology, 122*(3), 235–251.

Bekoff, M. (2001). Social play behavior: Cooperation, fairness, trust, and the evolution of morality. *Journal of Consciousness Studies, 8*(2), 81–90.

Bekoff, M., & Allen, C. (1998). Intentional communication and social play: How and why animals negotiate and agree to play. In M. Bekoff & J. A. Byers (Eds.), *Animal play: Evolutionary, comparative, and ecological perspectives* (pp. 97–114). Cambridge: Cambridge University Press.

Bekoff, M., & Byers, J. A. (1981). A critical reanalysis of the ontogeny of mammalian social and locomotor play: An ethological hornet's nest. In K. Immelmann, G. W. Barlow, L. Petrinovich, & M. Main (Eds.), *Behavioral development: The Bielefeld Interdisciplinary Project* (pp. 296–337). New York: Cambridge University Press.

Bender, C. E., Herzing, D. L., & Bjorklund, D. F. (2009). Evidence of teaching in Atlantic spotted dolphins (*Stenella frontalis*) by mother dolphins foraging in the presence of their calves. *Animal Cognition, 12,* 43–53.

Blomqvist, C., Mello, I., & Amundin, M. (2005). An acoustic play-fight signal in bottlenose dolphins (*Tursiops truncatus)* in human care. *Aquatic Mammals, 31*(2), 187–194.

Boran, J. R., & Heimlich, S. L. (1999). Social learning in cetaceans: Hunting, hearing, and hierarchies. In H. Box & K. Gibson (Eds.), *Mammalian social learning: Comparative and ecological perspectives* (pp. 281–307). Cambridge: Cambridge University Press.

Caldwell, M. C., & Caldwell, D. K. (1972). Vocal mimicry in the whistle mode by an Atlantic bottlenosed dolphin. *Cetology, 9,* 1–8.

Caldwell, M. C., & Caldwell, D. K. (1979). The whistle of the Atlantic bottlenosed dolphin (*Tursiops truncatus*): Ontogeny. In H. E. Winn & B. L. Olla (Eds.), *Behavior of marine animals* (pp. 369–401). New York: Plenum.

Call, J., & Carpenter, M. (2009). Monkeys like mimics. *Science, 325,* 824–825.

Caro, T. M., & Hauser, M. D. (1992). Is there teaching in nonhuman animals? *Quarterly Review of Biology, 67*(2), 151–174.

Caspi, A., Sugden, K., Moffitt, T. E., Taylor, A., Craig, I. W., Harrington, H., et al. (2003). Influence of life stress on depression: Moderation by a polymorphism in the 5-htt gene. *Science, 201*(5631), 386–389.

Chirighin, L. (1987). Mother-calf spatial relationships and calf development in the bottlenose dolphin (*Tursiops truncatus*). *Aquatic Mammals, 31*(1), 5–15.

Cockcroft, V. G., & Ross, G. J. B. (1990). Observations on the early development of captive bottlenose dolphin calf. In S. Leatherwood & R. R. Reeves (Eds.), *The bottlenose dolphin* (pp. 461–478). New York: Academic Press.

Coelho, A. M., & Bramblett, C. A. (1982). Social play in differentially reared infant and juvenile baboons (*Papio* sp.). *American Journal of Primatology, 3*(1–4), 153–160.

Connor, R. C., Wells, R. S., Mann, J., & Read, A. J. (2000). The bottlenose dolphin: Social relationships in a fission–fusion society. In J. Mann, R. C. Connor, P. L. Tyack, & H. Whitehead (Eds.), *Cetacean societies: Field studies of dolphins and whales* (pp. 91–126). Chicago: University of Chicago Press.

DeCasper, A. J., & Fifer, W. P. (1980). Of human bonding: Newborns prefer their mothers' voices. *Science, 208*(4448), 1174–1176.

Dudzinski, K. M. (1998). Contact behavior and signal exchange in Atlantic spotted dolphins (*Stenella frontalis*). *Aquatic Mammals, 24*(3), 129–142.

Dudzinski, K. M., Gregg, J. D., Ribic, C. A., & Kuczaj, S. A. (2009). A comparison of pectoral fin contact between two different wild dolphin populations. *Behavioural Processes, 80*, 182–190.

Duffy-Echevarria, E. E., Connor, R. C., & St. Aubin, D. J. (2008). Observations of strand-feeding behavior by bottlenose dolphins (*Tursiops truncatus*) in Bull Creek, South Carolina. *Marine Mammal Science, 24*(1), 202–206.

Dunbar, R. I. M. (2010). The social role of touch in humans and primates: Behavioral function and neurobiological mechanisms. *Neuroscience and Biobehavioral Reviews, 34*, 260–268.

Evans, W. E. (1967). Vocalizations among marine mammals. In W. N. Tavolga (Ed.), *Marine bioacoustics* (Vol. 2, pp. 159–186). Oxford: Pergamon Press.

Favaro, L., Gnone, G., & Pessani, D. (2013). Postnatal development of echolocation abilities in a bottlenose dolphin (*Tursiops truncatus*): Temporal organization. *Zoo Biology, 32*, 210–215.

Fearnbach, H., Durban, J., Parsons, K., & Claridge, D. (2012). Seasonality of calving and predation risk in bottlenose dolphins on Little Bahama Bank. *Marine Mammal Science, 28*(2), 402–411.

Feldman, R. (2007). Parent-infant synchrony: Biological foundations and developmental outcomes. *Current Directions in Psychological Science, 16*(6), 340–345.

Fellner, W., Bauer, G. B., & Harley, H. E. (2006). Cognitive implications of synchrony in dolphins: A review. *Aquatic Mammals, 32*(4), 511–516.

Fellner, W., Bauer, G. B., Stamper, S. A., Losch, B. A., & Dahood, A. (2013). The development of synchronous movement by bottlenose dolphins (*Tursiops truncatus*). *Marine Mammal Science, 29*(3), 203–225.

Frank, L. K. (1957). Tactile communication. *Genetic Psychology Monographs, 56*, 209–255.

Fripp, D., Owen, C., Quintana-Rizzo, E., Shapiro, A., Buckstaff, K., Jankowski, K., et al. (2005). Bottlenose dolphin (*Tursiops truncatus*) calves appear to model their signature whistles on the signature whistles of community members. *Animal Cognition, 8*, 17–26.

Gibson, Q. A., & Mann, J. (2008a). Early social development in wild bottlenose dolphins: Sex differences, individual variation, and maternal influence. *Animal Behaviour, 76*, 375–387.

Gibson, Q. A., & Mann, J. (2008b). The size, composition, and function of wild bottlenose dolphin (*Tursiops* sp.) mother-calf groups in Shark Bay, Australia. *Animal Behaviour, 76*, 389–405.

Gopnik, A., Meltzoff, A. N., & Kuhl, P. K. (1999). *The scientist in the crib: Minds, brains, and how children learn.* New York: Morrow.

Gubbins, C., McCowan, B., Lynn, S. K., Hooper, S., & Reiss, D. (1999). Mother-infant spatial relations in captive bottlenose dolphins, *Tursiops truncatus. Marine Mammal Science, 15*(3), 751–765.

Guinet, C. (1991). Intentional stranding apprenticeship and social play in killer whales (*Orcinus orca*). *Canadian Journal of Zoology, 69,* 2712–2716.

Held, S. D., & Špinka, M. (2011). Animal play and animal welfare. *Animal Behaviour, 81,* 891–899.

Herman, L. M., & Tavolga, W. N. (1980). The communication systems of cetaceans. In L. M. Herman (Ed.), *Cetacean behavior: Mechanisms and functions* (pp. 149–209). New York: Wiley Interscience.

Herzing, D. L. (1996). Vocalizations and associated underwater behavior of free-ranging Atlantic spotted dolphins, *Stenella frontalis,* and bottlenose dolphins, *Tursiops truncatus. Aquatic Mammals, 22*(2), 61–79.

Highfill, L. E., & Kuczaj, S. A. (2007). Do bottlenose dolphins (*Tursiops truncatus*) have distinct and stable personalities? *Aquatic Mammals, 33*(3), 380–389.

Highfill, L. E., & Kuczaj, S. A., II. (2010). How studies of wild and captive dolphins contribute to our understanding of individual differences. *International Journal of Comparative Psychology, 23,* 269–277.

Hill, H. M., Greer, T., Solangi, M., & Kuczaj, S. A. (2007). All mothers are not the same: Maternal styles in bottlenose dolphins (*Tursiops truncatus*). *International Journal of Comparative Psychology, 20,* 35–54.

Hoese, H. D. (1971). Dolphin feeding out of water in a salt marsh. *Journal of Mammalogy, 52*(1), 222–223.

Hoppitt, W. J. E., Brown, G. R., Kendal, R., Rendell, L., Thornton, A., Webster, M. M., et al. (2008). Lessons from animal teaching. *Trends in Ecology and Evolution, 23*(9), 486–493.

James, W. (1890). *The principles of psychology.* New York: Holt.

Janik, V. M., & Slater, P. J. B. (1997). Vocal learning in mammals. *Advances in the Study of Behavior, 26,* 59–99.

Kaplan, J. D., Lentell, B. J., & Lange, W. (2009). Possible evidence for infanticide among bottlenose dolphins (*Tursiops truncatus*) off St. Augustine, Florida. *Marine Mammal Science, 25*(4), 970–975.

Krützen, M., Mann, J., Heithaus, M. R., Connor, R. C., Bejder, L., & Sherwin, W. (2005). Cultural transmission of tool use in bottlenose dolphins. *Proceedings of the National Academy of Sciences of the United States of America, 102*(25), 8939–8943.

Kuczaj, S. A. (1983). *Crib speech and language play.* New York: Springer.

Kuczaj, S. A., II. (2014). Language learning in cetaceans. In P. Brooks, V. Kempe, & J. Golsoon (Eds.), *Encyclopedia of language development* (pp. 328–331). Thousand Oaks, CA: Sage.

Kuczaj, S. A., & Highfill, L. E. (2005). Dolphin play: Evidence for cooperation and culture? *Behavioral and Brain Sciences, 28*(5), 705–706.

Kuczaj, S. A., Highfill, L., & Byerly, H. (2012). The importance of considering context in the assessment of personality characteristics: Evidence from ratings of dolphin personality. *International Journal of Comparative Psychology, 25,* 309–329.

Kuczaj, S. A., & Horback, K. M. (2013). Play and emotions. In S. Watanabe & S. A. Kuczaj, II (Eds.), *Comparative perspectives on human and animal emotions* (pp. 87–112). Tokyo: Springer.

Kuczaj, S. A., & Makecha, R. (2008). The role of play in the evolution and ontogeny of contextually flexible communication. In U. Griebel & K. Oller (Eds.), *Evolution of communicative flexibility: Complexity, creativity, and adaptability in human and animal communication* (253–277). Cambridge, MA: MIT Press.

Kuczaj, S. A., Makecha, R., Trone, M., Paulos, R. D., & Ramos, J. A. (2006). Role of peers in cultural innovation and cultural transmission: Evidence from the play of dolphin calves. *International Journal of Comparative Psychology, 19,* 223–240.

Kuczaj, S. A., II, & Walker, R. T. (2012). Dolphin problem solving. In T. Zentall & E. Wasserman (Eds.), *Handbook of comparative cognition* (pp. 736–756). Oxford: Oxford University Press.

Kuczaj, S. A., & Yeater, D. B. (2006). Dolphin imitation: Who, what, when, and why? *Aquatic Mammals, 32*(4), 413–422.

Kuczaj, S. A., Yeater, D. B., & Highfill, L. E. (2012). How selective is social learning in dolphins? *International Journal of Comparative Psychology, 25,* 221–236.

Lakin, J. L., Chartrand, T. L., & Arkin, R. M. (2008). I am too just like you: Nonconscious mimicry as an automatic behavioral response to social exclusion. *Psychological Science, 19*(8), 816–822.

Levav, J., & Argo, J. J. (2010). Physical contact and financial risk taking. *Psychological Science, 21*(6), 804–810.

Lonsdorf, E. V. (2006). What is the role of mothers in the acquisition of termite-fishing behaviors in wild chimpanzees (*Pan troglodytes schweinfurthii*). *Animal Cognition, 9,* 36–46.

López, J. C., & López, D. (1985). Killer whales (*Orcinus orca*) of Patagonia and their behavior of intentional stranding while hunting nearshore. *Journal of Mammalogy, 66*(1), 181–183.

Lorenz, K. (1979). *The year of the greylag goose.* London: Eyre Methuen.

Maccoby, E. (1990). Gender and relationships: A developmental account. *American Psychologist, 45,* 513–520.

Maniscalco, J. M., Matkin, C. O., Maldini, D., Calkins, D. G., & Atkinson, S. (2007). Assessing killer whale predation on stellar sea lions from field observations in Kenai Fjords, Alaska. *Marine Mammal Science, 23*(2), 306–321.

Mann, J., & Barnett, H. (1999). Lethal tiger shark (*Galeocerdo cuvier*) attack on bottlenose dolphin (*Tursiops* sp.) calf: Defense and reactions by the mother. *Marine Mammal Science, 15*(2), 568–575.

Mann, J., Connor, R., Barre, L. M., & Heithaus, M. R. (2000). Female reproductive success in bottlenose dolphins (*Tursiops* sp.): Life history, habitat provisioning, and group size effects. *Behavioral Ecology, 11*, 210–219.

Mann, J., & Sargeant, B. (2003). Like mother, like calf: The ontogeny of foraging traditions in wild Indian Ocean bottlenose dolphins (*Tursiops* sp.). In D. M. Fragaszy & S. Perry (Eds.), *The biology of traditions: Models and evidence* (pp. 236–266). Cambridge: Cambridge University Press.

Mann, J., Sargeant, B. L., & Minor, M. (2007). Calf inspections of fish catches in bottlenose dolphins (*Tursiops* sp.): Opportunities for social learning? *Marine Mammal Science, 23*(1), 197–202.

Mann, J., & Smuts, B. B. (1998). Natal attraction: Allomaternal care and mother-infant separations in wild bottlenose dolphins. *Animal Behaviour, 55*, 1097–1113.

Mann, J., & Smuts, B. (1999). Behavioral development in wild bottlenose dolphin newborns (*Tursiops* sp.). *Behaviour, 136*(5), 529–566.

Mann, J., Stanton, M. A., Patterson, E. M., Bienenstock, E. J., & Singh, L. O. (2012). Social networks reveal cultural behavior in tool-using dolphins. *Nature Communications, 3*(980), 1–7.

Massen, J. J., & Koski, S. E. (2013). Chimps of a feather sit together: Chimpanzee friendships are based on homophily in personality. *Evolution and Human Behavior, 84*(3–5), 301–302.

McBride, A. F., & Hebb, D. O. (1948). Behavior of the captive bottle-nose dolphin, *Tursiops truncatus*. *Journal of Comparative and Physiological Psychology, 41*(2), 111–123.

McBride, A., & Kritzler, H. (1951). Observations on pregnancy, parturition, and post-natal behavior in the bottlenose dolphin. *Journal of Mammalogy, 32*, 251–266.

McCowan, B., & Reiss, D. (1993). Spontaneous vocal mimicry and production by bottlenose dolphins (*Tursiops truncatus*): Evidence for vocal learning. *Journal of Comparative Psychology, 107*(3), 301–312.

Mello, I., & Amundin, M. (2005). Whistle production pre- and post-partum in bottlenose dolphins (*Tursiops truncatus*) in human care. *Aquatic Mammals, 31*(2), 169–175.

Miles, J. A., & Herzing, D. L. (2003). Underwater analysis of the behavioral development of free-ranging Atlantic spotted dolphin (*Stenella frontalis*) calves (birth to 4 years of age). *Aquatic Mammals, 29*(3), 363–377.

Morisaka, T., Shinohara, M., & Taki, M. (2005a). Underwater sounds produced by neonatal bottlenose dolphins (*Tursiops truncatus*): I. Acoustic characteristics. *Aquatic Mammals, 31*(2), 248–257.

Morisaka, T., Shinohara, M., & Taki, M. (2005b). Underwater sounds produced by neonatal bottlenose dolphins (*Tursiops truncatus*): II. Potential function. *Aquatic Mammals, 31*(2), 258–265.

Noren, S. R., Biedenback, G., & Edwards, E. F. (2006). Ontogeny of swim performance and mechanics in bottlenose dolphins (*Tursiops truncatus*). *Journal of Experimental Biology, 209,* 4724–4731.

Palagi, E. (2008). Sharing the motivation to play: The use of signals in adult bonobos. *Animal Behaviour, 75,* 887–896.

Paukner, A., Suomi, S. J., Visalberghi, E., & Ferrari, P. F. (2009). Capuchin monkeys display affiliation toward humans who imitate them. *Science, 325,* 880–883.

Paulos, R. D., Trone, M., & Kuczaj, S. A., II. (2010). Play in wild and captive cetaceans. *International Journal of Comparative Psychology, 23,* 701–722.

Pellis, S. M., & Pellis, V. C. (2009). *The playful brain: Venturing to the limits of neuroscience.* Oxford: Oneworld.

Pellis, S. M., & Pellis, V. C. (2011). To whom the play signal is directed: A study of headshaking in black-handed spider monkeys (*Ateles geoffroyi*). *Journal of Comparative Psychology, 125*(1), 1–10.

Piaget, J. (1952). *The origins of intelligence in children.* New York: W. W. Norton.

Reid, K., Mann, J., Weiner, J. R., & Hecker, N. (1995). Infant development in two aquarium bottlenose dolphins. *Zoo Biology, 14,* 135–147.

Reiss, D. (1988). Observations on the development of echolocation in young bottlenose dolphins. In P. E. Nachtigall & P. W. B. Moore (Eds.), *Animal sonar* (pp. 121–127). New York: Springer.

Robinson, K. P. (2014). Agonistic intraspecific behavior in free-ranging bottlenose dolphins: Calf-directed aggression and infanticidal tendencies by adult males. *Marine Mammal Science, 30,* 381–388.

Saayman, G. S., & Tayler, C. K. (1979). The socioecology of humpback dolphins (*Sousa* sp.). In H. E. Winn & B. L. Olla (Eds.), *Behavior of marine animals* (Vol. 3, pp. 369–401). New York: Plenum.

Sakai, M., Hishii, T., Takeda, S., & Koshima, S. (2006). Flipper rubbing behaviors in wild bottlenose dolphins (*Tursiops aduncus*). *Marine Mammal Science, 22*(4), 966–978.

Sakai, M., Morisaka, T., Iwasaki, M., Yoshida, Y., Wakabayashi, I., Seko, A., et al. (2013). Mother-calf interactions and social behavior development in Commerson's dolphins (*Cephalorhynchus commersonii*). *Journal of Ethology, 31,* 305–313.

Sakai, M., Morisaka, T., Kogi, K., Hishii, T., & Koshima, S. (2010). Fine-scale analysis of synchronous breathing in wild Indo-Pacific bottlenose dolphins (*Tursiops aduncus*). *Behavioural Processes, 83,* 48–53.

Samuels, A., & Gifford, T. (1997). A quantitative assessment of dominance relations among bottlenose dolphins. *Marine Mammal Science, 13*(1), 70–99.

Sargeant, B. L., & Mann, J. (2009). Developmental evidence for foraging traditions in wild bottlenose dolphins. *Animal Behaviour, 27,* 715–721.

Sargeant, B. L., Mann, J., Berggren, P., & Krutzen, M. (2005). Specialization and development of beach hunting, a rare foraging behavior, by wild bottlenose dolphins (*Tursiops* sp.). *Canadian Journal of Zoology*, *83*(11), 1400–1410.

Scott, E. M., Mann, J., Watson-Capps, J. J., Sargeant, B. L., & Connor, R. C. (2005). Aggression in bottlenose dolphins: Evidence for sexual coercion, male-male competition, female tolerance through analysis of tooth-rake marks and behavior. *Behaviour*, *142*(1), 21–44.

Sewall, K. (2012). Vocal matching in animals. *American Scientist*, *100*, 306–315.

Silber, G. K., & Fertl, D. (1995). Intentional beaching by bottlenose dolphins (*Tursiops truncatus*) in the Colorado River Delta, Mexico. *Aquatic Mammals*, *21*, 183–186.

Spinka, M., Newberry, R. C., & Bekoff, M. (2001). Mammalian play: Training for the unexpected. *Quarterly Review of Biology*, *76*(2), 141–168.

Tamaki, N., Morisaka, T., & Taki, M. (2006). Does body contact contribute towards repairing relationships? The association between flipper-rubbing and aggressive behavior in captive bottlenose dolphins. *Behavioural Processes*, *73*, 209–215.

Tardin, R. H. O., Espécie, M. A., Lodi, L., & Simão, C. (2013). Parental care behavior in the Guiana dolphin, *Sotalia guianensis* (Cetacea: Delphinidae), in Ilha Grande Bay, southeastern Brazil. *Zoologia*, *30*(1), 15–23.

Tavolga, M. C. (1966). Behavior of the bottlenose dolphin (*Tursiops truncatus*): Social interactions in a captive colony. In K. S. Norris (Ed.), *Whales, dolphins, and porpoises* (pp. 718–730). Berkeley: University of California Press.

Tavolga, M. C., & Essapian, F. S. (1957). The behavior of the bottlenose dolphin *Tursiops truncatus*: Mating, pregnancy, parturition, and mother-infant behavior. *Zoologica*, *42*, 11–31.

Thompson, K. V. (1998). Self assessment in juvenile play. In M. Bekoff & J. A. Byers (Eds.), *Animal play: Evolutionary, comparative, and ecological perspectives* (pp. 183–204). Cambridge: Cambridge University Press.

Thorpe, W. H. (1963). *Learning and instinct in animals*. Cambridge, MA: Harvard University Press.

Tyack, P. L. (1997). Development and social functions of signature whistles in bottlenose dolphins, *Tursiops truncatus*. *Bioacoustics*, *8*, 21–46.

Tyack, P. L., & Sayigh, L. S. (1997). Vocal learning in cetaceans. In C. T. Snowdon & M. Hausberger (Eds.), *Social influences on vocal development* (pp. 208–233). New York: Cambridge University Press.

Van Baaren, R. B., Holland, R. W., Kawakami, K., & Van Knippenberg, A. (2004). Mimicry and prosocial behavior. *Psychological Science*, *15*(1), 71–74.

Visser, I. N., Smith, T. G., Bullock, I. D., Green, G. D., Carlsson, O. G. L., & Imberti, S. (2008). Antarctic peninsula killer whales (*Orcinus orca*) hunt seals and a penguin on floating ice. *Marine Mammal Science*, *24*(1), 225–234.

Weaver, A. (2003). Conflict and reconciliation in captive bottlenose dolphins, *Tursiops truncatus*. *Marine Mammal Science, 19*(4), 836–846.

Weir, J., Deutsch, S., & Pearson, H. C. (2010). Dusky dolphin calf rearing. In B. Würsig & M. Würsig (Eds.), *The dusky dolphin: Master acrobat off different shores* (pp. 177–193). Oxford: Academic Press.

Weir, R. H. (1962). *Language in the crib*. The Hague: Mouton.

Weiss, J. (2006). Foraging habitats and associated preferential foraging specializations of bottlenose dolphins (*Tursiops truncatus*) mother-calf pairs. *Aquatic Mammals, 32*(1), 10–19.

Wells, R. S. (1991). Bringing up baby. *Natural History, 100*(8), 56–62.

Welsh, L. S., & Herzing, D. L. (2008). Preferential association among kin exhibited in a population of Atlantic spotted dolphins (*Stenella frontalis*). *International Journal of Comparative Psychology, 21*, 1–11.

West, M. (1974). Social play in the domestic cat. *American Zoologist, 14*(1), 427–436.

Yeater, D. B., & Kuczaj, S. A., II. (2010). Observational learning in wild and captive dolphins. *International Journal of Comparative Psychology, 23*, 379–385.

IV Future Directions of Dolphin Research

9   The Cognitive Ecology of Dolphin Social Engagement

Christine M. Johnson

Investigating social cognition in dolphins is a complex, demanding task. Alien in senses, form, and habitat, dolphins are challenging to comprehend and difficult to study. As the previous chapters have made clear, dolphins are also sophisticated, innovative creatures whose cognition we have only just begun to understand. "Cognitive ecology" (Hutchins, 2010a, 2014) is a transformative new approach that is making exciting inroads in the study of human cognition (Enfield & Levinson, 2006; Goodwin, 2000, 2013; Hutchins, 1995b, 2001, 2010a, 2010b, 2014; Malafouris & Renfrew, 2010). Because cognitive ecology provides a way to study social cognition based on observations of naturally occurring behavior, some researchers working with nonhumans have adopted this approach as well (e.g., Byrne & Bates, 2014; Forster, 2002; Hirata & Matsuzawa, 2001; Johnson, 2001, 2010; King, 2004; Russon & Andrews, 2011). By outlining how this approach might be applied to the study of social interaction in dolphins, I aim to present a new set of conceptual and methodological tools for studying dolphin communication and cognition (see also Herzing, 2006; Johnson, 2010).

I begin with an overview of the basic tenets of cognitive ecology, describing the models and methods involved. I then illustrate how this approach can be used to investigate social cognition in four types of complex social engagements in dolphins. Note that while I focus here on social discourse in adults, the cognitive ecology approach also offers many advantages for the study of development (e.g., Rogoff, 1990; Russon, 2006; Hutchins & Johnson, 2009; de Barbaro, Johnson, & Deak, 2013). Its emphasis on system dynamics makes it a particularly good fit for addressing the unfolding of events. It allows us to study how individuals engage in real-world cognition—problem solving, learning, communicating—by giving us a way to make sense of the complexity and diversity of everyday behavior.

Cognitive Ecology

Traditional approaches in cognitive science take cognition as the set of abilities, possessed by an individual, that it brings to bear to solve the problems it encounters. The use of these abilities is presumed to involve the manipulation of mental representations and to correspond to specific neurological activity. While much has been, and will continue to be, learned from this perspective, it has its limitations, especially for the study of nonhumans. Since mental activity is not directly observable, these "individual ability"' approaches to animal cognition often start with postulations based on humans' experience of their own mental lives. This can prove problematic (Barrett & Würsig, 2014; Harley, 2013; Johnson, 2002). Particularly so, one would think, when studying animals like dolphins, whose ways of perceiving and acting in the world are so different from our own. This chapter focuses on an alternative view that suffers less from these difficulties, by enabling us to do a cognitive analysis of behavior itself.

From the perspective of cognitive ecology, cognition is not something you have but something you do. When animals (including humans) actively engage with the world—navigating around obstacles, searching for and obtaining food, negotiating with conspecifics, and so on—their behavior continually adapts to the constraints of their physical and social environments. Constrained as well by their own senses and range of responses, each species has its own way of "coming into coordination" with its world (Hutchins, 1995a, 1995b; Johnson, 2001). A hunting dolphin, for instance, using echolocation and pursuit, adaptively "comes into coordination" with its prey. Similarly, when dolphins mate, they must coordinate the trajectory, proximity, and orientation of their bodies to accomplish copulation. In cognitive ecology, such changes in coordination are documented and analyzed as visible incidents of cognitive activity. This approach provides a significant advantage for scientists studying nonhuman cognition, since it frees us from having to postulate unobserved mental events and allows us instead to employ systematic observations of engagement as the basis for understanding and comparing animal cognition.

Shifting to a view in which cognition is defined as "adaptive engagement with the world" brings along a host of theoretical and methodological consequences. Probably the foremost of these is that it demands that we treat the object of study *as a system* (Hutchins, 1995a, 2001, 2010a, 2010b). That is, the approach is called cognitive ecology because it recognizes that animals operate within a complex web of relationships, in a multifaceted setting. As a result, the cognitive ecologist must maintain a focus on the system as a whole. Just as a biologist studying a natural ecosystem will investigate how various system factors (like population density, predator–prey ratios, resource availability, etc.) co-constrain one another, and track transformations in these relation-ships over time, so too will the cognitive ecologist emphasize interaction and transfor-mation. When the system of interest is social cognition—as it is in this chapter—the resources that are transformed are those that enable animals to come into adaptive coordination with each other.

Information Flow

These social resources—the calls and gestures the animals produce, the attention they show to one another, the extent to which one animal's actions "fit" with another's, and so on—are all familiar aspects of what we typically think of as animal communication. In fact, from the view of cognitive ecology, communication and social cognition are virtually synonymous (Bateson, 1979). Both involve the coordination of what Good-win (2013) calls the "semiotic resources" that mediate meaningful engagement. But rather than characterizing these in terms of an exchange of signals between a sender and a receiver, requiring mental encoding and decoding, cognitive ecology directly observes transformations of the "media of information flow" within such a system.

Suppose, for example, that the cognitive system of interest was the brain. Research-ers could track information flow as changes in the electrochemical media that con-stitute the circuitry of that system, identifying the many factors that constrain its trajectory. In cognitive ecology, information flow is tracked not along neurological pathways but across changes in observable, embodied media, both within and between individuals. For instance, when an animal hears the call of another, it may turn to look at the source of that call. In this case, the cognitive ecologist would say that, within the animal-in-its-world system, information has been transformed across sensorimotor media, from audition to vision. Seen as a social system, when one animal calls and the other responds, the flow of information can be tracked in behavior from one to the next. As a result, the cognitive ecologist is primarily engaged in tracking how configu-rations of relevant media transform, within and across participants.

It is tracking such transformations that provides insights into which media matter—that is, which are salient and consequential, in a particular setting, for those animals. For example, a wandering infant, startled by the world, will repeatedly reorient its sensory modalities, and possibly vocalize, until it targets and then rapidly approaches and contacts its mother. The information flow in this adaptive event can readily be tracked in the infant's orientations and trajectories relative to the mother, as well as in the timing and types of her responses. This pattern would be species specific in the particular modalities engaged, the nature of the calls, the roles played by other family members, and so on. When an account of information flow also involves longer-term factors like rank or friendship, and especially when multiple individuals are involved, the configurations of relevant media can become quite complex. But if we adhere to some basic principles, interesting patterns can be revealed. In this way, the cognitive ecology approach not only allows us to study complex social cognition but provides us with a basis for comparing even quite disparate systems.

Embodied Media

By treating the study of cognition as a life science (Bateson, 1972, 1979; Rosch, Varela, & Thompson, 1991), cognitive ecology takes the notion of "embodied" cognition literally, using the attention, action, and arousal shown by interacting animals as its raw data (see also Barrett, 2011). When we see animals interacting with their surroundings—orienting, discriminating, engaging—we are directly observing cognition in action. Clearly, knowing something of the sensory systems and behavioral repertoire of a species is essential. For example, recognizing that dolphins can both vocalize and hear in what to humans are ultrasonic frequencies (Au, 1993), and elephants do so at infrasonic frequencies (Payne, Langbauer, & Thomas, 1986), would be critical to the species-appropriate study of information flow in those two systems. In contrast, other aspects of embodied cognition may cross such species boundaries. In most social animals, for instance, high arousal tends to involve increases in the pace and extent of the body movements observed, with related changes in vocal amplitude and other sound-making activities.

When one such sensorimotor-arousal system comes into coordination with another, *social* cognition occurs. During social engagement, relevant behaviors include the relative body positions, orientations, affective displays, gestures, vocalizations, and so on, that have long been the focus of ethological research. In fact, the methods of cognitive ecology have much in common with those of traditional ethology. Ethology is generally concerned with documenting a species' behavioral repertoire and understanding the ways in which factors like age, gender, rank, reproductive state, and so on, affect behavior. While such issues also come into play in cognitive ecology, it is more concerned with how adaptations are accomplished. For example, while ethology may be concerned with whether related animals are more likely to collaborate than nonrelated ones, cognitive ecology would further investigate which behaviors and senses mediate collaboration, and how the effectiveness of this coordination develops over time. As a result, one key difference between these approaches is that cognitive ecology includes the detailed microanalysis of engagement. A glance, a gesture, a turn to or from, are brief but often key components of social discourse. During social cognition, it is at this millisecond timescale that many critical adaptations occur.

Attentional behavior is a good case in point. Social attention consists of a shift in an animal's sensors such that it gains perceptual access to one or more conspecifics. Each overt act of attention—whether sniffing, turning to look, or echolocating—provides us with an observable cognitive event, critical to specifying information flow in that species. Social attention is often a joint activity, in which multiple animals concurrently monitor one another or direct their attention to a common target. Gaze following, for instance, is a widespread behavior among primates (Tomasello, Call, & Hare, 1998) and plays a major role in their social negotiations (Emery, 2000; Johnson & Karin-D'Arcy, 2006). Moreover, since acts of attention are often a preamble to interaction, they can become ritualized into displays that predict or deny engagement. Baboons, for example, can solicit affiliative engagement with a nonthreatening look, or avoid responding to a solicitation, or even to a threat, by directing their visual attention elsewhere—a tactic called "gaze aversion" (Chance & Jolly, 1970; Kummer, 1971).

In addition, attentional behavior can also be used as a general measure of "salience." Salience—the likelihood that a target will be attended—is a particularly useful cognitive metric here. Consider, for example, the case when one individual attends a target, thereby drawing the attention of others to it. In the mimicry literature, this is called "stimulus enhancement," and it is a common means by which animals come to converge in their behavior toward a particular target, such as food or other resources (Whiten & Ham, 1992). By recording micro-level changes in attention, the cognitive ecologist can track how the salience of a particular target moves through a group, and identify other actions that help to direct or deflect it. We can also see changes in salience over time within a given individual. For example, food is salient when one is hungry, less so when one is sated. Similarly, a female, previously ignored, may become salient to a male when a third party begins to court her. In general, changes of consequence—the appearance of a predator, the challenge of a competitor, the threat to an offspring—are more likely to be attended than less-important events would be. As a result, tracking salience can tell us a great deal about what matters to the participants in a given cognitive system.

Valences and Social Markets

Understanding "what matters" in a social group is necessary to an account of its cognitive ecology. If we are to define cognition as "adaptive engagement with the world," we must have a rubric for what counts as adaptive. Note that in this setting, "adaptive" is

not synonymous with "optimal," only with "advantageous in context." But identifying even the latter can be no small feat in socially complex animals. This is particularly
true when multiple animals with differing interests are involved. Fortunately, there are
a variety of behavioral patterns that we can use to get a handle on these slippery issues.

Long traditions, in both ethology and psychology, argue that an animal will tend
to move toward stimulus with a positive valence, and away from one with a negative
valence (e.g., Garcia & Koelling, 1996; Boyd, Robinson, & Fetterman, 2011). As a result,
measures like the proportion of moves *to* versus *from* a particular resource, especially
relative to the propensities of other participants, can inform us about the relevance
and changing value of that resource for that animal (e.g., Johnson & Oswald, 2001).
Similarly, the more effort an animal invests toward a particular end, or the stronger its
reaction to a diversion from that end, the greater the valence we can assign to it (see
Zahavi, 1977). Thus if an animal persists in its pursuit of a potential mate, despite many
obstacles, we can assume that that partner has a high positive valence for that animal.
And if it invests considerable effort in avoiding another individual, that individual can
be taken as having a high negative valence. Behaviors can be considered "adaptive,"
then, when they are consistent with maximizing positive and minimizing negative
valence.

Furthermore, we know that if a particular turn of events has a positive valence
for a given individual, then that animal will tend to act in ways that perpetuate
that state (Thorndike, 1911). In contrast, if the valence is negative, the individual is
likely to do something to destabilize that situation. For example, if having a partner
in a particular enterprise (such as grooming or cooperative foraging) has a positive
valence, then, other things being equal, animals with partners should tend to remain
with them, while animals without should be expected to put effort into altering
their solitary state, by either joining or displacing others. In this way, we can learn
what is important to the groups that we study. For example, data on reconciliation
after aggression suggest that dolphins (Samuels & Flaherty, 2000; Weaver, 2003), like
primates (Aureli & de Waal, 2000), value the reestablishment of peaceful relations
within their groups.

In some cases, assigning valence can be fairly straightforward. Prey, for example,
are undoubtedly a valuable commodity, and activity that generally results in their capture can be considered adaptive. High arousal often arises in critical survival contexts,
where behaviors related to attack and defense are performed. In dolphins, high-arousal
actions associated with biting and "hitting with sound"—such as head jerks, jaw claps,
and open mouths—are typical of negatively valenced interactions, whereas gentle contact or synchrony can be assigned more positive valences (McBride & Hebb, 1948;

Tavolga, 1966; Caldwell & Caldwell, 1972; Samuels & Gifford, 1997; see Herzing, chap. 6, this vol.). Shifting arousal—such as in the escalation or de-escalation of an engagement, often involving ritualized behaviors that mark the gradations—can also be informative. In primates, for example, an animal can display a relatively low-level yawn threat, or escalate to bristling pilo-erection, or move all the way to outright attack. When a dolphin escalates from a head bob to an "S posture" (Caldwell & Caldwell, 1972; Pryor, 1990), it alters the sensory access its adversary has, by filling more of its visual field (see also Norris & Dohl, 1980). Such changes in behavior and access can thus reveal how much is at stake for the participants in a given engagement.

But, especially in complex social creatures like dolphins, valence is often highly context dependent (Smith, 1977; Johnson, 1993) and can accrue to virtually any aspect of a negotiation. That is, social engagement is always situated within a particular "economy," and what operates as currency within that economy depends on the history of the group, as well as its current state. Barrett and Henzi (2006) have outlined, in their discussion of "biological markets," how the costs of such social commodities can vary with supply and demand (see also Noe & Hammerstein, 1994). In a baboon troop, for example, where animals pay the cost of grooming a mother to gain access to her infant, this cost increases when there are fewer infants in the troop (Henzi & Barrett, 2002). That is, an interested party must groom a lone mother for longer than she would if multiple mother-infant pairs were available. Note that while valences differ for different individuals, it is not necessary to attribute them as invisible "mental states." Instead they are observable states of a behavioral system, which can be tracked and analyzed within a given ecology.

Sociocognitive Complexity

Social cognition occurs, of course, among all animals that interact. But the way that dolphins are social is different from the way that, for example, fish are social. Certainly, information does flow through a fish school as they dodge and veer together, rapidly and jointly maintaining coordination. But dolphins organize into more variable subgroupings and show greater flexibility in how they interact (e.g., Connor, 2007; Lusseau & Newman, 2004; Norris & Dohl, 1980; Wells, 2003). Understanding the multiple factors that shape these complex interactions is the real challenge in dolphin cognitive research (see Johnson, 2010).

De Waal (1986) offers a definition of social complexity that may prove useful here. He suggests that a complex society is one in which power is not equal to rank. In many animal societies, rank is synonymous with power, that is, with one's capacity to acquire or control contested resources. But in de Waal's complex society, lower-ranking animals can band together to outcompete a higher-ranking one. In this sort of system, each individual needs to be concerned not only about its own relationships—for example, who it can dominate and who can dominate it—but also about the relationships between others. For example, a high-ranking animal needs to be on the alert for coalition building between subordinates, and low-ranking animals have new interests in courting coalitional support (Harcourt & de Waal, 1992). Thus the occurrence of polyadic interactions—such as those involving coalitions or other functional sub-groupings—is a hallmark of complex sociocognitive systems. Such a polyadic system is more cognitively demanding because it compounds the media that are relevant to monitor and manipulate—both for the animals and for the researcher (Johnson, 2001, 2010).

In what follows, we will examine social engagements that have been observed, or may be expected, in dolphins, each of which shows this polyadic complexity. By working our way through four types of engagement—collaboration, the use of social tools, social attention, and information brokering—we will see how cognitive ecology can be applied to learn more about the social cognition involved. In each case, we will review the embodied media and type of information flow displayed, situate these accounts within their social markets, and discuss the types of complexity that can arise.

Applying Cognitive Ecology to Complex Dolphin Engagements

Tracking Information Flow
Information flow, of the sort described in the previous sections, can be tracked in a number of ways (see table 9.1). Since information can be defined as "a difference that makes a difference" (Bateson, 1972), our unit of analysis in such a study is always an interaction. That is, when one animal makes a change in its behavior, only if that difference makes a difference in the behavior of the second animal can we observe information flow. These forms of information flow are not specific to dolphins but should be applicable to most social animals, as long as they are tracked in species-relevant media.

Contiguous Interactions
As in many other disciplines, cognitive ecologists rely heavily on the temporal or spatial contiguity of events. Behaviors that co-occur, or immediately follow, are often pertinent to one another. The antiphony of animal calling, the flight of the prey the

Table 9.1

Information flow during social engagement can be tracked through four types of interactions, distinguished by the relationships between the behaviors that embody them.

INTERACTION	RELATIONSHIP BETWEEN BEHAVIORS	
Contiguous	➡️◄	Proximal, Immediate
Congruent	〜➚ 〜➚	Synchronous, Imitative
Complementary	➡️◄◄	"Fit," Affordance
Ritualized	↗↓↗↓↗ ↓	Next Move in Conventional Sequence

moment the predator charges, the look toward a conspecific who quickly looks away, are all commonplace examples of this sort of information flow.

Congruent Interactions

In a congruent engagement, an animal witnesses the action of another and repeats it. Thus if one animal synchronizes with another or notices another's behavior and later imitates it, information flow can be assumed between the model and the mimic. Whether tracking pragmatic actions like obtaining food, attentional behavior like head turns, or signal exchanges like call matching, the cognitive ecologist observes the form of the action passing from one body to the next.

Complementary Interactions

Complementary activity (see Hutchins & Johnson, 2009) concerns the notion of "fit" (e.g., Hinde & Simpson, 1975), in which the activity of the second animal can, in some sense, be seen to "fit" with the first. For example, to accomplish mating, one party will tend to assume a certain posture that will make it easier for the other to assume the complementary posture. This evokes the Gibsonian notion of an "affordance" (Gibson, 1979), since, in a complementary interaction, the first action can be seen as affording— making it easier or more obvious to perform—the second. Complementarity can be simultaneous or sequential and can be used as a measure of confluent interest and of shared expertise.

Ritualized Interactions

Information flow can also depend on the normative routines within a particular species or group. As ethologists have often described (Tinbergen, 1952; Smith, 1977; Alcock, 1998), many social engagements, such as courtship, dominance, and play, can be ritualized, either phylogenetically or ontogenetically (e.g., Tomasello, Gust, & Frost, 1989). Once socialized within a community, participants can be assumed to be familiar with such routines, and so the performance of the "next move" in a familiar sequence also provides evidence of information flow. Not surprisingly, ritualized behaviors often take advantage of affordances that promote the coupling of action, of the sort previously described for congruent and complementary interactions (see Hutchins & Johnson, 2009).

In fact, it may often be the case that several of the foregoing types of information flow are observable in a given engagement. Moreover, especially when multiple parties are involved, relevant events can be separated in time, and links must be carefully substantiated. Collecting large quantities of data helps both to establish regularities and to recognize violations. For example, disruptions of information flow, like hesitancy, incongruity, and obstruction, can still be interpreted as adaptive, as long as one knows enough about habitual behavior and the valences of the relevant social economy.

Let us now consider how these forms of information flow may operate in complex polyadic engagements in dolphins.

Example 1: Collaboration

Dolphins have been observed to engage in multiple forms of collaboration. By working together to control schools of fish, individuals can feed more successfully than they might if foraging alone (e.g., Gazda, Connor, Edgar, & Cox, 2005; Würsig, 1986). Multiple dolphins can circle the prey, causing them to bunch tightly in defense, and then pin the school against the surface or drive it onto a beach. This limits the prey's escape options and makes it easier for dolphins to consume them. The emergence of this activity is an example of a cognitive adaptation arising through joint participation. Another form of collaboration occurs in coalitions of bottlenose dolphins, in which pairs or trios of adult males compete for, herd, and mate with select females (Connor, 2007). This collaboration can become particularly complex when coalitions band together, forming "superalliances," to compete with other such alliances (Connor, Heithaus, & Barre, 1999).

One important type of information flow that we might expect to see in such engagements is congruency. Synchronous displays, for example, are well documented between coalitional males (Connor, Smolker, & Richards, 1992), and there is some indication that synchronous calls may also be used during coalitional competitions (Herzing, 1996). Synchrony is a fundamental dolphin achievement, practiced from birth, that likely serves as a frequent medium of information flow (Fellner, Bauer, & Harley, 2006; Cusick & Herzing, 2014). While we know, from the lab, that dolphins are also excellent mimics (Tayler & Saayman, 1973; Herman, 2002; Kuczaj & Yeater, 2006), we have few data on how this ability functions for them in their daily social lives. With cognitive ecology's focus on unfolding processes, we can move beyond questions such as whether or how much imitation occurs to asking how it is used.

One hypothesis might be that imitation can serve as a solicitation—an invitation to join—as during the development of a social alliance (Johnson & Norris, 1994). Research with nonhuman primates reveals that, like humans, they recognize when they are being imitated (Paukner et al., 2005). Furthermore, humans become increasingly prosocial as a result of being imitated (van Baaren et al., 2004; Carpenter, Uebel, & Tomasello, 2013). If this is also the case for dolphins, we should be able to find changes in partner valence based on who duplicates whose behavior.

To investigate the role of congruence in collaboration, one might compare interactions between well-established allies and ones that are not as stable or long lasting. Might the long-term allies be more prone to coordinate their own actions first (e.g., synchronize their behavior) before taking on a joint effort like facing off with another coalition? Might the more novice collaborators invest more of their off time, outside interalliance conflicts, in imitation? As this behavior would align their attention to one another's bodies, it could provide relevant training for collaborative activities. Cantor and Whitehead (2013) suggest that similarity of behavior—what they call "conformism"—might be more likely within established social clusters than between them, further differentiating those subgroups. Thus, understanding how, when, and with whom imitation is used can also provide insights into the role that social cognition plays in structuring the social context.

If congruency affects not only feeding efficiency but also subgroup stability, then it seems likely that a social market for it could arise. As a result, we might ask if imitative effort is reciprocated. That is, we might predict that inequalities in such reciprocation could provoke shifts in arousal, aggression, or other indicators of partner valence between potential collaborators (see Schino & Aureli, 2009). Game theoretical models have argued that for reciprocity to become a stable strategy, "cheaters" must be identified and sanctioned (Trivers, 1971; Maynard Smith, 1982; see also Connor & Norris, 1982). This could be put to an empirical test by doing microanalyses of solicitations for collaboration, comparing successful and unsuccessful trajectories, as well as any retaliation against errant participants.

Another pertinent aspect of collaboration is the division of labor and the emergence of tactical roles. Among bottlenose dolphins off Cedar Key, Florida, one animal serves as the "driver," herding a fish school toward a tight-knit line of collaborators, who serve as a "barrier" against which the fish can be pinned (Gazda et al., 2005). Such collaboration provides an excellent example of how apropos—and in fact necessary—a systems perspective is; the cognitive accomplishment here could not be captured by the behavior of any one participant alone. Only so long as the barrier dolphins are doing their part are the driver's actions meaningful and effective, and vice versa. Clearly, a relevant type of information flow here is complementarity, as the two types of behavior "fit" together and only together create a functional system.

The two collaborative groups studied by Gazda et al. (2005) also showed distinct differences in the nature and effectiveness of their routines. The more successful group had a long history of being sighted together, suggesting they had had ample opportunity to practice working as a team (see Anderson & Franks, 2001). We could examine how the micro-level behavior of these two groups differs. Are the practices of the more successful cadre more energetically efficient? Expert performance in many domains is characterized by a "smoothness" of execution. In this case, that smoothness could be a function of a particularly tight contiguity between the actions of the participants, as well as a sustained complementarity by the driver, and congruency between the barrier dolphins.

Similarly, might the less-effective group be more prone to "bobbling" their discourse—generating false starts, missed cues, vacillations, and so on? Breaks in information flow—violations like hesitancy and incongruency—offer a direct measure of the relative stability and proficiency of a cognitive adaptation. How susceptible a discourse is to disruption is another measure of the robustness of the cognition that accomplishes it (Hutchins, 2006, 2010b). Such disruptions can include errors by participants, and so how experts versus novices recover from such errors is also of interest (Hayashi, Raymond, & Sidnell, 2013). Plus, longitudinally, one could explore the development of such expertise, and thus what behaviors, specifically, the animals are learning to do, and the role their co-constrained situation plays in that process.

One way to think about complexity during collaboration is in terms of multitasking. Consider participation in a "superalliance" engagement (Connor, Heithaus, & Barre, 1999). Each animal in such an event is simultaneously participating in a coalitional undertaking, an intercoalitional collaboration, a cross-coalitional competition, and courtship. Moreover, inasmuch as these animals live in a fission–fusion society (Connor et al., 2000) but still need to maintain such relationships, they must also spend time tending to different relationships intermittently, and shifting between roles depending on current subgroup membership, requiring a more protracted form of multitasking. A society with a myriad of social roles—each with its distinctive behavioral patterns—is necessarily complex. The number of roles with which an individual shows facility provides one measure of its cognitive flexibility—a hallmark of "intelligent" behavior.

Example 2: Social Tool Use

Another form of role-based, polyadic engagement has been called "social tool" use (Chance & Jolly, 1970; Byrne, 1995). In this triadic interaction, one animal, the "user," engages another as a "tool," to influence its relationship with a third, the "target." For example, a user may position a tool between it and an aggressive target, as a "buffer" against the threat. Alternatively, the user may groom a "passport"—a close associate of an attractive target—so as to get closer to the target itself. In the earlier case of adolescent female baboons grooming mothers to gain access to their infants, the mother would be the adolescent user's "passport" to the target infant. In some schemes, even the recruitment of an ally against a common adversary can be considered a case of social tool use. More typically, however, the term "social tool" implies that the user is, in some way, exploiting the tool, serving primarily the user's own, rather than their joint, interests.

Because this sort of exploitation is connoted as "insincere" and "Machiavellian," it has engendered discussions of topics like whether engagement with the tool involves premeditation, or whether the target is necessarily represented by the user as being duped (Byrne & Whiten, 1988). But these are not the key questions facing the cognitive ecologist. Because of our focus on cognitive activity, we are more concerned with the complexity of the social configurations that are established in these engagements, and the cognitive demands this complexity puts on participation.

Social tool use arises when the current interaction between the user and the target has a negative valence for the user. If an aggressive target threatens a user, or an attractive target ignores the user, the user will act to destabilize this interaction, to either end or change it. What makes this a polyadic (and thus complex) engagement is that the user accomplishes this by initiating a routine with a third party, the tool. Furthermore, this routine must be incompatible with the current user-target trajectory. Thus, in a "buffer" engagement, in response to a threat from the target, the user initiates an affiliative interaction with a receptive tool, especially one toward which the target is unlikely to aggress. As a result, if the target were to carry through on its threat, doing so would also affect the valence of the target's relationship with the tool. If this poses too high a cost for the target, the threat is de-escalated, and the cumulative information flow is diverted from an aggressive to an affiliative routine. In the case of a "passport," the initial negative valence for the user of being prevented from approaching a desirable target is mitigated through its friendly interaction with the tool. In the baboon example (Barrett & Henzi, 2006), as the mother becomes increasingly relaxed by being groomed, she likewise relaxes her constraints on the user's access to her infant.

While bottlenose dolphins have been proposed as likely candidates in which to find social tool use (Connor & Mann, 2006)—in part because of their embedded coalitional behavior—no study to date has directly investigated exploitative engagements in these animals. The communicative repertoire of dolphins is, of course, quite different from that of primates, but some of the same measures could be used. Certainly the complementarity involved in the solicitation of a partner, the violation of complementarity that constitutes a rejection, along with ritualized behaviors such as petting or threat, should be readily observable. In addition, as discussed earlier, characterizing an engagement as a social tool also requires evidence of a "conflict of interest" between the user and target.

As discussed in the introduction, to/from patterns can be used to identify both shared and conflicting interests. For example, in a study of social tool use in bonobos, researchers compared triadic social tool interactions with interactions between a dyad with a third party nearby (Johnson & Oswald, 2001). In their microanalysis of these engagements, the researchers recorded every time one of the three animals turned its head or body to or from another. In the dyads, the to/from patterns tended to be identical. That is, if one turned to the other, the other would also turn to it, and if one turned from, the other also turned from. Plus, there was no such contingency between the dyad and the third party. This was interpreted as only the pair having converging interests in their current engagement.

These findings contrasted starkly with the triadic social tool engagements. In those engagements, the user and target had consistently opposite patterns. In a buffer, for example, when the target turned to the user, the user predictably turned away; and if the user turned to the target, the target also turned from the user. This was taken as evidence of conflicting interests. In fact, each type of social tool use studied had its own distinctive conflicting pattern between the user and target. Furthermore, the insincerity often attributed to the user regarding its engagement with the tool was reflected in a much less tight contingency between their turns than was seen in the normal dyads. It would be most interesting to conduct such analyses of dolphin interactions to see if similar patterns emerge.

Example 3: Social Attention

The emergence of sophisticated social attention, it is argued, most likely occurs in species with embedded social networks, who must therefore track and coordinate multiple levels of social engagement (Grove, Pearce, & Dunbar, 2012). Many dolphins, such as bottlenose (Wells, Irvine, & Scott, 1980; Connor, 2007), spinner (Norris & Dohl, 1980), dusky (Würsig, Würsig, & Cipriano, 1989), and spotted dolphins (Herzing, 2011), show complex, embedded sociality (see Würsig & Pearson, chap. 4, this vol.). Laboratory research reveals that they can use human attentional behavior, such as pointing or head turns (Pack & Herman, 2006; although see Tomonaga, Uwano, Ogura, & Saito, 2010), and may even direct the attention of others (Xitco, Gory, & Kuczaj, 2001, 2004). As such, these dolphins provide a promising model by which to test hypotheses concerning the competitive control and cooperative sharing of information through social attention.

The media of information flow in social attention are the behaviors that achieve a change in access to others. In dolphins, this will mostly involve visual and acoustic access. Dolphins' eyes are laterally placed, each having a panoramic view of its own visual hemisphere (e.g., Madsen & Herman, 1980; see Hanke & Erdsack, chap. 3, this vol.). Recent research suggests that bottlenose dolphins may show lateralization in their use of one eye over another for certain types of targets (e.g., Yaman et al., 2003; Thieltges et al., 2011), although little is known yet about the role this may play during social engagement. While dolphins have omnidirectional hearing, their optimal listening position is with head directed at the target (see Cranford, Amundin, & Krysl, chap. 2, this vol.). Their narrow echolocation beam likewise points directly forward, and evidence suggests that they can adjust the depth and direction of their "acoustic gaze" (Wisniewska et al., 2012). They also have a limited binocular visual field directly ahead and below, which is probably involved when approaching prey or conspecifics. As a result, a shift in head orientation toward another animal, especially in conjunction with echolocation, can be a telling act of social attention in these animals (see Gregg, Dudzinski, & Smith, 2008).

Information flow might also involve congruency, such as when one animal notices another attending a target and shifts its own attention to that target as well. Such "gaze following" has been observed in a wide variety of nonhuman primates (Tomasello, Call, & Hare, 1998), as well as in human-dog interactions (Kaminski, Brauer, Call, & Tomasello, 2009). Developmentally, gaze following provides a means of learning about what matters in a shared environment, especially in association with displays of arousal and affect. Dolphins can engage in such shared attention not only by looking at and ensonifying the same target but also by passively "listening in" to the returning echoes of another's echolocation (Xitco & Roitblat, 1996). While this practice affords less information about the target than if they had ensonified it themselves, such eavesdropping may provide important social information about the nature and extent of an echolocator's involvement (see Gotz, Verfuss, & Schnitzler, 2006).

Situating observations of attention within a social market can help identify the factors that affect salience in that system. Long ago, Chance (1967) suggested that rank within a monkey troop is, in part, expressed through patterns of social monitoring. He suggested, and later data support, that subordinates would be more likely to monitor dominants than vice versa (e.g., McNelis & Boatright-Horowitz, 1998; for a review, see Johnson & Karin-D'Arcy, 2006). Presumably, the potential consequences of the dominant's actions are of greater valence to the subordinate than vice versa, making the subordinate more wary and watchful. In many cases, however, the distribution of social attention in a group is more complicated than Chance's rank-based model would predict. As discussed earlier, if two subordinates engage in bonding behavior, which might lead to their coalitional activity against a third dominant animal, that dominant is likely to find such an event highly salient. This would be observable in how the dominant animal monitors and potentially intervenes in such an engagement (Kummer, 1971; de Waal, 1982; Ferreira, Izar, & Lee, 2006). Only by situating this event within the current social economy can such salience be understood.

As ever, the cognition involved here becomes increasingly complex when social attention occurs during polyadic engagement. For example, displays of attention can offer additional options for social tool use. In an "alibi," for instance, the user can direct its own attention to a tool, as a way of avoiding engagement with an unwelcome target (Byrne, 1995; Johnson, 2010). Similarly, when performing what has been called a "distraction display" (Whiten & Byrne, 1988), the user can redirect the attention of the target, through attention following, to the tool, thus distracting the target from the user or its resources. For example, a subordinate chimpanzee who has discovered a treat that a dominant may be liable to take from him can make a big display of attending a third party (the tool). Following his gaze, the dominant may then become engaged with the tool, freeing the user to enjoy the treat unmolested (e.g., Goodall, 1986; Hirata & Matsuzawa, 2001). In fact, the sort of to/from analysis discussed in the earlier section on social tools often includes attentional moves to or from another.

Under polyadic conditions, social attention might also become embedded—as when A attends B attending C attending D, and so on. It would be interesting to see how many such embeddings are observable in dolphins. To what extent do they modify their behavior based on who is, or is not, present to observe them? Do they suppress behavior that might provoke negative consequences from a particular audience? Might they even display surprise or heightened interest about a phenomenon already known to them, as a means of maintaining the appearance of ignorance? By tracking social attention through a polyadic marketplace, we may begin to find answers to such intriguing questions.

Example 4: Information Brokering

An individual who provides information about a third party, or about other events that it witnessed but that its audience did not, is an "information broker." One form of information brokering found in a number of species is alarm calling. The individual making the alarm presumably has access to information about a predator that its audience does not—until they hear the call. Various avian and primate species, for example, have distinct calls associated with particular predators and particular evasive responses (Seyfarth, Cheney, & Marler, 1980; Evans, Evans, & Marler, 1993). But alarm calls are "broadcast behaviors" and, as such, are directed to the group at large, making them more like dyadic interactions, rather than truly polyadic ones. Given the subgroup structure of polyadic engagement, information brokering is liable to become more complex (Lusseau, 2007). This should be particularly true in fission–fusion societies, such as those seen in many dolphin groups (see Würsig & Pearson, chap. 4, this vol.), where, as individuals alter their subgroup structure, they have differential access to important social information.

One focus of research on information brokering has been the study of apprenticeship (Vygotsky, 1978; Rogoff, 1990; Russon, 2006). The information brokering between teacher and learner is foundational to human cognition. Unlike in nonhuman primates, there is some evidence that odontocetes may also engage in teaching. In particular, killer whales (Guinet & Bouvier, 1995) and Atlantic spotted dolphins (Bender, Herzing, & Bjorklund, 2009) have been observed to modify their foraging behavior, in the presence of an attentive less-experienced conspecific, in a way that decreases their own foraging efficiency but makes the relevant moves more salient to the observing animal (see Caro & Hauser, 1992). Presumably this increase in salience would lead to better reproductions of the behavior by the novice. Thus, through the tutor's congruent demonstration, information about its group's foraging practices flows through this system, helping the novice's performance become more adaptive.

Information brokering also arises in network analyses of dolphin social structure. In these analyses, clusters of association within a dolphin community can be identified, and animals at the intersection of such clusters—animals that are relatively common associates of both groups—can be seen as likely brokers of information (Lusseau & Newman, 2004; see also Coussi-Korbel & Fragaszy, 1995). Such a broker might learn an activity in one group and then perform it in another group that does not yet practice that behavior. In this setting, we could observe the information flow from one group to the next as the members of the second group produce congruent imitations of the broker's activity, as well as the complementary actions afforded by the broker playing his part. In this way, the two groups and the broker together accomplish the cultural transmission of that activity. Evidence continues to accumulate for the occurrence of culturally distinctive practices in cetaceans (e.g., Rendell & Whitehead, 2001; Mann & Sargeant, 2003; Allen, Bejder, & Krutzen, 2011), but we still know little about how such practices emerge (see Kuczaj & Winship, chap. 8, this vol.).

One important way that humans broker information is through practices that "make reference"—that is, direct the attention of others to current, absent, or even imaginary events. Probably the most species specific of these practices organizes vocalizations into syntactical combinations of arbitrary symbols. The arbitrariness of human words (there is nothing doglike about *dog*) opens up a huge range of vocal activities to a species like ourselves, who can master a large vocabulary and rules for producing novel but still informative combinations. But using symbols is not the only way to provide others with access to events displaced in space or time.

Humans also broker information by making "iconic reference." In iconic reference, the relation between a communicative act and its referent is not arbitrary; instead the act is congruent with its referent. Consider how much a human in a foreign land can communicate, even if she does not speak the language. Using pantomime and vocal mimicry, a human can request and gain information and even tell a story, effectively recounting events that its audience does not know. Humans also incorporate these skills during normal language use—including mimicking the words, gestures, and facial expressions of others—as a way of passing on information about them. Thus iconic reference plays a vital role in information brokering in humans.

In experimental work, scientists have attempted to determine if dolphins can communicate arbitrary information to one another (Bastian, 1968; Zanin, Markov, & Sidorva, 1990). In these inconclusive studies, the experimenter provided visual or auditory information—such as a flashing light—to one dolphin, who was then required to somehow vocally pass on this information to another dolphin. Due to issues involving training protocols, it was never clear whether the dolphins could accomplish this task, but their performance was suggestive enough to make pursuing the question further worthwhile.

A more tractable way to address the question of whether dolphins engage in referential communication may be to track a group's use of their signature whistles. We have long known that many dolphin species produce signatures—individual-specific whistles that they may use to identify themselves to others (Caldwell & Caldwell, 1965; Harley, 2008; Sayigh, Esch, Wells, & Janik, 2007; see Lammers & Oswald, chap. 5, this vol.). But while a signature whistle is defined as the contour most often produced by a particular animal—let's call it the "signatore"—a fair number of times this call is also made by others in the group. At times, this appears to be in answer to the signatore producing its own whistle—a practice known as "call matching" (Janik, 2000; Nakahara & Miyazaki, 2011; Watwood, Tyack, & Wells, 2004). But at other times, signatures are produced in the absence of the signatore (e.g., Watwood, Owen, Tyack, & Wells, 2005; personal observation).

While we still know little about how dolphins use and respond to each other's signatures, research in the lab shows that dolphins can differentiate between familiar versus unfamiliar signatures (Bruck, 2013; Caldwell & Caldwell, 1972; see Harley, 2008). Surely, in a long-term social group, it is reasonable to assume that the animals have had sufficient exposure to the regularities of signature use, such that the correlations between signatores and their whistles are well established. Consider, too, that these animals are documented to be vocal mimics (Richards, Wolz, & Herman, 1984; Reiss & McCowan, 1993). A mimic could make reference to an absent animal by reproducing a portion of the group's common experience of that animal, that is, its signature. The congruence between the whistle made by the signatore itself, and the one made by the nonsignatore, would allow information about the signatore to flow in this system. This, then, would also be a case of iconic reference.

Furthermore, with an interest in discourse, we would also want to track how the audience embodies transformations of such information. Might the call pass from animal to animal to animal? Might the use of its call alter the salience of the signatore, or the readiness of the audience to respond to it, should it reappear? Perhaps hearing an absent animal's whistle would prompt others to seek it out? And what if the signatore itself heard the call—would that change its likelihood of approaching the caller? In addition, if the animal using the signature alters it in ways that connote it, or adds information besides the signatore's identity—such as who it was with, what mood it was in, what it was doing, and so on—would audience members act in ways consistent with having witnessed such events? Nonsignatore whistle use could also tell us something about the cognitive accomplishments of the informant, such as tracking who witnessed what, and making strategic use of privileged information. However the dolphins' complicated communication system operates, using cognitive ecology may give us the best hope for understanding it.

Conclusion

As intriguing as such speculations may be, it is clear that there is much work to be done if we are to gain answers to the sorts of questions posed in this chapter. Fortunately, the technology for making multimodal recordings of dolphin engagement is more accessible and advanced than ever. It would be low impact and relatively inexpensive, for example, to equip any of the many facilities currently housing dolphins with multiple hydrophones and underwater video cameras. In the wild, habituated groups tolerate audio-video recording by humans swimming with the animals (Herzing, 2011), and new developments in aerial cameras can capture interactions even in nonhabituated animals, or in less clear or hospitable seas (Nowacek, 2002). As long as the data that are collected are multimodal and multiparty and are analyzed at multiple timescales, from frame-by-frame microanalysis to long-term history, we can situate events in their social economy. In this way, we can investigate their cognitive ecology and track information flow through these sociocognitive systems. Imagine what we might discover!

References

Alcock, J. (1998). *Animal behavior: An evolutionary approach*. Sunderland: Sinauer Associates.

Allen, S. J., Bejder, L., & Krutzen, M. (2011). Why do Indo-Pacific bottlenose dolphins carry conch shells (*Turbinella* sp.) in Shark Bay, Western Australia? *Marine Mammal Science, 27*, 449–454.

Anderson, C., & Franks, N. R. (2001). Teams in animal societies. *Behavioral Ecology, 12*, 534–540.

Au, W. W. L. (1993). *The sonar of dolphins*. New York: Springer.

Aureli, F., & de Waal, F. B. M. (2000). *Natural conflict resolution*. Berkeley: University of California Press.

Barrett, L. (2011). *Beyond the brain: How body and environment shape animal and human minds*. Princeton, NJ: Princeton University Press.

Barrett, L., & Henzi, S. P. (2006). Monkeys, markets, and minds: Biological markets and primate sociality. In P. M. Kappeler & C. P. van Schaik (Eds.), *Cooperation in primates and humans: Mechanisms and evolution* (pp. 209–232). New York: Springer.

Barrett, L., & Würsig, B. (2014). Why dolphins are not aquatic apes. *Animal Behavior and Cognition, 1*(1), 1–18.

Bastian, J. (1968). The transmission of arbitrary environmental information between bottlenose dolphins. In R. G. Busnel (Ed.), *Animal sonar systems* (pp. 803–874). Jouy-en-Josas, France: Laboratoire de Physiologie Acoustique.

Bateson, G. (1972). *Steps to an ecology of mind*. New York: Ballantine.

Bateson, G. (1979). *Mind and nature: A necessary unity*. New York: Dutton.

Bender, C. E., Herzing, D. L., & Bjorklund, D. F. (2009). Evidence of teaching in Atlantic spotted dolphins (*Stenella frontalis*) by mother dolphins foraging in the presence of their calves. *Animal Cognition, 12*, 43–53.

Boyd, R. L., Robinson, M. D., & Fetterman, A. K. (2011). Miller (1944) revisited: Movement times in relation to approach and avoidance conflicts. *Journal of Experimental Social Psychology, 47*, 1192–1197.

Bruck, J. N. (2013). Decades-long social memory in bottlenose dolphins. *Proceedings of the Royal Society B: Biological Sciences, 280*, 1–6.

Byrne, R. W. (1995). *The thinking ape: Evolutionary origins of intelligence*. Oxford: Oxford University Press.

Byrne, R. W., & Bates, L. A. (2014). Cognition in the wild: Exploring animal minds with observational evidence. *Biology Letters, 7*, 619–622.

Byrne, R. W., & Whiten, A. (1988). *Machiavellian intelligence*. Oxford: Oxford University Press.

Caldwell, D. K., & Caldwell, M. C. (1972). *The world of the bottlenose dolphin*. Philadelphia, PA: Lippincott.

Caldwell, M. C., & Caldwell, D. K. (1965). Individualized whistle contours in bottlenose dolphins (*Tursiops truncatus*). *Nature, 207*, 434–435.

Cantor, M., & Whitehead, H. (2013). The interplay between social networks and culture. *Philosophical Transactions of the Royal Society of London B: Biological Sciences, 368*, 1–10.

Caro, T. M., & Hauser, M. D. (1992). Is there teaching in nonhuman animals? *Quarterly Review of Biology, 67*, 151–174.

Carpenter, M., Uebel, J., & Tomasello, M. (2013). Being mimicked increases prosocial behavior in 18-month-old infants. *Child Development, 84*, 1511–1518.

Chance, M. R. A. (1967). Attention structure as the basis of primate rank orders. *Man, 2*, 503–518.

Chance, M. R. A., & Jolly, C. (1970). *Social groups of monkeys, apes, and men*. New York: Dutton.

Connor, R. C. (2007). Complex alliance relationships in bottlenose dolphins and a consideration of selective environments for extreme brain size evolution in mammals. *Philosophical Transactions of the Royal Society B: Biological Sciences, 362*, 587–602.

Connor, R. C., Heithaus, R. M., & Barre, L. M. (1999). Super-alliance of bottlenose dolphins. *Nature, 371*, 571–572.

Connor, R. C., & Mann, J. (2006). Social cognition in the wild: Machiavellian dolphins? In S. Hurley & M. Nudd (Eds.), *Rational animals* (pp. 329–367). Cambridge: Oxford University Press.

Connor, R. C., & Norris, K. S. (1982). Are dolphins reciprocal altruists? *American Naturalist, 119*, 358–374.

Connor, R. C., Smolker, R. A., & Richards, A. F. (1992). Two levels of alliance formation among male bottlenose dolphins (*Tursiops* sp.). *Proceedings of the National Academy of Sciences of the United States of America, 89*, 987–990.

Connor, R. C., Wells, R., Mann, J., & Read, A. (2000). The bottlenose dolphin: Social relationships in a fission-fusion society. In J. Mann, R. Connor, P. Tyack, & H. Whitehead (Eds.), *Cetacean societies: Field studies of whales and dolphins* (pp. 9–126). Chicago: University of Chicago Press.

Coussi-Korbel, S., & Fragaszy, D. (1995). On the relation between social dynamics and social learning. *Animal Behaviour, 50*, 1441–1453.

Cusick, J. A., & Herzing, D. L. (2014). The dynamics of aggression: How individual and group factors affect the long-term interspecific aggression between two sympatric species of dolphin. *Ethology, 120*, 287–303.

de Barbaro, K., Johnson, C. M., & Deak, G. O. (2013). Twelve-month "social revolution" emerges from mother-infant sensorimotor coordination: A longitudinal investigation. *Human Development, 56*, 223–248.

de Waal, F. B. M. (1982). *Chimpanzee politics*. New York: Harper & Row.

de Waal, F. B. M. (1986). Dynamics of social relationships. In B. B. Smuts, D. L. Cheney, R. M. Seyfarth, R. W. Wrangham, & T. T. Struhsaker (Eds.), *Primate societies* (pp. 421–430). Chicago: University of Chicago Press.

Emery, N. J. (2000). The eyes have it: The neuro-ethology, function, and evolution of social gaze. *Neuroscience and Biobehavioral Reviews, 24*, 581–604.

Enfield, N. J., & Levinson, S. C. (2006). *Roots of human sociality: Culture, cognition, and interaction*. New York: Berg.

Evans, C. S., Evans, L., & Marler, P. (1993). On the meaning of alarm calls: Functional reference in an avian vocal system. *Animal Behaviour, 46*, 23–38.

Fellner, W., Bauer, G. B., & Harley, H. E. (2006). Cognitive implications of synchrony in dolphins: A review. *Aquatic Mammals, 32*(4), 511–516.

Ferreira, R. G., Izar, P., & Lee, P. C. (2006). Exchange, affiliation and protective interventions in semifree-ranging brown Capuchin monkeys (*Cebus apella*). *American Journal of Primatology, 68*, 765–776.

Forster, D. (2002). Consort turnovers as distributed cognition in olive baboons: A distributed approach to mind. In M. Bekoff, C. Allen, & G. M. Burghardt (Eds.), *The cognitive animal: Empirical and theoretical perspectives on animal cognition* (pp. 163–171). Cambridge, MA: MIT Press.

Garcia, J., & , Koelling, R. A. (1996). Relation of cue to consequence in avoidance learning. In L. D. Houck and L. C. Drickamer (Eds.), *Foundations of animal behavior: Classic papers with commentaries* (pp. 374–375). Chicago: University of Chicago Press.

Gazda, S. K., Connor, R. C., Edgar, R. K., & Cox, F. (2005). Division of labor with role specialization in group hunting bottlenose dolphins (*Tursiops truncatus*) off Cedar Key, Florida. *Proceedings of the Royal Society B: Biological Sciences, 227*, 135–140.

Gibson, J. J. (1979). *The ecological approach to visual perception*. Boston: Houghton Mifflin.

Goodall, J. (1986). *The chimpanzees of Gombe*. Cambridge, MA: Harvard University Press.

Goodwin, C. (2000). Action and embodiment within situated human interaction. *Journal of Pragmatics, 32*, 1489–1522.

Goodwin, C. (2013). The cooperative, transformative organization of human action and knowledge. *Journal of Pragmatics, 46*, 8–23.

Gotz, T., Verfuss, U. K., & Schnitzler, H. U. (2006). Eavesdropping in wild rough-toothed dolphins (*Steno bredanensis*)? *Biology Letters, 2*(1), 5–7.

Grove, M., Pearce, E., & Dunbar, R. I. M. (2012). Fission-fusion and the evolution of hominin social systems. *Journal of Human Evolution, 62*, 191–200.

Gregg, J. D., Dudzinski, K. M., & Smith, H. V. (2008). 3D MASC: A method for estimating relative head angle and spatial distance of dolphins from underwater video footage. *Animal Behaviour, 75*, 1181–1186.

Guinet, C., & Bouvier, J. (1995). Development of intentional stranding techniques in killer whale (*Orcinus orca*) calves at Crozet archipelago. *Canadian Journal of Zoology, 73*, 27–33.

Harcourt, A., & de Waal, F. B. M. (1992). *Coalitions and alliances in humans and other animals*. New York: Oxford University Press.

Harley, H. E. (2008). Whistle discrimination and categorization by the Atlantic bottlenose dolphin (*Tursiops truncatus*): A review of the signature whistle framework and a perceptual test. *Behavioural Processes, 77*, 243–268.

Harley, H. E. (2013). Consciousness in dolphins? A review of recent evidence. *Journal of Comparative Physiology, 199*, 565–582.

Hayashi, M., Raymond, G., & Sidnell, J. (2013). *Conversational repair and human understanding*. Cambridge: Cambridge University Press.

Henzi, S. P., & Barrett, L. (2002). Infants as a commodity in a baboon market. *Animal Behaviour, 63*, 915–921.

Herman, L. M. (2002). Vocal, social, and self-imitation by bottlenosed dolphins. In C. Nehaniv & K. Dautenhahn (Eds.), *Imitation in animals and artifacts* (pp. 63–108). Cambridge, MA: MIT Press.

Herzing, D. L. (1996). Underwater behavioral observations and associated vocalizations of free-ranging Atlantic spotted dolphins, *Stenella frontalis*, and bottlenose dolphins, *Tursiops truncatus*. *Aquatic Mammals, 22*, 61–79.

Herzing, D. L. (2006). The currency of cognition: Assessing tools, techniques and media for complex behavioral analyses. *Aquatic Mammals, 32*(4), 544–553.

Herzing, D. L. (2011). *Dolphin diaries*. New York: St. Martin's Press.

Hinde, R. A., & Simpson, M. J. A. (1975). Qualities of mother-infant relationships in monkeys. In R. Porter & M. O'Connor (Eds.), *CIBA Foundation Symposium 33: Parent-infant interaction* (pp. 39–67). Amsterdam: Elsevier.

Hirata, S., & Matsuzawa, T. (2001). Tactics to obtain a hidden food item in chimpanzee pairs (*Pan troglodytes*). *Animal Cognition, 4*, 285–295.

Hutchins, E. (1995a). *Cognition in the wild*. Cambridge, MA: MIT Press.

Hutchins, E. (1995b). How a cockpit remembers its speed. *Cognitive Science, 19*, 265–288.

Hutchins, E. (2001). Distributed cognition. In N. J. Smelser & P. B. Baltes (Eds.), *The International Encyclopedia of the Social and Behavioral Sciences* (4th ed., pp. 2068–2072). New York: Elsevier Science.

Hutchins, E. (2006). The distributed cognition perspective on human interaction. In N. J. Enfield & S. C. Levinson (Eds.), *Roots of human sociality: Culture, cognition and interaction* (pp. 375–398). Oxford: Berg.

Hutchins, E. (2010a). Cognitive ecology. *Topics in Cognitive Science, 2*(4), 705–715.

Hutchins, E. (2010b). Enaction, imagination, and insight. In J. Stewart, O. Gapenne, & E. A. Di Paolo (Eds.), *Enaction: Toward a new paradigm for cognitive science* (pp. 425–450). Cambridge, MA: MIT Press.

Hutchins, E. (2014). The cultural ecosystem of human cognition. *Philosophical Psychology, 27*(1), 34–50.

Hutchins, E., & Johnson, C. M. (2009). Modeling the emergence of language as an embodied collective cognitive activity. *Topics in Cognitive Science, 1*, 523–546.

Janik, V. M. (2000). Whistle matching in wild bottlenose dolphins (*Tursiops truncatus*). *Science, 289*, 1355–1357.

Johnson, C. M. (1993). Animal communication via coordinated cognitive systems. In P. P. G. Bateson, N. Thompson, & P. Klopfer (Eds.), *Perspectives in ethology: Vol. 10. Variability in behavior* (pp. 187–207). New York: Plenum.

Johnson, C. M. (2001). Distributed primate cognition: A review. *Animal Cognition, 3*(4), 167–183.

Johnson, C. M. (2002). The Vygotskian advantage in cognitive modeling: Participation precedes and thus prefigures understanding. *Behavioral and Brain Sciences, 25*, 628–629.

Johnson, C. M. (2010). Observing cognitive complexity in primates and cetaceans. *International Journal of Comparative Psychology, 23*, 587–624.

Johnson, C. M., & Karin-D'Arcy, M. R. (2006). Social attention in primates: A behavioral review. *Aquatic Mammals, 32*, 423–442.

Johnson, C. M., & Norris, K. S. (1994). Social behavior. In K. S. Norris, B. Würsig, R. S. Wells, & M. Würsig (Eds.), *The Hawaiian spinner dolphin* (pp. 243–286). Berkeley: University of California Press.

Johnson, C. M., & Oswald, T. M. (2001). Distributed cognition in apes. In J. D. Moore & K. Stenning (Eds.), *Proceedings of the Twenty-third Annual Conference of the Cognitive Science Society* (pp. 453–458). University of Edinburgh, Scotland: Human Communication Research Centre.

Kaminski, J., Brauer, J., Call, J., & Tomasello, M. (2009). Domestic dogs are sensitive to a human's perspective. *Behaviour, 146*, 979–998.

King, B. J. (2004). *The dynamic dance: Nonvocal communication in African great apes.* Cambridge, MA: Harvard University Press.

Kuczaj, S. A., & Yeater, D. B. (2006). Dolphin imitation: Who, what, when, and why? *Aquatic Mammals, 32*(4), 413–422.

Kummer, H. (1971). *Primate societies.* Arlington Heights, IL: Harlan Davidson.

Lusseau, D. (2007). Evidence for social role in a dolphin social network. *Evolutionary Ecology, 21*, 357–366.

Lusseau, D., & Newman, M. E. J. (2004). Identifying the role that animals play in their social networks. *Proceedings of the Royal Society B: Biological Sciences, 271*, S477–S481.

Madsen, C. J., & Herman, L. M. (1980). Social and ecological correlates of cetacean vision and visual appearance. In L. M. Herman (Ed.), *Cetacean behavior: Mechanisms and functions* (pp. 101–147). New York: Wiley Interscience.

Malafouris, L., & Renfrew, C. (2010). *The cognitive life of things: Recasting the boundaries of the mind.* Cambridge: McDonald Institute Monographs.

Mann, J., & Sargeant, B. (2003). Like mother, like calf: The ontogeny of foraging traditions in wild Indian Ocean bottlenose dolphins (*Tursiops* spp.). In D. M. Fragaszy & S. Perry (Eds.), *The biology of traditions: Models and evidence* (pp. 236–266). Cambridge: Cambridge University Press.

Maynard Smith, J. (1982). *Evolution and the theory of games.* Cambridge: Cambridge University Press.

McBride, A. F., & Hebb, D. O. (1948). Behavior of the captive bottlenose dolphin, *Tursiops truncatus. Journal of Comparative and Physiological Psychology, 41*, 111–123.

McNelis, N. L., & Boatright-Horowitz, S. L. (1998). Social monitoring in a primate group: The relationship between visual attention and hierarchical ranks. *Animal Cognition, 1*(1), 65–69.

Nakahara, F., & Miyazaki, N. (2011). Vocal exchanges of signature whistles in bottlenose dolphins (*Tursiops truncatus*). *Journal of Ethology, 29*, 309–320.

Noe, R., & Hammerstein, P. (1994). Biological markets: Supply and demand determine the effect of partner choice in cooperation, mutualism, and mating. *Behavioral Ecology and Sociobiology, 35*, 1–11.

Norris, K. S., & Dohl, T. (1980). The structure and function of cetacean schools. In L. M. Herman (Ed.), *Cetacean behavior: Mechanisms and functions* (pp. 211–262). New York: Wiley Interscience.

Nowacek, D. (2002). Sequential foraging behaviour of bottlenose dolphins, *Tursiops truncatus*, in Sarasota Bay, FL. *Behaviour, 139*(9), 1125–1145.

Pack, A. A., & Herman, L. M. (2006). Dolphin social cognition and social attention: Our current understanding. *Aquatic Mammals, 32*, 443–460.

Paukner, A., Anderson, J. R., Borelli, E., Visalberghi, E., & Ferrari, P. F. (2005). Macaques (*Macaca nemestrina*) recognize when they are being imitated. *Biology Letters, 1*(2), 219–222.

Payne, K., Langbauer, W. R., Jr., & Thomas, E. (1986). Infrasonic calls of the Asian elephant (*Elephas maximus*). *Behavioral Ecology and Sociobiology, 18*, 297–301.

Pryor, K. (1990). Non-acoustic communication in small cetaceans: Glance, touch, position, gesture, and bubbles. *NATO ASI Series, 196*, 537–544.

Reiss, D., & McCowan, B. (1993). Spontaneous vocal mimicry and production by bottlenose dolphins (*Tursiops truncatus*): Evidence for vocal learning. *Journal of Comparative Psychology, 107*, 301–312.

Rendell, L., & Whitehead, H. (2001). Culture in whales and dolphins. *Behavioral and Brain Sciences, 24*, 309–382.

Richards, D. G., Wolz, J. P., & Herman, L. M. (1984). Vocal mimicry of computer-generated sounds and vocal labeling of objects by a bottlenosed dolphin, *Tursiops truncatus*. *Journal of Comparative Psychology, 98*, 10–28.

Rogoff, B. (1990). *Apprenticeship in thinking: Cognitive development in social context*. Oxford: Oxford University Press.

Rosch, E., Varela, F., & Thompson, E. F. (1991). *The embodied mind*. Cambridge, MA: MIT Press.

Russon, A. E. (2006). Acquisition of complex foraging skills in juvenile and adolescent orangutans (*Pongo pygmaeus*): Developmental influences. *Aquatic Mammals, 32*, 500–510.

Russon, A., & Andrews, K. (2011). Orangutan pantomime: Elaborating the message. *Biology Letters, 7*, 627–630.

Samuels, A., & Flaherty, C. (2000). Peaceful conflict resolution in the sea? In F. Aureli & F. B. M. de Waal (Eds.), *Natural conflict resolution* (pp. 229–231). Berkeley: University of California Press.

Samuels, A., & Gifford, T. (1997). A quantitative assessment of dominance relations among bottlenose dolphins. *Marine Mammal Science, 13*, 70–99.

Sayigh, L. S., Esch, H. C., Wells, R. S., & Janik, V. M. (2007). Facts about signature whistles of bottlenose dolphins (*Tursiops truncatus*). *Animal Behaviour, 74*, 1631–1642.

Schino, G., & Aureli, F. (2009). Reciprocal altruism in primates: Partner choice, cognition and emotions. *Advances in the Study of Behavior, 39*, 45–69.

Seyfarth, R. M., Cheney, D. L., & Marler, P. (1980). Vervet monkey alarm calls: Semantic communication in a free-ranging primate. *Animal Behaviour, 28*(4), 1070–1094.

Smith, W. J. (1977). *The behavior of communicating: An ethological approach.* Cambridge, MA: Harvard University Press.

Tavolga, M. C. (1966). Behavior of the bottlenose dolphin (*Tursiops truncatus*): Social interactions in a captive colony. In K. S. Norris (Ed.), *Whales, dolphins, and porpoises* (pp. 718–730). Berkeley: University of California Press.

Tayler, C. K., & Saayman, G. S. (1973). Imitative behaviour by Indian Ocean bottlenose dolphins (*Tursiops aduncus*) in captivity. *Behaviour, 46,* 286–298.

Thieltges, H., Lemasson, A., Kuczaj, S., Boeye, M., & Blois-Heulin, C. (2011). Visual laterality in dolphins when looking at (un)familiar humans. *Animal Cognition, 14*(2), 303–308.

Thorndike, E. L. (1911). *Animal intelligence: Experimental studies.* New York: MacMillan.

Tinbergen, N. (1952). "Derived" activities: Their causation, biological significance, origin, and emancipation during evolution. *Quarterly Review of Biology, 27,* 1–32.

Tomasello, M., Call, J., & Hare, B. (1998). Five primate species follow the visual gaze of conspecifics. *Animal Behaviour, 55,* 1063–1069.

Tomasello, M., Gust, D., & Frost, G. T. (1989). The development of gestural communication in young chimpanzees: A follow-up. *Primates, 30,* 35–50.

Tomonaga, M., Uwano, Y., Ogura, S., & Saito, T. (2010). Bottlenose dolphins' (*Tursiops truncatus*) theory of mind as demonstrated by responses to their trainers' attentional state. *International Journal of Comparative Psychology, 23,* 386–400.

Trivers, R. L. (1971). The evolution of reciprocal altruism. *Quarterly Review of Biology, 46,* 35–57.

van Baaren, R. B., Holland, R. W., Kawakami, K., & van Knippenberg, A. (2004). Mimicry and prosocial behavior. *Psychological Science, 15,* 71–74.

Vygotsky, L. S. (1978). *Mind in society: The development of higher psychological processes.* Cambridge, MA: Harvard University Press.

Watwood, S. L., Owen, E. C. G., Tyack, P. L., & Wells, R. S. (2005). Signature whistle use by temporarily restrained and free-swimming bottlenose dolphins, *Tursiops truncatus. Animal Behaviour, 69,* 1373–1386.

Watwood, S. L., Tyack, P. L., & Wells, R. S. (2004). Whistle sharing in paired male bottlenose dolphins, *Tursiops truncatus. Behavioral Ecology and Sociobiology, 55,* 531–543.

Weaver, A. (2003). Conflict and reconciliation in captive bottlenose dolphins, *Tursiops truncatus. Marine Mammal Science, 19*(4), 836–846.

Wells, R. S. (2003). Dolphin social complexity: Lessons from long-term study and life history. In F. B. M. de Waal & P. L. Tyack (Eds.), *Animal social complexity: Intelligence, culture, and individualized societies* (pp. 32–56). Cambridge, MA: Harvard University Press.

Wells, R. S., Irvine, A. B., & Scott, M. D. (1980). The social ecology of inshore odontocetes. In L. M. Herman (Ed.), *Cetacean behavior: Mechanisms and functions* (pp. 263–318). New York: Wiley Interscience.

Whiten, A., & Byrne, R. W. (1988). Tactical deception in primates. *Behavioral and Brain Sciences, 11*, 233–273.

Whiten, A., & Ham, R. (1992). On the nature and evolution of imitation in the animal kingdom. In P. J. B. Slater, J. S. Rosenblatt, C. Beer, & M. Milkinski (Eds.), *Advances in the study of behavior* (Vol. 21, pp. 239–283). New York: Academic Press.

Wisniewska, D. M., Johnson, M., Beedholm, K., Wahlberg, M., & Madsen, P. T. (2012). Acoustic gaze adjustments during active target selection in echolocating porpoises. *Journal of Experimental Biology, 215*, 4358–4373.

Würsig, B. (1986). Delphinid foraging strategies. In R. J. Schusterman, J. A. Thomas, & F. G. Wood (Eds.), *Dolphin cognition and behavior* (pp. 347–359). Hillsdale, NJ: Erlbaum.

Würsig, B., Würsig, M., & Cipriano, F. (1989). Dolphins in different worlds. *Oceanus, 32*, 71–75.

Xitco, M. J., Gory, J. D., & Kuczaj, S. A. (2001). Spontaneous pointing by bottlenose dolphins (*Tursiops truncatus*). *Animal Cognition, 4*, 115–123.

Xitco, M. J., Jr., Gory, J. D., & Kuczaj, S. A., II. (2004). Dolphin pointing is linked to the attentional behavior of a receiver. *Animal Cognition, 7*, 231–238.

Xitco, M. J., & Roitblat, H. L. (1996). Object recognition through eavesdropping: Passive echolocation in bottlenose dolphins. *Animal Learning and Behavior, 24*, 355–365.

Yaman, S., von Fersen, L., Dehnhardt, G., & Güntürkün, O. (2003). Visual lateralization in the bottlenose dolphin (*Tursiops truncatus*): Evidence for a population asymmetry? *Behavioural Brain Research, 142*, 109–114.

Zahavi, A. (1977). The testing of a bond. *Animal Behaviour, 25*, 246–247.

Zanin, A. V., Markov, V. I., & Sidorva, I. E. (1990). The ability of bottlenose dolphins, *Tursiops truncatus*, to report arbitrary information. In J. Thomas & R. Katelein (Eds.), *Sensory abilities in cetaceans* (pp. 685–696). New York: Plenum Press.

10   Whales, Dolphins, and Ethics: A Primer

Thomas I. White

One of the most important features of science is that major discoveries regularly raise important ethical questions. This is especially true with research about cetaceans, because the discoveries by marine mammal scientists over the last fifty years have made it clear that whales and dolphins share traits once believed to be unique to humans: self-awareness, abstract thought, the ability to solve problems by planning ahead, understanding linguistically sophisticated concepts such as syntax, and the formation of cultural communities (Herman, 1984; Reiss & Marino, 2001; Norris, Würsig, Wells, & Würsig, 1991).

Accordingly, humanity faces a number of profound questions: What are the ethical implications of the fact that whales and dolphins demonstrate such intellectual and emotional sophistication? Which ethical standards should be used in evaluating how humans treat them? When looked at through this lens, which human behaviors are ethically problematic? How do we change our behavior to improve the situation?

Engaging with these questions, however, poses a special challenge for marine mammal scientists. The scientific disciplines employ methodologies that emphasize the careful collection, cataloging, and description of empirical data. By contrast, ethical considerations are essentially conceptual and normative. Ethical analyses begin with the facts related to the actions under investigation, but the primary point of an ethical analysis is to conclude what those facts tell us about the ethical acceptability or unacceptability of the actions under investigation.

The fundamental challenge for marine mammal scientists who want to explore the ethical implications of what marine mammal science has discovered about whales and dolphins is to move from the description of facts about whales and dolphins to the evaluation of what those facts say about human behavior toward cetaceans. A simple way of putting this is that the task is to move from *is* to *ought*—that is, to move from what we know about various cognitive and affective capacities of whales and dolphins and the impact of human actions on these beings (what *is* the case) to a conclusion about whether such actions are ethically acceptable, that is, whether humans should behave toward cetaceans in this fashion (what *ought* to be the case). This entails becoming familiar with the philosophical literature regarding ethics, in general, and environmental ethics, in particular, and acquiring the appropriate technical skills and intellectual perspective for engaging in conceptual discussion and analysis.

My essay aims to serve as a brief introduction to ethics for marine mammal scientists interested in discussing the moral status of practices such as: dolphin drive hunts; commercial or scientific whaling; the deliberate or preventable harm to cetaceans resulting from certain human fishing practices; and the use of captive cetaceans for entertainment, education, military purposes, therapy for human medical conditions, or scientific research. I begin by explaining the basic elements and appropriate procedure for an ethical analysis (fundamental ideas such as *moral standing, moral rights*, and *flourishing*) and briefly describe their application to ethical issues connected with human treatment of whales and dolphins.

What Is Ethics?

Ethics is one of a host of ways we use to evaluate human actions. Nonethical evaluations include whether or not an action is legal, profitable, aesthetically pleasing, well executed, novel, humorous, consistent with the rules of a particular activity (e.g., baseball or chess), in agreement with the traditions of a particular group (sorority, village, community, religion), and the like. The number of nonethical perspectives we use to evaluate actions is almost limitless.

Despite the many differences that may surface among philosophers in debates about ethics, there is a consensus that, at the least, ethical judgments do not rest on standards that might be subjective, arbitrary, irrational, contradictory, or internally inconsistent (e.g., law, religion, social or cultural norms or traditions, individual conscience or emotions). The goal is to base ethical judgments on objective standards with, as much as possible, universal validity.

Accordingly, the most basic goal of an ethical evaluation can be seen as determining whether or not the action in question is consistent with the *well-being* of those affected by that action. Does it increase or decrease their ability to live a successful life? Does it support or restrict their growth and development? Does it promote or undermine their key interests?

Given the complexities of the practical world, it is, of course, regularly the case that an action will promote the interest of one group at the expense of another. Hence the classic ethical dilemmas: Do we have reason to favor one group over the other? Is the best solution one in which each group must compromise? How important is the amount or type of good or harm produced? Are some harms or actions never ethically defensible?

Moral Standing

The first step in determining the ethical character of an action or resolving ethical dilemmas is to determine whether all the parties involved have *moral standing*. Put prosaically, "Who 'counts'?" Whose interests deserve to be taken into account? Who can be harmed or benefited in an ethically significant way?

Traditionally, only some animals have been seen as having a claim for moral standing. (The capacity of animals to feel pain and their vulnerability to be killed are typically seen as sufficient conditions to grant moral standing.) However, thinkers such as Christopher Stone have raised the possibility of a more expansive understanding of the concept with his provocative essay "Do Trees Have Standing?" (Stone, 1972).

Fortunately, the most pressing ethical issues involving whales and dolphins center on a clash between humans and cetaceans. And since both groups have sophisticated intellectual and emotional abilities that make them vulnerable to a wide range of benefits and harms, there should be no question that both groups have moral standing.

Having moral standing, we might say, gets a biological family, order, or species only "in the door." It does not mean that all beings with moral standing deserve *the same* protections. For example, publicly shaming an innocent human for serious wrongdoing that he or she did *not* do could lead to substantial harm to that individual. But it is difficult to believe that verbally abusing a cow in the same way would compromise the well-being of that mammal. Accordingly, the capacities, needs, and traits of a species will determine the type of consideration any member of that species should be entitled to. (I discuss the relevant features of whales and dolphins below.)

The issue of moral standing also raises the question of whether it should extend to a group or to individuals. The risk of extinction, for example, may entitle a species to moral standing, but this does not necessarily extend to individual members of that species. If an identifiable population or community is threatened, but not the species, members of the community may enjoy moral standing—but only in their capacity as members of the specific group.

All individual humans, of course, are seen as having moral standing. The special combination of advanced cognitive and affective capacities that distinguishes us creates a uniqueness for each individual, which we see as having intrinsic value. These capacities also produce a distinctive vulnerability to pain and suffering. Because humans experience life as self-aware individuals with sophisticated intellectual and emotional abilities (the capacity to plan and control behavior, to form significant emotional relationships, to recall past events, and the like), we are vulnerable to a greater range of harms than beings who lack these abilities. We can suffer from not simply physical pain but complex emotional pain such as traumatic memories, fear in the present, dread regarding the future, and so on.

Discussions about the welfare of whales and dolphins have traditionally been limited to whether or not a species is threatened with extinction, a specific population is threatened, and the like. From this perspective, groups of cetaceans, not individuals, have moral standing. However, research that demonstrates advanced cognitive abilities among dolphins—particularly self-awareness (Reiss & Marino, 2001)—offers evidence for the claim that *individual* cetaceans should be regarded as having moral standing. Self-awareness makes possible a sense of self-identity and creates the individual uniqueness that humans prize so highly in ourselves. The rich inner world resulting from a combination of self-awareness and sophisticated intellectual and emotional abilities carries with it a significant vulnerability to affective as well as physical harm that parallels the risk among individual humans.

It is critical to recognize that if we recognize individual whales and dolphins as having moral standing on the basis of key traits they share with humans (self-awareness, ability to control their actions, intellectual and emotional abilities advanced enough to produce a rich inner life that includes associated vulnerabilities), in an ethical dispute, our respective species would appear to have *equal* moral standing.

Ethical Standards

Having identified who is entitled to moral standing, the next question is which standard should be applied. What should determine the ethical character of the action under question? In the case of a clash of interests, what is the most ethically appropriate resolution to such a conflict?

In the two-thousand-year history of ethics, the two most important competing standards come from *teleological* and *deontological* approaches to ethics. A teleological approach argues that all actions are morally neutral, and their ethical character is determined by the consequences of the actions. The best-known example is the utilitarianism of Jeremy Bentham and John Stuart Mill. In his *Introduction to the Principles of Morals and Legislation*, Bentham (1789/1948) writes:

Nature has placed mankind under the governance of two sovereign masters, *pain* and *pleasure*. It is for them alone to point out what we ought to do. ... The standard of right and wrong ... [is]

fastened to their throne. … By utility is meant that property in any object whereby it tends to produce benefit, advantage, pleasure, good, or happiness (all this in the present case comes to the same thing) or (what comes again to the same thing) to prevent the happening of mischief, pain, evil, or unhappiness to the party whose interest is considered: if that party be the community in general, then the happiness of the community: if a particular individual, then the happiness of that individual. (pp. 1–3)

This approach regularly surfaces as "cost-benefit analysis" in contemporary economic and political discussions and is embraced by many as a practical and commonsense ethical standard. If the benefits outweigh the costs, then the action or policy is good. When viewed from a democratic perspective, this approach often endorses actions that benefit the majority over the minority.

By contrast, a deontological perspective rejects the importance of consequences and argues that actions have intrinsic moral properties. Such a perspective is best represented by Immanuel Kant (1785/1981), who writes: "Everything has either a price or a dignity. Whatever has a price can be replaced by something else as its equivalent; on the other hand, whatever is above all price, and therefore admits of no equivalent, has a dignity. … Skill and diligence in work have a market value; … but fidelity in promises and benevolence on principle … have intrinsic worth" (p. 434). Kant's central moral principle is the "categorical imperative": "Act in such a way that you treat humanity, whether in your own person or in the person of any other, always at the same time as an end and never simply as a means" (p. 429).

This approach offsets the weakness of utilitarianism, which can produce calculations that justify ethically problematic actions because of the tangible benefits they produce. Our repudiation of slavery, human experimentation, and the like are classic cases where the very action (treating persons as property or in some other way showing no respect for the dignity of the human person) is considered so morally offensive that any tangible benefits that may result are considered irrelevant.

Despite the robust disagreements among ethical theorists about which of these approaches is correct, in the more practical domain of applied ethics, a more productive approach uses both perspectives. (A deontological approach corrects for the risk of utilitarianism producing situations where "the ends justify the means." A teleological approach, with its focus on particular situations, softens the moral rigidity and narrow focus that can come from examining only the actions involved.) Each can serve as, we might say, a lens that reveals different features of the ethical issue at hand or the actions being evaluated.

Such a practical, eclectic approach gives two major strengths to any ethical analysis. First, it allows us to identify the most fundamental issues in any ethical analysis:

• When we look at the consequences of any actions under question, is anyone with moral standing harmed?

• Are these harms offset by an appropriate amount and/or type of benefits?

• Setting the consequences aside and examining the actions themselves, do they treat all parties appropriately, that is, in a way that is consistent with the respect and dignity they are due? Are any of the actions so ethically indefensible as to trump the tangible benefit?

Second, combining both teleological and deontological approaches creates a more objective, complete, and stricter ethical standard than either perspective alone. It also derails the temptation to select one's ethical perspective according to whichever one will advance one's personal interest. Therefore, for either an action under study or a resolution of a clash between parties to be ethically defensible, it must be the case that there is a proper mix of benefits versus harms *and* that all parties are treated appropriately.

Ethical Standards, Humans, Cetaceans, and Flourishing

Given the difficulties of resolving ethical disputes among humans or determining the ethical character of actions performed by different individuals, for different reasons in different circumstances, it is obvious that the apparent simplicity of these three questions belies the complexities connected with most ethical issues. It should come as no surprise, then, that when we inject different species into the mix—both profoundly similar to, and fundamentally different from, humans—ethical discussions become geometrically much more complex. And one of the most important challenges is how to conduct such an inquiry in a neutral, objective way so that we do not—even unconsciously—tilt the analysis in a direction that automatically favors one species over the other.

Accordingly, the best approach should be *species specific* and grounded as much as possible in facts. For the purposes of this introduction to ethics, then, the concept of the "flourishing" of a being (and its relationship to the concept of moral rights) forms an appropriate foundation for an ethical standard.

The most important thinker representing this perspective is Martha Nussbaum (2006), who advances a "capabilities approach" to animal ethics. Reflecting a deontological perspective, Nussbaum (2011) takes as "a fundamental ethical starting point … that we must respect each individual sentient being as an end in itself, not a mere means to the ends of others" (p. 237). But the more tangible part of her theory is the idea that "each creature has a characteristic set of capabilities, or capacities for functioning, distinctive of that species, and that those more rudimentary capacities need support from the material and social environment if the animal is to *flourish* in its characteristic way" (p. 237; italics mine).

The assumption underlying this perspective is that animals have evolved in such a way that a certain set of conditions must be met for them to grow, develop, and acquire the traits, skills, and dispositions necessary to have a satisfying and successful life *as a member of that species*. The environment in which a species evolved, the challenges it faced, the resulting adaptations, and the features that came to distinguish that species all determined these conditions.

For example, for humans to flourish, we require: physical and emotional health and safety; absence of pain and suffering; protection when we are young or infirm; freedom of choice, equality, justice, and so on; treatment consistent with appropriate respect for our dignity as autonomous individuals; opportunity to learn what we need to know to navigate a social group's culture; access to meaningful emotional relationships; and rest.

It is, of course, possible for humans to tolerate situations that lack many of these conditions. However, we do not flourish in such circumstances. Because of the nature of the cloth from which we are cut, a life characterized by, for example, being prevented from acquiring the skills necessary to make a living, being discriminated against or enslaved, or being prevented from having meaningful relationships would be unsatisfying at a deep and fundamental level. No human in these circumstances could develop the sense of autonomy, safety, and control that would let him or her lead a successful and satisfying life.

Indeed, the conditions for flourishing are so important to humans that we enshrine them as rights—specifically, human rights. (These are moral rights, which proceed from our mere membership in the species, as opposed to legal rights or political rights, which require some sort of action by some outside party.) The best-known statement is the United Nations' Declaration of Human Rights. This is essentially a list of what any member of our species would categorically *need* in order to have the possibility of living a successful and satisfying life. That is, we say that humans have a right to these conditions because we need them to flourish.

The conditions necessary for flourishing, then, actually become the foundation of ethics. That is, from this perspective, to say that an action is *ethically positive* is to say that it *promotes the flourishing of those involved*. To say that an action is *ethically negative* is to say that it *prevents or undermines it*. Ethical disputes, then, are understood most simply as clashes over competing basic needs in situations where there is no obvious way for all involved to have them met—at least not in the way that competing individuals initially desire.

In an ethical clash between human and cetacean interests, then, a fundamental question is: What are the conditions for *cetacean* flourishing? For the purposes of this essay, the most significant attempt to detail these conditions is the "Declaration of Rights for Cetaceans: Whales and Dolphins" (Brakes & Simmonds, 2011). In this document, the conditions identified as necessary for whales and dolphins to flourish include: life; freedom of movement and residence within their natural environment; freedom from cruel treatment, removal from their natural environment, or treatment as property; and cultures free of disruption.

This perspective is supported by what has been learned about cetacean intellectual, emotional, and social sophistication in the last half century's research. These discoveries feature: self-awareness (Reiss & Marino, 2001); the structural sophistication of the dolphin brain (Marino, 2002; Morgane, Jacobs, & Galaburda, 1986; Ridgway, 1986); the ability to understand artificial human languages, representations of reality, and human pointing and gazing behavior (Herman, 1984; Herman & Pack, 1999; Herman, Morrel-Samuels, & Pack, 1990; Herman, Pack, & Morrel-Samuels, 1993; Herman, Richards, & Wolz, 1984); dolphins' abilities to plan (Kuczaj, Gory, & Xitco, 1998); and cetacean social intelligence (Connor & Peterson, 1994; Herzing, 2000, 2011; Norris, 1991; Norris, Würsig, Wells, & Würsig, 1991; Reynolds, Wells, & Eide, 2000; Smolker, 2001). Especially important are the discoveries of cetacean culture (Rendell & Whitehead, 2001). Particularly significant in this regard are: the ongoing studies of the Pacific Northwest orcas by a variety of scientists; Denise Herzing's long-term research on Atlantic spotted dolphins (Herzing, 2011); and Hal Whitehead's work on culture in sperm whales (Whitehead, 2011).

As in the case of human rights, cetacean rights are moral (not legal) rights. Similarly, as Nussbaum argued earlier, the ethical requirements include *both* the material conditions that have to be met for whales and dolphins to grow and develop in a way that gives them a reasonable opportunity to live satisfying and successful lives *and* being treated with appropriate respect for their dignity as individuals with moral standing. Like humans, cetaceans can be harmed not only by physical abuse but also by treatment inconsistent with their dignity.

Examination of the list of cetacean rights asserted in the Declaration reveals the same duality noted earlier regarding the ethical constraints on human behavior toward whales and dolphins. Respect for the *intrinsic worth* and *dignity* of individual cetaceans is reflected in prohibitions against treating them as property, constraining their movements, disrupting their cultures, and removing them from a natural environment.

These prohibitions are based on the idea that whales and dolphins have the capacity for free, autonomous behavior and that, as is the case with humans, any interference with someone's free choice would be ethically unacceptable. The possibility of *tangible harm* is referenced in the need for protection against cruel treatment and other actions that contain the risk of harm.

Humans, Whales, Dolphins, and Ethics

In light of this discussion of moral standing, moral rights, and the conditions needed for flourishing, the ethical character of some human treatment of whales and dolphins would appear ethically questionable.

Inasmuch as the first condition to be able to flourish is to be alive, the most problematic human practice would be the deliberate killing of whales and dolphins in drive hunts or in "scientific whaling," and the preventable deaths and injuries of cetaceans produced by certain human fishing practices and military exercises.

The more debatable issue of captivity brings with it a series of questions. First, is it possible for any captive facility to provide the conditions necessary for the flourishing of the whales and dolphin who live there? Seen from this perspective, the central issue is not the life span or even the physical condition of cetaceans in captivity. The basic question is whether captivity can provide the sort of stimulation needed for normal growth and development.

It is regularly argued that the tangible benefits of captivity—through entertainment, education, and research—outweigh any harm. However, this claim must overcome the objection raised by the moral standing of whales and dolphins as individuals, which implies that buying and selling cetaceans—or anything that amounts to treating them as property—would be intrinsically wrong.

Of course, this ethically problematic aspect of captivity is complicated by the practical problem of what to do with the cetaceans currently captive, even if the facilities holding them agreed to release them. Some cetaceans might be able to be trained to make the transition to living in the wild, but others might not. What is the best course of action in such a situation?

Implications for Future Research

Greater Familiarity with Normative Traditions
As noted at the outset, this essay is firmly grounded in the idea that the ethical implications of the scientific research on whales and dolphins become evident only when

viewed through the lens of philosophical concepts such as moral standing, moral rights, and flourishing. One of the most important needs in future cetacean research, then, is for the descriptive methodology of science to be supplemented by perspectives from intellectual approaches that specialize in normative judgments. Future marine scientists must become as adept at ethical analysis as data analysis. They must acquire a thorough understanding of the methodology, intellectual perspectives, and relevant literature of fields like philosophy and environmental ethics. Failure to do so will produce the disappointing situation of scientists not fully understanding the ethical dimensions of their own research. And this will obviously slow the pace of improving the treatment of cetaceans by humans.

Key Areas for Future Inquiry

I have suggested that an appropriate standard for evaluating the ethical character of human treatment of whales and dolphins is the set of necessary conditions required for individual cetaceans to flourish in their natural habitat. Unfortunately, in comparison to the large amount we understand about what humans need to flourish and to experience a sense of well-being, we actually know relatively little about the necessary conditions for the growth and development of all facets—physical, emotional, social—of cetaceans. Given such ignorance, it is likely that various human behaviors currently harm whales and dolphins in ways that are both unintended and preventable. If whales and dolphins have the right to be protected from harm and to be treated with appropriate respect as individuals, this implies a duty on the part of researchers to orient their investigations in ways that advance the goal of raising the ethical character of human treatment of whales and dolphins.

Greater Importance of Research in the Wild

The importance of the research that has been done on captive cetaceans in the past cannot be underestimated. However, the very effectiveness of this research raises the question of whether ongoing captive research is ethically defensible. In particular, what has been discovered about the cognitive and affective sophistication of dolphins calls into question the practice of treating them as property. More importantly, research in the wild, which has uncovered the social complexity of whales and dolphins and revealed the existence of cetacean cultures, raises the possibility that captive research could be insufficient for establishing appropriate standards for the treatment of dolphins by humans. From an ethical perspective, field studies could provide more relevant results than research done in captivity for determining species-appropriate standards. A variety of areas come to mind where specific research efforts in various disciplines could make significant contributions.

Reducing Harm to Individuals

Ship Strikes

A significant number of whales are struck by vessels each year in practically every ocean on the planet. Substantial research in this area has already led to some progress in reducing the number of strikes. However, more work is needed to improve our understanding of whale behavior that seems to increase their risk of being struck, as well as to develop possible technologies that could be used to warn them off.

Military Sonar Testing

Ongoing disputes about sonar testing by the military frequently involve claims and counterclaims about whether or not such tests harm cetaceans, and if so, how serious that harm is. While it would obviously be unethical to conduct research that would subject live cetacean test subjects to different types and levels of sonar to determine at what point they are seriously harmed, ongoing research on whales that may have died as a result of such tests could yield important results that could be used to alter or stop such sonar testing.

Aboriginal Whaling

The killing of whales in connection with human aboriginal cultures is a particularly contentious issue because any objection to such practices can sound like cultural imperialism. At the same time, as is clear from the defense of human slavery as "our peculiar institution" used in the United States before the Civil War, and long-standing and revered patterns of discrimination against women and nonwhites in modern America, cultural traditions are not necessarily ethically defensible. While it seems unlikely that the research of marine scientists could affect these practices, this would be a fertile ground for anthropologists. Ideally, a better understanding of the cultural and economic factors that drive such practices could lead to discovering ways that the aboriginal communities involved might be willing to choose to end the practices without seriously compromising their cultural values or way of life.

Reducing Harm to Groups

Disruption of Cetacean Cultures

From an ethical perspective, research into the structure and dynamics of cetacean cultures—and their fragility—is unquestionably an important area for the future. As the case of North Atlantic right whales has shown, human actions can unintentionally remove cultural knowledge from a cetacean community to such an extent that the long-term existence of the community is put at risk (Whitehead, Rendell, Osborne, & Würsig, 2004). Hopefully, more research in this area will lower that risk. Central questions include the following: How is cultural information stored, retrieved, and passed on from generation to generation? Do various whales in a community have distinct responsibilities in preserving certain knowledge critical to the welfare of the group? Can cultural information be transmitted from community to community? Which human behaviors disrupt critical cultural processes?

A Sanctuary for Dolphins

An especially pressing ethical issue is that there is currently no appropriate home for dolphins who should be "retired" from military service or performing at entertainment facilities. If, in the future, captivity is banned in countries that currently allow it, the scale of the problem will increase. While sanctuaries for elephants, chimpanzees, and other nonhuman animals exist, there are none for cetaceans. Research into every aspect of a cetacean sanctuary—possible locations, necessary conditions, financial support—is needed before one can actually be established.

Final Remarks

By providing a brief overview of the elements of ethical analysis, I have attempted to demonstrate that a full understanding of the ethical issues related to the treatment of whales and dolphins by humans requires a multidisciplinary approach—specifically, a methodology that integrates scientific findings with their philosophical implications. It is important to recognize, however, that once the central questions in an ethical dilemma have been identified and all the relevant evidence surveyed, the next step is to construct an argument that advances a specific position about the ethical character of the actions in question and attempts to defend that argument against likely objections.

It is beyond the scope of this essay to proceed to this next step, but the reader is welcome to consider extended arguments that I have offered for the idea that much human treatment of dolphins is ethically indefensible because dolphins are nonhuman persons (White, 2007, 2011, 2013; White & Herzing, 1998).

References

Bentham, J. (1948). *Principles of morals and legislation*. New York: Hafner. (Original work published 1789.)

Brakes, P., & Simmonds, M. P. (2011). Thinking whales and dolphins. In P. Brakes & M. P. Simmonds (Eds.), *Whales and dolphins: Cognition, culture, conservation and human perceptions* (pp. 207–214). London: Earthscan.

Connor, R. C., & Peterson, D. M. (1994). *The lives of whales and dolphins.* New York: Henry Holt.

Herman, L. M. (1984). Cognition and language competencies of bottlenosed dolphins. In R. J. Schusterman, J. A. Thomas, & F. G. Wood (Eds.), *Dolphin cognition and behavior: A behavioral approach* (pp. 221–252). Hillsdale, NJ: Erlbaum.

Herman, L. M., Morrel-Samuels, P., & Pack, A. A. (1990). Bottlenosed dolphin and human recognition of veridical and degraded video displays of an artificial gestural language. *Journal of Experimental Psychology: General, 119*(2), 215.

Herman, L. M., & Pack, A. A. (1999). Dolphins (*Tursiops truncatus*) comprehend the referential character of the human pointing gesture. *Journal of Comparative Psychology, 113*(4), 347.

Herman, L. M., Pack, A. A., & Morrel-Samuels, P. (1993). Representational and conceptual skills of dolphins. In H. L. Roitblat, L. M. Herman, & P. E. Nachtigall (Eds.), *Language and communication: Comparative perspectives* (pp. 403–442). Hillsdale, NJ: Erlbaum.

Herman, L. M., Richards, D. G., & Wolz, J. P. (1984). Comprehension of sentences by bottlenosed dolphins. *Cognition, 16*, 129–219.

Herzing, D. L. (2000). A trail of grief. In M. Bekoff (Ed.), *The smile of a dolphin: Remarkable accounts of animal emotions* (pp. 138–139). New York: Discovery Books.

Herzing, D. L. (2011). *Dolphin diaries: My 25 years with spotted dolphins in the Bahamas.* New York: St. Martin's Press.

Kant, I. (1981). *Grounding of the metaphysics of morals* (J. W. Ellington, Trans.). Indianapolis, IN: Hackett. (Original work published 1785.)

Kuczaj, S. A., Gory, J. D., & Xitco, M. J. (1998). Using programs to solve problems: Imitation versus insight. *Behavioral and Brain Sciences, 21*(05), 695–696.

Marino, L. (2002). Convergence of complex cognitive abilities in cetaceans and primates. *Brain, Behavior, and Evolution, 59*, 21–32.

Morgane, P. J., Jacobs, M. S., & Galaburda, A. (1986). Evolutionary morphology of the dolphin brain. In R. J. Schusterman, J. A. Thomas, & F. G. Wood (Eds.), *Dolphin cognition and behavior: A comparative approach* (pp. 5–30). Hillsdale, NJ: Erlbaum.

Norris, K. S. (1991). *Dolphin days: The life and times of the spinner dolphin.* New York: W. W. Norton.

Norris, K. S., Würsig, B., Wells, R., & Würsig, M. (1991). *The Hawaiian spinner dolphin.* Berkeley, CA: University of California Press.

Nussbaum, M. C. (2006). *Frontiers of justice: Disability, nationality, species membership.* Cambridge, MA: Harvard University Press.

Nussbaum, M. C. (2011). The capabilities approach and animal entitlements. In T. L. Beauchamp & R. G. Frey (Eds.), *The Oxford handbook of animal ethics* (pp. 228–254). Oxford: Oxford University Press.

Pryor, K., & Norris, K. S. (1991). *Dolphin societies: Discoveries and puzzles.* Berkeley: University of California Press.

Reiss, D., & Marino, L. (2001). Mirror self-recognition in the bottlenose dolphin: A case of cognitive convergence. *Proceedings of the National Academy of Sciences of the United States of America, 98*(10), 5937–5942.

Rendell, L., & Whitehead, H. (2001). Cetacean culture: Still afloat after the first naval engagement of the culture wars. *Behavioral and Brain Sciences, 24*, 360–373.

Reynolds, J. E., Wells, R. S., & Eide, S. D. (2000). *The bottlenose dolphin: Biology and conservation.* Gainesville, FL: University Press of Florida.

Ridgway, S. (1986). Physiological observations on dolphin brains. In R. J. Schusterman, J. A. Thomas, & F. G. Wood (Eds.), *Dolphin cognition and behavior: A comparative approach* (pp. 31–60). Hillsdale, NJ: Erlbaum.

Smolker, R. (2001). *To touch a wild dolphin.* New York: Doubleday.

Stone, C. D. (1972). Should trees have standing? Toward legal rights for natural objects. *Southern California Law Review, 45*, 450–487.

White, T. I. (2007). *In defense of dolphins: The new moral frontier.* Oxford: Blackwell.

White, T. I. (2011). What is it like to be a dolphin? In P. Brakes & M. P. Simmonds (Eds.), *Whales and dolphins: Cognition, culture, conservation and human perceptions* (pp. 188–206). London: Earthscan.

White, T. I. (2013). Humans and dolphins: An exploration of anthropocentrism in applied environmental ethics. *Journal of Animal Ethics, 3*(1), 85–99.

White, T. I., & Herzing, D. L. (1998). Dolphins and the question of personhood. *Etica e Animali, 9*, 64–84.

Whitehead, H. (2011). The cultures of whales and dolphins. In P. Brakes & M. P. Simmonds (Eds.), *Whales and dolphins: Cognition, culture, conservation and human perceptions* (pp. 149–168). London: Earthscan.

Whitehead, H., Rendell, L., Osborne, R. W., & Würsig, B. (2004). Culture and conservation of non-humans with reference to whales and dolphins: Review and new directions. *Biological Conservation, 120*, 435.

11 Visions of the Future

Whitlow Au, John Ford, Louis M. Herman, Paul Nachtigall, Sam Ridgway, Jeanette Thomas, Randall Wells, Hal Whitehead, Christine M. Johnson, and Denise L. Herzing

Johnson and Herzing: One of the explicit goals of this book is to help prepare those interested in studying dolphin communication and cognition to take the field in new directions. In any scientific discipline, the future of research depends on what went before. This is due, in part, to the fact that each new finding generates more questions than answers. Thus it occurred to us (the editors) that the people likely to have the most useful insights into the future might be those responsible for the preponderance of its past. As a result, we have turned to several luminaries in the field—researchers whose long-term, seminal work has provided much of the basis for the preceding chapters—for their ideas about where the field might, and should, go. In particular, we asked each to briefly describe the three questions he or she would most like to see answered in future research. Their responses were indeed insightful, intriguing, and sometimes surprising. And we agree: we want to know the answers to *all* these questions!

For most people working in the fields of dolphin communication and cognition, these folks need no introduction. But for those new to any aspect of these fields—and to acknowledge our appreciation for their many contributions—we will briefly introduce, and say a word or two about, each of these founders.

Dr. Whitlow Au literally "wrote the book" on dolphin sonar (Au, 1993). His long career of studying auditory processes, signal processing, and echolocation in dolphins has involved conducting psychophysical testing, electrophysiological measurements, underwater acoustics measurements, computer modeling of auditory systems, and artificial neural network computations (e.g., Au, Floyd, & Haun, 1978; Au, Moore, & Pawloski, 1988; Au & Benoit-Bird, 2003; Au & Ou, 2014). Chief Scientist in the Marine Mammal Research Program at the University of Hawaii, he has worked with animals both in captivity, at the Naval Undersea Centers in San Diego and Hawaii (e.g., Au & Pawloski, 1992), and in the wild (e.g., Au, Ford, Horne & Newman-Allman, 2004) and has been a leading figure in animal bioacoustics (Au & Hastings, 2008).

Dr. John Ford has been studying killer whales off the coast of British Columbia since the 1980s, with a focus on underwater communication (Ford, 1989, 1991). Through identifying and tracking long-standing matrilineal pods, he and his collaborators discovered that these animals develop group-specific dialects composed of stereotyped burst pulse calls. Over several decades, Ford and his colleagues have shed much light on related aspects of behavior and communication in killer whales in the eastern North Pacific (e.g., Deecke, Barrett-Lennard, Spong, & Ford, 2010; Riesch, Ford, & Thomsen, 2006), including the identification of vocal "clans," groups with overlapping repertoires that may represent distinct lineages within the resident population. Plus, their work on dialect change has contributed significantly to our understanding of cultural transmission in cetaceans (e.g., Deecke, Ford, & Spong, 2000; Yurk, Barrett-Lennard, Ford, & Matkin, 2002) and sets the standard for how systematic field studies can inform cognitive research.

Dr. Louis Herman was the founder and director of the Kewalo Basin Marine Mammal Laboratory at the University of Hawaii. For decades, Kewalo Basin was the premier research laboratory for studies of bottlenose dolphin cognition and allied topics (see summaries in Herman, 1980, 2006). Both Herman's early work on comprehension of artificial gestural and acoustic languages (Herman, Richards, & Wolz, 1984) and his more recent work on social cognitive issues, including imitation (Herman, 2002), social attention (Pack & Herman, 2006), and self-awareness (Herman, 2012), have not only yielded numerous insights into the cognitive abilities of dolphins but also provided critical comparisons for research being done with other species (e.g., Johnson & Herzing, 2006). Herman also pioneered field studies of humpback whales, producing the first descriptive study and first aerial surveys of the species in Hawaiian waters (Herman & Antinoja, 1977), as well as later studies of social organization (e.g., Baker & Herman, 1984; Craig & Herman, 1997), life histories (Herman et al., 2011), and singers and song (Mobley, Herman, & Frankel, 1988; Herman et al., 2013).

Dr. Paul Nachtigall is the director of the Marine Mammal Research Program at the University of Hawaii's Hawaii Institute of Marine Biology and has been involved in a wide range of studies on a variety of cetacean species. Focusing primarily on sensory and perceptual processes—including taste, vision, hearing, and echolocation—in dolphins and small whales (Nachtigall, Murchison, & Au, 1978; Nachtigall & Moore, 1988; Nachtigall, 2008), he has also contributed to studies of cognition (e.g., Roitblat, Penner, & Nachtigall, 1991; see Roitblat, Herman, & Nachtigall, 1993). His research on dolphin hearing includes studies of sound localization, the use of echolocation in fishnet avoidance, and how hearing processes adapt while the animals echolocate (e.g., Nachtigall, Lien, Au, & Read, 1995; Nachtigall & Supin, 2008). His work has involved a variety of approaches, including the use of evoked auditory potential hearing tests, the development of low-frequency audiometrics, and computational modeling of their discrimination abilities (e.g., Au & Nachtigall, 1995). In addition, along with his colleague Alexander Supin, Dr. Nachtigall has developed a way to measure hearing *while* an animal is echolocating (Supin, Nachtigall, Pawloski & Au, 2003), to better understand its natural function.

Dr. Sam Ridgway (a veterinarian, DVM; and neurobiologist, PhD) has often been called "the father of marine mammal medicine," beginning with his breakthrough work on dolphin physiology and medicine (Ridgway & Johnston, 1966; Ridgway & McCormick, 1967; Ridgway, Scronce, & Kanwisher, 1969) and decades of research on dolphin health. His popular book about the first ten years of his work, *The Dolphin Doctor* (1987), is a classic, and still in demand. One of the founders of the Navy Marine Mammal program in 1961, he has had extensive experience working with dolphins in captive and open-ocean settings (e.g., Ridgway, Scronce, & Kanwisher, 1969; Ridgway & Howard, 1979). A world authority on the dolphin brain (e.g., Ridgway, 1990), he has developed and adapted technologies for neurological studies, including evoked potentials, PET scans, and MRI (e.g., Ridgway et al., 2006). After a career that spans more than fifty years, he remains at the cutting edge of dolphin research, especially involving the relationship between brains, audition, and vocalization (e.g., Ridgway et al., 2009; Ridgway, 2011; Ridgway & Hanson, 2014; Ridgway, Moore, Carder, & Romano, 2014).

Dr. Jeanette Thomas, professor of biology emerita at Western Illinois University, has studied both marine and terrestrial mammals. Equally interested in captive and field behavior, she has supervised many research projects at places like the Shedd Aquarium, Brookfield Zoo, Lincoln Park Zoo, Hubbs-SeaWorld Research Institute, and the Columbus Zoo (e.g., Tremel et al., 1998; Therrien, Thomas, Therrien, & Stacey, 2012; Loeding, Thomas, Bernier, & Santymire, 2011; Stansbury, Thomas, et al., 2014), as well as participating in diverse field projects ranging from studies of Antarctic seals, to Alaskan sea otters, to Mediterranean monk seals (e.g., Thomas & Kuechle, 1982; Stirling & Thomas, 2003; Williams, Kastelein, Davis, & Thomas, 1988; Munoz et al., 2011). Her research interests include studying sensory behavior, using bioacoustics as a population assessment tool, evaluating the effects of human-made noise on animals, and the behavioral enrichment of captive animals (e.g., Wartzok, Erbe, Getz, & Thomas, 2012). Her compilation, with Ronald Kastelein (1990), entitled *Sensory Abilities in Cetaceans* is, even after all these years, an important sourcebook; and her more recent volume, with Cynthia Moss and Marianne Vater (2004), *Echolocation in Bats and Dolphins*, shows again the many insights to be gained from comparative research. As a longtime editor of *Aquatic Mammals*, she helped structure and promote our field and is in a unique position to evaluate and predict its future (see Thomas, 2014).

Dr. Randall Wells is the director of the Chicago Zoological Society's Sarasota Dolphin Research Program, which, after forty-four years of research, is the world's longest-running study of a dolphin population (Wells, Scott, & Irvine, 1987; Wells, 2014). Based at Mote Marine Laboratory in Sarasota, Florida, Dr. Wells and colleagues have produced a wealth of information on the social behavior, ecology, and factors in the conservation of bottlenose dolphins. Long-term observational studies have resulted in the identification of more than 5,100 individual dolphins along Florida's central west coast. Occasional capture–release projects have provided life history, genetic, health, body condition, and environmental contaminant data for many of the roughly 160 resident dolphins of Sarasota Bay, a community that spans up to five concurrent generations and includes individuals up to sixty-four years of age (Wells & Scott, 1990; Duffield & Wells, 2002; Wells, 2014). The wide variety of research efforts in Sarasota Bay and surrounding waters has produced data on everything from social structure, mating system, reproductive success, and communication (e.g., Wells, 2003; Sayigh, Esch, Wells, & Janik, 2007) to the impact of human activities on dolphin health and behavior (e.g., Wells et al., 2005, 2008, 2013). The well-studied resident dolphin community in Sarasota Bay is used as a reference population for NOAA's comparative investigations of potential impacts on bottlenose dolphins of incidents such as the *Deepwater Horizon* oil spill and PCB pollution at EPA Superfund sites (e.g. Schwacke et al., 2011, 2013).

Dr. Hal Whitehead teaches at the Department of Biology at Dalhousie University in Nova Scotia, but much of his career has been spent living and working at sea, observing and recording the great whales. While he has studied various species, the bulk of his work has focused on the behavior, social structure, population biology, and conservation of sperm whales (Whitehead, 1996, 1998, 2010; Whitehead & Weilgart, 1991; Whitehead et al., 2012; see Whitehead, 2003). His decades of studying behavior in the wild have led to valuable texts on methods for the analysis of social structure (Whitehead, 2008a, 2008b; Whitehead & Van Parijs, 2010), and his development of SOCPROG, an open-source MATLAB program, provides a highly useful and relevant platform for researchers doing quantitative analyses of behavior (Whitehead, 2009). He has also been a major proponent of our recognition of cetacean culture (Rendell

& Whitehead, 2001; Whitehead & Lusseau, 2012) and has helped advance our understanding of the relationship between cultural and genetic evolution (Whitehead, Richerson, & Boyd, 2002).

The questions posed by these researchers have been organized into the following general categories of inquiry: perception, vocalizations, development, cognition, problem solving in the wild, conservation and management, and useful technologies.

Perception

Nachtigall: How do dolphins perceive the world?

We say, "Do you see what I mean?" because we rely primarily on vision as a way to make sense of the world. Dolphins may rely on echolocation as the primary sensory system. If they spoke like humans, they might say, "Do you hear what I mean?" Humans can discriminate the differences in positions of sounds down to one degree if the two sounds are on the horizontal plane, but if they are switched to the vertical plane, the discrimination ability falls to 8 degrees. Renaud and Popper, in 1975, showed that when dolphin clicks are used as sound sources, bottlenose dolphins discriminate sound sources of less than one degree in both the horizontal and the vertical planes. That is a remarkable ability, but how is it done? Recent work by Kristen Taylor (2013) showed that hearing thresholds varied by frequency. A dolphin heard whistle frequencies best when they came from the side, and it heard clicks best when they came from directly in front. What sort of three-dimensional world do dolphins organize with sound, and what processes do they use? Is there anything in their auditory system similar to human stereopsis, and could it be enhanced by movement, as it is in human vision (Paradis et al., 2000)? How do dolphins perceive a world sensed with sound?

Herman: How are the senses of vision and audition related to each other?

Studies have shown that these senses are closely intertwined, in that a dolphin can match arbitrary objects or shapes with great accuracy across these senses (e.g., Pack, Herman, & Hoffmann-Kuhnt, 2004; Harley, Putman, & Roitblat, 2003). These studies argue that echolocation yields a representation of the ensonified image that is closely related to the image created through vision. The findings also argue that objects are directly perceived through echolocation, as they are through vision, without a need to correlate sound spectra and objects through experience and learning. Further study of this remarkable finding is needed, such as an examination of how changes in aspects of echoic or visual targets affect cross-modal matching.

Thomas: How do marine mammals adjust to the fact that the velocity of sound in water can change, depending on the temperature, salinity, and depth profile at their location?

The velocity or speed of sound depends on the temperature, salinity, and pressure (or depth) of the medium. Sounds propagated in air are not very affected by salinity or pressure (unless at very high altitudes) but are affected slightly by air temperature. The velocity of sound in air is fairly constant, at about 344 meters per second. Humans and terrestrial animals rely on this rather constant speed of sound to determine the distance between them and another sound source.

In contrast, the velocity of sound in water is about 4.5 times faster (~1,500 meters per second) than the same sound in air. As a result, marine mammals can communicate over long distances in water. Salinity is fairly constant in the ocean, with the exception of highly saline areas, like the Red Sea, or polar areas, where freshwater runoff may cause local areas of lower salinity. In polar regions of the ocean, the water temperature is isothermal, so the sound velocity varies little from the water surface to the bottom. In contrast, in other oceanic areas, the water temperature varies with depth and creates sound velocity contours. Marine mammals can dive several times per day over a wide range of depths and thus encounter varying sound velocity contours. Somehow they must be able to adjust to determine distances between themselves and other objects or sound sources. This includes adjusting both communication sounds and echolocation in species that have that ability.

Au: Can dolphins perceive time separation pitch (TSP)?

When two short broadband signals that are highly correlated and separated in time by a short interval are played to human subjects, many can perceive a pitch that is the inverse of the time separating the two broadband pulses. Noise with a rippled spectrum will also produce time separation pitch in humans. That is, many human subjects can match a tone signal to frequency separating the ripple spectrum of the noise. An experiment was done with *Tursiops truncatus* in which the animal demonstrated an ability to discriminate rippled noise from nonrippled noise (Tarakanov, Pletenko & Supin, 1996). However, this demonstration merely showed that it might be possible for dolphins to perceive TSP. Further research is required to determine if this ability, which comes into play, for instance, when humans listen to an echolocation signal with an increasing repetition rate, is also relevant to a dolphin's perception of such a sequence.

Thomas: What strategies does a social group of odontocetes use when echolocating, to avoid masking each other's signals, and to use energy for echolocation most effectively?

Echolocating mammals—bats and dolphins—are often highly social. Bats emerge in large groups at night to feed, and some dolphins feed in a large group. If all individuals in the group echolocated at the same time, it is likely they would "jam" or mask each other's signals, and this would waste energy. Xitco and Roitblat (1996) demonstrated at Epcot Center that an "eavesdropping" bottlenose dolphin could listen to the outgoing signals and echoes of another dolphin actively involved in a target detection task and properly choose the correct match. Thus it is possible for one member of a social group to listen to outgoing signals and returning echoes from other animals and know the important information about the target: what it is, how far away, and whether it is stationary or moving toward or away from the listener. As a result, it is not necessary for all pod members to echolocate all the time for information transfer. The interesting question is, how do group members decide who echolocates and how often?

Vocalizations

Au: How are the acoustic signals emitted by the dolphin produced?

While we do know some things about this process (see Cranford, Amundin, & Krysl, chap. 2, this vol.), many other questions must still be answered. For example, what is the role of the various structures such as the air sacs, rostrum, and melon in the formation of the biosonar signal and beam? Current research seems to suggest that the biosonar beam is formed even before the signal reaches the melon as it travels outward from the phonic lips into the water, and the melon is not really a focusing structure but functions more as a collimator and as an acoustic impedance matching device. Recent research findings also suggest that biosonar clicks are produced mainly from the right phonic lip, and whistles are produced mainly by the left phonic lip, even in porpoise species that do not typically produce whistles (Madsen, Lammers, Wisniewska, & Beedholm, 2013). So one obvious question regarding porpoises is, why is the left phonic lip still present and has not evolved away? Finally, in 1978, we showed that the beam pattern of single biosonar clicks of a stationary dolphin can vary in width in both the vertical and the horizontal planes (Au, Floyd, & Haun, 1978). Recently, Danish scientists found the same phenomenon in free-swimming animals (Wahlberg et al., 2011). Now we must ask: How is the beam pattern being manipulated? Is the geometry of the different air sacs being manipulated? Is the melon geometry being manipulated? Are the sounds being manipulated at the level of the phonic lips?

Wells: What is the role of vocalizations in social organization?

With few exceptions, the activities of resident bottlenose dolphins occur within their community ranges (Wells, 2003). The animals live in a fission–fusion society, where social associates can change frequently as individuals move through the community range in relatively small groups (Wells et al., 1987; Connor, Wells, Mann, & Read, 2000). The community range near Sarasota, Florida, is an estuary, where murky waters largely preclude visual communication signals and create a situation where the use of acoustic signals is more effective. What role does acoustic communication play in bringing individuals together, coordinating or guiding their activities, and maintaining or ending associations? Over the decades, we have learned that (1) individually specific signature whistles can be identified for each resident dolphin, (2) they are formed within the first few months of life, (3) signature whistles serve as names for individuals, and (4) they are used as contact calls (Caldwell, Caldwell, & Tyack, 1990; Tyack, 1997; Janik, Sayigh, & Wells, 2006; Sayigh et al., 2007). Individuals produce a wide variety of whistles in addition to their signature whistles. What information is conveyed in the myriad other whistles produced by these animals?

Ridgway: What about those mysterious "burst pulse" sounds?

Studies of natural dolphin sound have focused on whistles and echolocation clicks. Burst pulse sounds deserve more study. There is much evidence in the literature for food- or feeding-associated calls from cetaceans. In the first group of dolphins at Marineland of Florida, Wood (1953) described mewing and rasping feeding sounds. After taking a fish, a dolphin emitted a "brush of pulses" (Norris, Prescott, Asa-Dorian, & Perkins, 1961). Borrowing terminology from the bat literature, researchers have sometimes called these sounds "feeding buzzes." Dolphins continue their buzzes even after taking a fish. During learning, the after part often moves forward, occurring immediately after the trainer's bridge signal promising future reward. This has been called a "victory squeal" (Branstetter et al., 2012; Ridgway et al., 2014). Such dolphin reward-associated calls deserve more study.

Development

Ridgway: What can we learn about fetal development?

Asymmetry of the skull, and of brain development, is a remarkable feature of odontocetes. Do different parts of the brain develop at different rates, and how does such development affect cognition? To what extent is phylogeny repeated in ontogeny? The ancestors of cetaceans had a large, wide cerebellum, larger than the cerebrum (Edinger,

1955). Does the dolphin embryo show the cerebellum leading in brain development? The dolphin embryo has olfactory bulbs. Sleptsov (1939) proposed that the right bulb regressed first, creating the asymmetry seen in the adult skull. We might learn about brain and skull development and the ontogeny of dolphin cognition from fetal ultrasound. Ultrasonography (Smith et al., 2013) is relatively low in cost, does not harm the fetus or mother, and can be used to study the physiology of the fetus in real time. Because of its use in human medicine, we can expect continuing advances in this technology.

Together with studies of development of the fetus, we would also like to know more about its capability for hearing. Ocean sound should transmit right through the animal's body, and the fetus should be able to hear at 25 or 28 weeks, as the human fetus can. Physiological responses of the fetus such as heart rate acceleration or bradycardia can reveal how hearing develops. Scientists might learn how sound from the mother and from the group is perceived and ultimately helps shape both perception and communication in these animals.

Ford: How do young killer whales learn their calls?

Vocal learning is a rare ability in mammals, but it appears to be common, if not ubiquitous, in dolphins and other cetaceans. Just as vocal learning allowed languages to evolve in humans, so it has been critical to the development of advanced communication systems in cetaceans. Signature whistles in dolphins, group-specific dialects in killer whales and sperm whales, and the constantly changing songs of male humpback whales have all arisen because of the ability of cetaceans to imitate and learn the sounds of conspecifics. Although studies of captive dolphins and killer whales have provided some insights, how and when vocal learning takes place in the wild is poorly known. Because young killer whales only acquire the repertoire of their mother and other close kin, they must learn selectively and ignore the calls of neighboring groups that are often within close proximity and highly audible. It is possible that some call learning takes place before birth, as the developing fetus would be exposed to the calls of its mother and close kin in utero because of the efficient propagation of sound from seawater into tissue. Because killer whales occasionally mimic the calls of unrelated groups, it is apparent that they know and are capable of producing calls outside their natal repertoire but normally do not do so. However, we do have evidence of horizontal transmission of structural variations in calls shared by closely related matrilines, suggesting that call learning occurs not strictly through vertical transmission from mother to offspring. Finally, we do not yet know if the acquisition of call repertoires depends completely on learning or whether some basic components are inherited and thus constrain the extent of vocal variation. While these questions are specific to killer whales, answers would surely contribute to our overall understanding of delphinid communication systems.

Whitehead: How much do dolphins teach?

When studying social learning, we usually focus on changes in the behavior of the learner, but the model animals, from whom the learners learn, can also affect the learning process. If they change their behavior in a way that is costly to themselves but improves the performance of the learner, then this can be called teaching (Caro & Hauser, 1992). Teaching is also linked to cognition, in that some concepts of teaching require theory of mind (Premack & Premack, 1996), as well as to communication, which is a vital element of the teaching process. However, from an evolutionary perspective, the strongest link may be with sociality, as teaching is a form of altruism (Thornton & Raihani, 2008). Thus teaching can be a fundamental element in the structuring of societies, and it is for modern humans. Teaching is extremely difficult to study in the wild, especially when experiments are impossible or difficult. Conducting experiments with wild dolphins is particularly tricky, and even in captivity there are many challenges. A key obstacle when testing hypotheses about teaching is measuring changes in the behavior of the teacher and learner, with the changes in the learner occurring over long timescales. But despite these formidable difficulties, we have some evidence for teaching by dolphins from nonexperimental observations (Bender et al., 2009). A thorough cross-species investigation of teaching by dolphins would be fascinating.

Cognition

Nachtigall: How are echolocation and cognition blended to solve real-world dolphin problems?

In studying echolocation and how outgoing clicks and hearing interact, we have found that changing hearing sensation levels depend a great deal on making sure that the echoes are received at a hearable and constant level (Nachtigall & Supin, 2008; Supin & Nachtigall, 2013). An "automatic gain control" to keep echo levels constant as echoes return from a fast-moving fish while the dolphin is also moving may take a great deal of neural effort. This may involve a beautiful neural system that might be thought of as cognitively simple. But the study of the interaction of hearing and clicks while echolocating also promises a fascinating look at how echolocation and higher levels of cognition, such as expectation, work. The animal may well have to concentrate and pay attention to particular targets while making extremely fine adjustments in a dynamic, rapid, and changing environment. The initial work on attention and echo-location with fixed targets by Penner (1988) showed how important attention is. If the animal expected the targets to be in one place, then the characteristics of the clicks and the performance of the animal differed substantially. The new work by Kloepper (2013) is similarly fascinating. The overall cognitive questions of the relationships between attention, echolocation, and perception are just beginning to be answered. The descriptive changes in outgoing signals, combined with the animal's movements, provide objective, quantifiable, fine-grained observations of a sensory system under cognitive control. If those data were then combined with an examination of the changes in hearing during echolocation, we could learn much about the acoustic functioning of the dolphin brain.

Herman: How can we better understand a dolphin's awareness of others and of self?
The proficiency of dolphins for imitating the motor behaviors and actions of other dolphins or of humans—either viewed live, or on a television screen, or observed though echolocation—implies physical and social awareness of others. For example, the work on imitating human motor behaviors (Herman, 2002) implies an awareness of the actions of others, as well as an awareness of the different body plan of humans and how to relate that plan to the dolphin's own body image. However, questions are still open about the awareness of intentions or beliefs of others, sometimes categorized as questions about a "theory of mind." Although some investigative effort has been made on this topic, further, well-controlled, and conceptually richer studies are needed for this challenging topic.

The study of self-awareness in animals, including dolphins, has primarily been limited to investigations of mirror self-recognition (MSR). But, as noted by Gordon Gallup, the developer of the MSR test, there is more to self-awareness that simply recognizing yourself in a mirror (see Gallup, 1983) . Herman (2012) summarized studies revealing that dolphins were consciously aware of their own recent behaviors, repeating them or not repeating them on request, and were also aware of their own body and body parts, understanding symbolic references to those parts and requests to use them in novel ways. It is important to expand on this idea that self-awareness is multifaceted, to continue to go beyond MSR, and to develop a richer understanding of the range and depth of self-awareness demonstrable in the dolphins.

Whitehead: How do dolphins learn socially?
Social learning, learning that is influenced by observation of, or interaction with, another animal or its products (Heyes, 1994), is the basis of culture. It comes in a

variety of forms, ranging from stimulus enhancement and social facilitation to imitation and emulation (Hoppitt & Laland, 2008). The form of social learning by which information is transferred from one animal to another is interesting in its own right, because of its links with aspects of cognition such as theory of mind, as well as through its effects on the characteristics of the resultant culture. For instance, imitation is particularly likely to lead to accumulating cultures in which the forms of behavior, as well as potentially their products, become progressively more complex. We know very little of the types of social learning used by dolphins—it is likely that they use different types in different circumstances—but it would be fascinating to find out.

Herman: Is there "episodic memory" in dolphins?

Episodic memory entails memory of the *what*, *where*, and *when* of one's own past behaviors or experiences and is sometimes referred to as "mental time travel" (Tulving, 1972; Suddendorf & Busby, 2009). While some scientists claim that episodic memory is unique to humans, Schwartz and Evans (2001) have tentatively suggested that great apes may be capable of demonstrating episodic memory, at least for the *what* and *where* of events, but they caution that "there have been no definitive tests to determine if they can retrieve the 'when' component of an event" (p. 81). Similarly, the dolphin's demonstrated ability to repeat its own behaviors, including successfully repeating actions taken to objects, appears to meet with the *what* and *where* components of episodic memory (Mercado et al., 1998, 1999; also see Herman, 2012). Studies that also address the *when* component would thus be important and valuable. There may be many approaches to this problem. One possibility might be to teach the dolphin the concepts of "first" and "last," a type of serial-order concept. With that concept well established, one might then, for example, ask the dolphin to place object A in bin 1 and, later, to place object B in bin 2 (for numerous examples of dolphins performing this type of relational task, see Herman, Richards, & Woltz, 1984). Then, using the trained concepts of first and last, one would ask the dolphin to fetch "first" or fetch "last." The duration of time between caching A and caching B would be a variable, as would the interval after B is cached. This time variation would help to address questions of where information is stored: long-term memory, short-term memory, or both. Long-term memory storage would eliminate alternative interpretations of successful behavior based on residual or lingering physiological effects.

Problem Solving in the Wild

Au: What acoustic skills might be required for foraging at sea?

Several species of odontocetes—specifically pilot whales, Risso's dolphins, sperm whales, and all species of beaked whales—forage at depths that vary from about 400 to more than 1,000 meters. Stomach content analyses of these deep-diving odontocetes indicate that various species of squids are the major prey items, along with a variety of demersal fish species as secondary prey items. The question is, how do these odontocetes know that appropriate prey items are present? This question must take into account that foraging dives might last between one-half and one hour, and a considerable amount of energy is expended diving to deep depths. Therefore each foraging dive must be successful, not with just one prey item but with many prey items. The prey items are probably too deep to be sensed from the surface by echolocating odontocetes, since the target strength of squids is very low, and the prey is relatively far away. In the Sea of Cortez, Humboldt squid migrate with the mesopelagic layer toward the surface at night, but sperm whales still feed on the deep ones. There are also preferred foraging locations that change seasonally, thus complicating the whole food-gathering process.

Thomas: In areas of the ocean where there is a warm upper mixed layer, do marine mammals intentionally swim into the upper layer to listen or to communicate over long distances and position themselves below the mixed layer to acoustically hide?

Propagation of marine mammal sounds in water can vary greatly by region of the world. In polar areas, the water column is isothermal, and there is little variation in water temperature from the surface to the bottom. In contrast, equatorial areas often have an upper mixed layer of warm water lying above deeper, cooler water. The sound propagation in the upper mixed layer can be cylindrical, and acoustic signals will travel over longer distances than in deeper water, where spherical spreading is present. The water's surface is a reflective boundary that causes sound to bend downward, and the layer between the upper mixed layer and deeper, cooler water is reflective, causing sound to bounce upward. This creates a ducting effect of sound within the upper mixed layer, and sounds propagate further there.

These features were demonstrated to me empirically during a couple of survey studies on cetacean sightings and acoustics in the eastern tropical Pacific using a towed array of hydrophones. During these studies, my colleagues and I experienced three conditions: (a) a single delphinid or large herd of delphinids would be next to the survey ship, and their underwater whistles and clicks were clearly heard; (b) delphinids were clearly visible near the survey ship, but no whistles or clicks were detected by the towed array; and (c) delphinid whistles and clicks were clearly detected by the towed array, but no dolphins were in sight. After thorough tests of the hydrophone array and recording system verified that the equipment was operating properly, we believed something else was going on, so we started to collect CTD (conductivity, temperature, and depth) measurements at survey sights and then varied the depth of the towed array to be within the upper duct or below the upper duct. If the array was within the upper mixed layer, then sounds were readily detected, even when dolphins were out of visual range. If the array was below the duct, then dolphin sounds were either not detected or highly attenuated, even when they were obviously seen nearby the ship.

These observations led us to wonder if delphinids actively use this upper mixed layer to their advantage. A delphinid wanting to communicate over a long distance could swim within the duct. In contrast, a delphinid wanting to acoustically hide, for instance, from a predator, could swim below the duct and perhaps once in a while pop up into the layer to listen for possible prey.

Wells: How are community ranges established and maintained?

The apparently continuous distribution of bottlenose dolphins along coastlines and through bays, sounds, and estuaries is deceptive with regard to social structure. Research on ranging and social association patterns, genetics, and stable isotopes has demonstrated that considerable structuring exists, with many of these dolphins living in multidecadal, multigenerational communities with slightly overlapping boundaries and well-established patterns of social interactions (Wells et al., 1987; Duffield & Wells, 2002; Sellas, Wells, & Rosel, 2005; Urian, Hofmann, Wells, & Read, 2009; Barros et al., 2010; Wells, 2014). How do these dolphins perceive the limits of their community ranges? What factors define these limits? What role does dolphin communication play in establishing or maintaining these communities? As climate disruption proceeds, what effects will physiological and environmental stresses have on community structure—at what point will social barriers dissolve or be reshaped to accommodate the possible need for shifts in ranges or habitat use (Wells, 2010)?

Ford: What are the origin and function of "acoustic clans" in killer whales?

How different clans of fish-eating "resident" killer whales came to exist sympatrically and be part of the same social network has long been a mystery. Although pods belonging to a clan share much of their call repertoire, pods in different clans have no calls in common and thus sound quite distinct even to an untrained human ear. Pods from different clans commonly mix and travel together, with no clear relationship between clan affiliation and association patterns. Clans have been described in fish-eating killer whale populations in British Columbia, southern Alaska (Yurk et al., 2002), and

Kamchatka, Russia (Filatova et al., 2007). Although it is likely that clan members have dialects that are culturally inherited and reflect their common heritage, since no calls are shared between clans in the same population, their origin is uncertain. Genetic studies have revealed some differences in microsatellite DNA between clans, but no significant differences in mitochondrial DNA haplotypes, which would be expected if clans represent distantly related maternal lineages. If clans do have a recent common ancestor, their unique call repertoires are unlikely to have developed through a process of gradual change, as intermediate forms of at least some calls would be expected. It is possible that dialects can evolve through a process of punctuated equilibrium, with periods of relative stability interrupted by periods of rapid and major repertoire change. More than forty years of acoustic monitoring have so far not revealed anything but minor changes in dialects. Alternatively, perhaps the unique call repertoires of clans arose in isolation and came together in the same region through independent founding events in the past. Whatever their origin, clan members are remarkably faithful to their traditional call repertoires and resistant to cultural diffusion from associating groups with different dialects. Hopefully, future research will shed new light on how and why these unusual systems of dialects have evolved.

Whitehead: What is the scope of dolphin culture?

How much of dolphin behavior is cultural? We have convincing data for a few wild dolphin behaviors being cultural: sponging in Shark Bay (Krützen et al., 2005) and tail walking (Whitehead & Rendell, 2014). But inferring culture in the behavior of wild dolphins is difficult. We need to develop better methodologies for studying social learning and culture (see Hoppitt & Laland, 2013), and then to apply them to a wide range of behavior in a wide range of dolphin species. The behaviors studied should go beyond foraging, communication, and play, the three areas where we now know a little. We should also study the potential for culture in very fine-scale behavior, such as patterns of body movement, as well as in structuring overarching ways of life—conservative cultures, explorative cultures, pacific cultures, and so on. These studies should go well beyond the bottlenose dolphin, focusing also on oceanic species, and those with unusual lifestyles, such as Risso's dolphins and rough-toothed dolphins.

Conservation and Management

Wells: How do human activities affect dolphin life?

Anthropogenic noise is believed to adversely impact dolphins in a variety of settings around the world. In Sarasota Bay, Florida, the resident bottlenose dolphins have been exposed to loud noises of human origin for about 100 years, since the introduction of engine-powered vessels to the area. Since then they have had to contend with increasing numbers of power boats (tens of thousands registered within the community range), along with noise from dredge operations and marine construction. To what extent do human activities in these coastal waters interfere with effective dolphin communication, and what are the impacts of this interference? We have learned that they make short-term changes to their dive patterns, group spread, and whistle communications when power boats approach (Nowacek, Wells, & Solow, 2001; Buckstaff, 2004), and to their habitat use during major marine construction projects (Buckstaff, Wells, Gannon, & Nowacek, 2013). To what extent are these short-term changes occurring in response to the need for the animals to communicate in spite of the disturbance? What are the long-term impacts on health, survival, and reproductive success of repeated short-term responses to disturbances that interfere with communication? Is there a threshold of exposure level, duration, or frequency of anthropogenic noise and disturbance beyond which the animals will vacate their long-term community range?

Thomas: Are there acoustic features about sounds produced by marine mammals that could indicate stress or a health problem?
The acoustic features of laboratory mice and rat vocalizations are so well understood that researchers in the biomedical and pharmaceutical fields use subtle changes in the types of vocalizations used and acoustic properties of their sounds to determine whether an animal is stressed, positively stimulated (as in testing addictive drugs), or suffering from a health problem. Sam Ridgway (1983) and Moore and Ridgway (1996) have published some research on this topic using captive animals. In addition, one of my graduate students has examined this topic. Sara (Crowell) Therrien et al. (2012) recorded underwater sounds from a bottlenose dolphin at Brookfield Zoo during a baseline period and opportunistically before it died. The contour of the animal's whistle structure changed during its last few days of life. Given that marine mammals encounter a variety of potential anthropogenic stressors, it would be useful to study and determine whether the types of sounds used by a marine mammal or specific features of their sounds could be used as an indication of stress, suffering from a health problem, or even recovering from a health problem.

Useful Technologies

Ridgway: How might contemporary technology be used to study dolphin brains?
I have some experience with two potential windows on the dolphin mind. These are imaging technology (like MRIs) and dolphin event-related potentials (ERP). Brain waves

recorded from the scalp reveal an ERP elicited during decisions (Squires, Squires, & Hillyard, 1975). Sometimes called the "aha" response in human studies, dolphins also showed an event-related brain potential (Woods, Ridgway, Carder & Bullock, 1986). ERPs offer particularly fertile areas for scientific discovery on cetacean cognition, since the technique could be applied to even large species such as stranded sperm or beaked whales. Dolphins have a multitiered cerebral cortex with an extensive extra lobe, the paralimbic lobe. This lobe does not occur in mammals other than cetaceans. Because it is inaccessible, there are no functional studies of the paralimbic cortex. Imaging offers an opportunity to study activity in this unique part of the dolphin brain. Such studies might help reveal functions of the paralimbic cortex, insular cortex, basal ganglia, cerebellum, and other areas (Ridgway, 1990; Ridgway et al., 2006). We can compare all these windows between humans and other mammals. Such comparisons are in short supply in dolphin cognition studies.

Nachtigall: How might technology help us to determine who is making what calls?
Peter Tyack (1993) started by attempting to put "voca-lights" on dolphins that would illuminate when they made individual whistles, so that he could study social behavior in a group of bottlenose dolphins. His idea was to identify who was whistling when the whistle occurred; it was the beginning of the ideas for the development of DTAGs. We now have DTAGs, and they are used highly effectively for measuring behavior, diving, and the effects of sounds on wild, free-swimming animals. Perhaps it would be valuable to take a step back to a more fine-grained examination of social behavior and whistling, where investigators examine which animal makes which whistle and how they interact. Each animal in the group could now be tagged. It seems the technology now exists, by using tags, or even DTAG-like tags, without all the bells and whistles. Great technology has been developed. It could now be used to examine basic social behavior and sound production. The studies could be started in clear-water laboratory situations and advance to the open sea with wild animal groups. Gathering evidence for studying the bases of cognition during social interactions via whistle analysis seems, finally, to be achievable.

Herzing and Johnson: As evidenced by the preceding chapters, our authors, too, have insights into possible futures for the field. To aid in a quick review, we have summarized the main questions on the minds of the book's contributors in table 11.1. One might think of this, in combination with the questions in this chapter, as a guide to the future for the graduate student or active researcher. We look forward to many of *you* exploring this vast array of unanswered areas in dolphin communication and cognition.

Table 11.1
Future areas of inquiry suggested by the chapter authors.

Chapter	First Area of Inquiry	Second Area of Inquiry	Third Area of Inquiry
Chapter 1: Brain Lori Marino	Understand brain evolution better and in a comparative context	Incorporate the cognitive complexity of species when applying conservation measures	Develop a new ethic of respect for coexistence with humans, including a change in human behavior toward cetaceans
Chapter 2: Vocal Anatomy Ted W. Cranford, Mats Amundin, and Petr Krysl	Use computational tools to understand bioacoustic pathways	Figure out the directionality of sound reception, according to frequency, in the dolphin head	Study sound production and reception with nonstationary dolphins in their natural habitats
Chapter 3: Sensory Systems Wolf Hanke and Nicola Erdsack	Study how dolphin senses aid in navigation, object recognition, and general perception of their world	Assess the presence of other senses, including magnetic and electric	Explore dolphin sensory systems in the context of a model for their unique cognition and psychology
Chapter 4: Social Ecology Bernd Würsig and Heidi C. Pearson	Continue basic collection of observations in the wild, on the surface and underwater	Develop underwater sound localization techniques to determine details of communication	Combine aerial observation techniques using new technology with sound localization techniques to obtain the big picture of communication
Chapter 5: Vocalizations Marc O. Lammers and Julie N. Oswald	Match behavioral observations with sound data using advanced technology	Assess perceptual abilities of dolphins, specifically the function and perception of burst pulse sounds during communication	Expand communication studies to include ecological variables for offshore as well as coastal dolphins to determine communication constraints of environments and social structure variation
Chapter 6: Multimodal Communication Denise L. Herzing	Develop advanced technology to localize vocalizer and correlate with sounds	Use keyboards and cognitive interfaces to enhance our understanding of communication	Use pattern recognition software to determine fundamental units of information or potential structure
Chapter 7: Cognition Adam A. Pack	Explore long-term memory issues, especially around past actions and relationships with conspecifics	Explore goal-directed imitation abilities, especially those beyond motor imitation	Explore dolphin's inferential reasoning of social knowledge

Table 11.1 (continued)

Chapter	First Area of Inquiry	Second Area of Inquiry	Third Area of Inquiry
Chapter 8: Development Stan A. Kuczaj II and Kelley A. Winship	Study when young calves develop their understanding of meaning, friendship, conspecific recognition, personality, and synchrony	Apply both observational and experimental techniques, used in human infant studies, to the study of calf behavior	Study calf behavior in the wild relative to social behavior, play, and communication
Chapter 9: Social Cognition Christine M. Johnson	Apply new cognitive frameworks, including cognitive ecology to dolphin research	Use multimodal signal recording to monitor larger dynamics	Enhance data collection with new technology on micro and macro scales
Chapter 10: Ethics Thomas I. White	Apply ethical norms and standards to researchers and research protocols	Do more research in the wild to determine what it takes for dolphins to flourish both as individuals and in groups	Reduce harm to individuals and cetacean cultures caused by whaling, the military, etc., and work toward a sanctuary for nonreleasable dolphins

References

Au, W. W. L. (1993). *The sonar of dolphins*. New York: Springer.

Au, W. W. L., & Benoit-Bird, K. (2003). Automatic gain control in the echolocation system of dolphin. *Nature, 423*, 861–863.

Au, W. W. L., Floyd, R. W., & Haun, J. E. (1978). Propagation of dolphin echolocation signals. *Journal of the Acoustical Society of America, 64*, 411–412.

Au, W. W. L., Ford, J. K. B., Horne, J. K., & Newman-Allman, K. A. (2004). Echolocation signals of free-ranging killer whales (*Orcinus orca*) and modeling of foraging for chinook salmon (*Oncorhynchus tshawytscha*). *Journal of the Acoustical Society of America, 56*, 1280–1290.

Au, W. W. L., & Hastings, M. C. (2008). *Principles of marine bioacoustics*. New York: Springer.

Au, W. W. L., Moore, P. W. B., & Pawloski, D. A. (1988). Detection of complex echoes in noise by an echolocating dolphin. *Journal of the Acoustical Society of America, 83*, 662–668.

Au, W. W. L., & Nachtigall, P. E. (1995). Artificial neural network modeling of dolphin echolocation. In R. A. Kastelein, J. A. Thomas, & P. E. Nachtigall (Eds.), *Sensory systems of aquatic mammals* (pp. 183–201). London: Plenum Press.

Au, W. W. L., & Ou, H. (2014). Temporal signal processing of dolphin biosonar echoes from salmon prey. *Journal of the Acoustical Society America, Express Letter, 136*, EL67–71.

Au, W. W. L., & Pawloski, D. (1992). Cylinder wall thickness difference discrimination by an echolocating Atlantic bottlenose dolphin. *Journal of Comparative Physiology A, 172*, 41–47.

Baker, C. S., & Herman, L. M. (1984). Aggressive behavior between humpback whales (*Megaptera novaeangliae*) wintering in Hawaiian waters. *Canadian Journal of Zoology, 62*, 1922–1937.

Barros, N. B., Ostrom, P., Stricker, C., & Wells, R. S. (2010). Stable isotopes differentiate bottlenose dolphins off west central Florida. *Marine Mammal Science, 26*, 324–336.

Bender, C., Herzing, D., & Bjorklund, D. (2009). Evidence of teaching in Atlantic spotted dolphins (*Stenella frontalis*) by mother dolphins foraging in the presence of their calves. *Animal Cognition, 12*, 43–53.

Branstetter, B. K., Moore, P. W., Finneran, J. J., Tormey, M. N., & Aihara, H. (2012). Directional properties of bottlenose dolphin (*Tursiops truncatus*) clicks, burst-pulse, and whistle sounds. *Journal of the Acoustical Society of America, 131*, 1613–1621.

Buckstaff, K. C. (2004). Effects of watercraft noise on the acoustic behavior of bottlenose dolphins, *Tursiops truncatus*, in Sarasota Bay, Florida. *Marine Mammal Science, 20*, 709–725.

Buckstaff, K. C., Wells, R. S., Gannon, J. G., & Nowacek, D. P. (2013). Responses of bottlenose dolphins (*Tursiops truncatus*) to construction and demolition of coastal marine structures. *Aquatic Mammals, 39*, 174–186.

Caldwell, M. C., Caldwell, D. K., & Tyack, P. L. (1990). Review of the signature-whistle hypothesis for the Atlantic bottlenose dolphin, *Tursiops truncatus*. In S. Leatherwood & R. R. Reeves (Eds.), *The bottlenose dolphin* (pp. 199–234). San Diego: Academic Press.

Caro, T. M., & Hauser, M. D. (1992). Is there teaching in non-human animals? *Quarterly Review of Biology, 67*, 151–174.

Connor, R. C., Wells, R. S., Mann, J., & Read, A. J. (2000). The bottlenose dolphin, *Tursiops* spp.: Social relationships in a fission–fusion society. In J. Mann, R. C. Connor, P. L. Tyack, & H. Whitehead (Eds.), *Cetacean societies: Field studies of dolphins and whales* (pp. 91–126). Chicago: University of Chicago Press.

Craig, A. S., & Herman, L. M. (1997). Sex differences in site fidelity and migration of humpback whales (*Megaptera novaeangliae*) to the Hawaiian Islands. *Canadian Journal of Zoology, 75*, 1923–1933.

Deecke, V. B., Barrett-Lennard, L. G., Spong, P., & Ford, J. K. B. (2010). The structure of stereotyped calls reflects kinship and social affiliation in resident killer whales (*Orcinus orca*). *Naturwissenschaften, 97*, 513–518.

Deecke, V. B., Ford, J. K. B., & Spong, P. (2000). Dialect change in resident killer whales (*Orcinus orca*): Implications for vocal learning and cultural transmission. *Animal Behaviour, 60*, 629–638.

Duffield, D. A., & Wells, R. S. (2002). The molecular profile of a resident community of bottlenose dolphins, *Tursiops truncatus*. In C. J. Pfeiffer (Ed.), *Molecular and cell biology of marine mammals* (pp. 3–11). Melbourne, FL: Krieger.

Edinger, T. (1955). Hearing and smell in cetacean history. *Mschr Psychiatry Neurology, 129,* 37–58.

Filatova, O. A., Fedutin, I. D., Burdin, A. M., & Hoyt, E. (2007). The structure of the discrete call repertoire of killer whales *Orcinus orca* from southeast Kamchatka. *Bioacoustics, 16,* 261–280.

Ford, J. K. B. (1989). Acoustic behaviour of resident killer whales (*Orcinus orca*) off Vancouver Island, British Columbia. *Canadian Journal of Zoology, 67,* 727–745.

Ford, J. K. B. (1991). Vocal traditions among resident killer whales (*Orcinus orca*) in coastal waters of British Columbia, Canada. *Canadian Journal of Zoology, 69,* 1454–1483.

Gallup, G. G., Jr. (1983). Toward a comparative psychology of mind. In R. L. Mellgren (Ed.), *Animal cognition and behaviour* (pp. 473–510). Amsterdam: North-Holland.

Harley, H. E., Putman, E. A., & Roitblat, H. L. (2003). Bottlenose dolphins perceive object features through echolocation. *Nature, 424,* 667–669.

Herman, L. M. (Ed.). (1980). *Cetacean behavior: Mechanisms and functions.* New York: Wiley Interscience.

Herman, L. M. (2002). Vocal, social, and self-imitation by bottlenosed dolphins. In C. Nehaniv & K. Dautenhahn (Eds.), *Imitation in animals and artifacts* (pp. 63–108). Cambridge, MA: MIT Press.

Herman, L. M. (2006). Intelligence and rational behaviour in the bottlenosed dolphin. In S. Hurley & M. Nudds (Eds.), *Rational animals?* (pp. 439–467). Oxford: Oxford University Press.

Herman, L. M. (2012). Body and self in dolphins. *Consciousness and Cognition, 21,* 526–545.

Herman, L. M., & Antinoja, R. C. (1977). Humpback whales in the Hawaiian breeding waters: Population and pod characteristics. *Scientific Reports of the Whales Research Institute (Tokyo), 29,* 59–85.

Herman, L. M., Pack, A. A., Rose, K., Craig, A., Herman, E. Y. K., & Milette, A. (2011). Resightings of humpback whales in Hawaiian waters over spans of 10–32 years: Site fidelity, sex ratios, calving rates, female demographics, and the dynamics of social and behavioral roles of individuals. *Marine Mammal Science, 27,* 736–768.

Herman, L. M., Pack, A. A., Spitz, S. S., Herman, E. Y. K., Rose, K., Hakala, S., et al. (2013). Humpback whale song: Who sings? *Behavioral Ecology and Sociobiology, 67,* 1653–1663.

Herman, L. M., Richards, D. G., & Wolz, J. P. (1984). Comprehension of sentences by the bottlenosed dolphin. *Cognition, 16,* 129–219.

Heyes, C. M. (1994). Social learning in animals: Categories and mechanisms. *Biological Reviews of the Cambridge Philosophical Society, 69,* 207–231.

Hoppitt, W., & Laland, K. N. (2008). Social processes influencing learning in animals: A review of the evidence. *Advances in the Study of Behavior, 38,* 105–165.

Hoppitt, W. H., & Laland, K. N. (2013). *Social learning: An introduction to mechanisms, methods, and models*. Princeton: Princeton University Press.

Janik, V., Sayigh, L. S., & Wells, R. S. (2006). Signature whistle shape conveys identity information to bottlenose dolphins. *Proceedings of the National Academy of Sciences of the United States of America, 103*, 8293–8297.

Johnson, C. M., & Herzing, D. L. (2006). Primate, cetacean, and pinniped cognition compared: An introduction. *Aquatic Mammals, 32*, 409–412.

Kloepper, L. N. (2013). Using array technology to understand dynamics of echolocation in bats and toothed whales. *Journal of Post Doctoral Research, 1*(4): 1–8.

Krützen, M., Mann, J., Heithaus, M. R., Connor, R. C., Bejder, L., & Sherwin, W. B. (2005). Cultural transmission of tool use in bottlenose dolphins. *Proceedings of the National Academy of Sciences of the United States of America, 102*, 8939–8943.

Loeding, E., Thomas, J. A., Bernier, D., & Santymire, R. (2011). Using fecal hormonal and behavioral analysis to evaluate the introduction of two sable antelope at Lincoln Park Zoo. *Journal of Applied Animal Welfare Science, 14*, 220–246.

Madsen, C. J., & Herman, L. M. (1980). Social and ecological correlates of cetacean vision and visual appearance. In L. M. Herman (Ed.), *Cetacean behavior: Mechanisms and functions* (pp. 101–147). New York: Wiley Interscience.

Madsen, P. T., Lammers, M., Wisniewska, D., & Beedholm, K. (2013). Nasal sound production in echolocating delphinids (*Tursiops truncatus* and *Pseudorca crassidens*) is dynamic, but unilateral: Clicking on the right side and whistling on the left side. *Journal of Experimental Biology, 216*, 4091–4102.

Mercado, E., Murray, S. O., Uyeyama, W. K., Pack, A. A., & Herman, L. M. (1998). Memory for recent actions in the bottlenose dolphin (*Tursiops truncatus*): Repetition of arbitrary behaviours using an abstract rule. *Animal Learning and Behavior, 26*, 210–218.

Mercado, E., Uyeyama, W. K., Pack, A. A., & Herman, L. M. (1999). Memory for action events in the bottlenose dolphin. *Animal Cognition, 2*, 17–25.

Mobley, J. R., Jr., Herman, L. M., & Frankel, A. S. (1988). Responses of wintering humpback whales (*Megaptera novaeangliae*) to playback of recordings of winter and summer vocalizations and synthetic sound. *Behavioral Ecology and Sociobiology, 23*, 211–223.

Moore, S. E., & Ridgway, S. H. (1996). Patterns of dolphin sound production and ovulation. *Aquatic Mammals, 22*(3), 175—4.

Munoz, G., Karamanlidis, K., Dendrinos, P., & Thomas, J. A. (2011). Aerial vocalizations by wild and rehabilitating Mediterranean monk seals (*Monachus monachus*) in Greece. *Aquatic Mammals, 37*(3), 262–279.

Nachtigall, P. E. (2008). Recent directions in odontocete cetacean hearing. *Bioacoustics, 17*, 82–85.

Nachtigall, P. E., Lien, J., Au, W. W. L., & Read, A. J. (1995). *Harbour porpoises: Laboratory studies to reduce bycatch*. Woerden, the Netherlands: De Spil.

Nachtigall, P. E., Mooney, T. A., Taylor, K. A., & Yuen, M. L. (2007). Hearing and auditory evoked potential methods applied to odontocete cetaceans. *Aquatic Mammals, 33*(1), 6–13.

Nachtigall, P. E., & Moore, P. W. B. (1988). *Animal sonar: Processes and performance*. New York: Plenum Press.

Nachtigall, P. E., Murchison, A. E., & Au, W. (1978). Discrimination of solid cylinders and cubes by a blindfolded echolocating bottlenose dolphin (*Tursiops truncatus*). *Journal of the Acoustical Society of America, 64*, S87.

Nachtigall, P. E., & Supin, A. Y. (2008). A false killer whale adjusts its hearing when it echolocates. *Journal of Experimental Biology, 211*, 1714–1718.

Norris, K. S., Prescott, J. H., Asa-Dorian, P. V., & Perkins, P. (1961). An experimental demonstration of echolocation behavior in the porpoise *Tursiops truncatus* (Montagu). *Biological Bulletin, 120*, 163–176.

Nowacek, S. M., Wells, R. S., & Solow, A. R. (2001). Short-term effects of boat traffic on bottlenose dolphins, *Tursiops truncatus*, in Sarasota Bay, Florida. *Marine Mammal Science, 17*, 673–688.

Pack, A. A., & Herman, L. M. (2006). Dolphin social cognition and joint attention: Our current understanding. *Aquatic Mammals, 32*, 443–460.

Pack, A. A., Herman, L. M., & Hoffmann-Kuhnt, M. (2004). Dolphin echolocation shape perception: From sound to object. In J. Thomas, C. Moss, & M. Vater (Eds.), *Echolocation in bats and dolphins* (pp. 288–308). Chicago: University of Chicago Press.

Paradis, A. L., Cornilleau-Peres, V., Droulez, J., Van de Moortele, P. F., Berthoz, A., Le Bihan, D., et al. (2000). Visual perception of motion and 3-D structure from motion: An fMRI study. *Cerebral Cortex, 10*, 772–783.

Penner, R. H. (1988). Attention and detection in dolphin echolocation. In P. E. Nachtigall & P. W. B. Moore Animal (Eds.), *Sonar: Processes and performance* (pp. 707–713). New York: Plenum.

Premack, D., & Premack, A. J. (1996). Why animals lack pedagogy and some cultures have more of it than others. In D. R. Olson & N. Torrance (Eds.), *Handbook of education and human development: New models of learning, teaching, and schooling* (pp. 302–323). Oxford: Blackwell.

Ralston, J. V., & Herman, L. M. (1989). Dolphin auditory perception. In J. R. Dooling & S. H. Hulse (Eds.), *The comparative psychology of audition: Perceiving complex sounds* (pp. 295–328). Hillsdale, NJ: Erlbaum.

Renaud, D. L., & Popper, A. N. (1975). Sound localization by the bottlenose porpoise (*Tursiops truncatus*). *Journal of Experimental Biology, 63*, 569–585.

Rendell, L., & Whitehead, H. (2001). Culture in whales and dolphins. *Behavioral and Brain Sciences, 24*, 309–382.

Ridgway, S. H. (1983). Dolphin hearing and sound production in health and illness. In R. R. Fay and G. Gourevitch (Eds.), *Hearing and other senses: Presentation in honor of E. G. Weller* (pp. 223–254). Groton, CT: Amphora Press.

Ridgway, S. H. (1987). *The dolphin doctor: A pioneering veterinarian remembers the extra-ordinary dolphin that inspired his career.* New York: Ballentine Books.

Ridgway, S. H. (1990). The central nervous system of the bottlenose dolphin. In S. Leatherwood & R. Reeves (Eds.), *The bottlenose dolphin* (pp. 69–97). New York: Academic Press.

Ridgway, S. H. (2011). Neural time and movement time in choice of whistle or pulse burst responses to different auditory stimuli by dolphins. *Journal of the Acoustical Society of America, 129,* 1073–1080.

Ridgway, S. H., Houser, D., Finneran, J., Carder, D., Keogh, M., Van Bonn, W., et al. (2006). Functional imaging of dolphin brain metabolism and blood flow. *Journal of Experimental Biology, 209,* 2902–2910.

Ridgway, S. H., Keogh, M., Carder, D., Finneran, J., Kamolnick, T., Todd, M., et al. (2009). Dolphins maintain cognitive performance during 72 to 120 hours of continuous auditory vigilance. *Journal of Experimental Biology, 212,* 1519–1527.

Ridgway, S. H., Carder, D. A., & Romano, T. A. (1991). The victory squeal of dolphins and white whales on the surface and at 100 m or more in depth. *Journal of the Acoustical Society of America, 90*(4), 2335.

Ridgway, S., & Hanson, A. (2014). Sperm whales and killer whales with the largest brains of all toothed whales show extreme differences in cerebellum. *Brain, Behavior and Evolution, 83(4),* 1–9. doi:10.1159/000360519.

Ridgway, S. H., & Howard, R. (1979). Dolphin lung collapse and intramuscular circulation during free diving: Evidence from nitrogen washout. *Science, 206,* 1182–1183.

Ridgway, S. H., & Johnston, D. G. (1966). Blood oxygen and ecology of porpoises of three genera. *Science, 151*(3709), 456–458.

Ridgway, S. H., & McCormick, J. G. (1967). Anesthetization of porpoises for major surgery. *Science, 158*(3800), 510–512.

Ridgway, S., Moore, P. W., Carder, D. A., & Romano, T. A. (2014). Forward shift of feeding buzz components of dolphins and belugas during associative learning reveals a likely connection to reward expectation, pleasure, and brain dopamine activation. *Journal of Experimental Biology, 217,* 2910–2919.

Ridgway, S. H., Scronce, B. L., & Kanwisher, J. (1969). Respiration and deep diving in the bottlenose porpoise. *Science, 166,* 1651–1654.

Riesch, R., Ford, J. K. B., & Thomsen, F. (2006). Stability and group specificity of stereotyped whistles in resident killer whales, *Orcinus orca,* off British Columbia. *Animal Behaviour, 71,* 79–91.

Roitblat, H. L., Herman, L. M., & Nachtigall, P. E. (1993). *Language and communication: Comparative perspectives*. Hillsdale, NJ: Erlbaum.

Roitblat, H. L., Penner, R. A., & Nachtigall, P. E. (1991). Attention and decision making in echolocation matching-to-sample by a bottlenose dolphin (*Tursiops truncatus*): The microstructure of decision making. In J. A. Thomas & R. A. Kastelein (Eds.), *Sensory abilities of cetaceans: Laboratory and field evidence* (pp. 665–667). London: Plenum Press.

Sayigh, L. S., Esch, H. C., Wells, R. S., & Janik, V. M. (2007). Facts about signature whistles of bottlenose dolphins (*Tursiops truncatus*). *Animal Behaviour, 74*, 1631–1642.

Schwacke, L. H., Zolman, E. S., Balmer, B. C., De Guise, S., George, R. C., Hoquet, J., ... , & Rowles, T. K. (2011). Anemia, hypothyroidism, and immune suppression associated with polychlorinated biphenyl exposure in bottlenose dolphins (*Tursiops truncatus*). *Proceedings of the Royal Society B: Biological Sciences, 279*, 48–57.

Schwacke, L. H., Smith, C. R., Townsend, F. I., Wells, R. S., Hart, L. B., Balmer, B. C., et al. (2013). Health of common bottlenose dolphins (*Tursiops truncatus*) in the Gulf of Mexico following the *Deepwater Horizon* oil spill. *Environmental Science and Technology, 48*, 93–103.

Schwartz, B. L., & Evans, S. (2001). Episodic memory in primates. *American Journal of Primatology, 55*(2), 71–85.

Sellas, A. B., Wells, R. S., & Rosel, P. E. (2005). Mitochondrial and nuclear DNA analyses reveal fine scale geographic structure in bottlenose dolphins (*Tursiops truncatus*) in the Gulf of Mexico. *Conservation Genetics, 6*, 715–728.

Sleptsov, M. M. (1939). On the problem of the asymmetry of the skull in *odontoceti*. *Zoologicheskyt Zhurnal, 18*, 367–386.

Smith, C. R., Jensen, E. D., Blankenship, B. A., Greenberg, M., D'Agostini, D. A., Pretorius, D. H., et al. (2013). Fetal omphalocele in a common bottlenose dolphin (*Tursiops truncatus*). *Journal of Zoo and Wildlife Medicine, 44*, 87–92.

Squires, N. K., Squires, K. C., & Hillyard, S. A. (1975). Two varieties of long-latency positive waves evoked by unpredictable auditory stimuli in man. *Electroencephalography and Clinical Neurophysiology, 38*, 387–401.

Stansbury, A. L., Thomas, J. A., Stalf, C. E., Murphy, L. D., Lombardi, D., Carpenter, J., et al. (2014). Behavioural audiogram of two arctic foxes (*Vulpes lagopus*). *Polar Biology*. doi:10.1007/s00300-014-1446-5.

Stirling, I., & Thomas, J. A. (2003). Relationships between underwater vocalizations and mating systems in phocid seals. *Aquatic Mammals, 29*(2), 227–246.

Suddendorf, T., & Busby, J. (2009). Making decisions with the future in mind: Developmental and comparative identification of mental time travel. *Learning and Motivation, 36*, 110–125.

Supin, A. Y., Nachtigall, P. E., Pawloski, J. L., & Au, W. W. L. (2003). Evoked potential recording during echolocation in a false killer whale (*Pseudorca crassidens*). *Journal of the Acoustical Society of America, 113*(5), 2408–2411.

Supin, A. Ya, & Nachtigall, P. E. (2013). Gain control in the sonar of odontocetes. *Journal of Comparative Physiology, A, 199* (6), 471–478.

Tarakanov, M. B., Pletenko, M. G., & Supin, A. Y. (1996). Frequency resolving power of the dolphin's hearing measured by rippled noise. *Aquatic Mammals, 22* (3), 141–152.

Taylor, K. A. (2013). *Directional hearing and a head-related transfer function (HRTF) of a bottlenose dolphin (Tursiops truncatus)*. PhD thesis, University of Hawaii at Manoa.

Therrien, S. E., Thomas, J. A., Therrien, R. E., & Stacey, R. (2012). Time of day and social change effects on underwater sound production by bottlenose dolphins (*Tursiops truncatus*) at the Brookfield Zoo. *Aquatic Mammals, 38*(1), 65–75. doi:10.1578/AM.38.1.2012.65.

Thomas, J. A. (2014). Editing *Aquatic Mammals* (1999–2009): The times, they were a changing. *Aquatic Mammals, 40*(1), 3.

Thomas, J. A., & Kastelein, R. A. (1990). *Sensory abilities of cetaceans: Laboratory and field evidence*. New York: Plenum Press.

Thomas, J. A., & Kuechle, V. (1982). Quantitative analysis of the underwater repertoire of the Weddell seal (*Leptonychotes weddellii*). *Journal of the Acoustical Society of America, 72*, 1730–1738.

Thomas, J. A., Moss, C. F., & Vater, M. (2004). *Echolocation in bats and dolphins*. Chicago: University of Chicago Press.

Thornton, A., & Raihani, N. J. (2008). The evolution of teaching. *Animal Behaviour, 75*, 1823–1836.

Tremel, D. P., Thomas, J. A., Ramirez, K. T., Dye, G. S., Bachman, W. A., Orban, A. N., et al. (1998). Underwater hearing sensitivity of a Pacific white-sided dolphin, *Lagenorhynchus obliquidens*. *Aquatic Mammals, 24*(2), 63–69.

Tulving, E. (1972). Episodic and semantic memory. In E. Tulving & W. Donaldson (Eds.), *Organization of memory*. New York: Academic Press.

Tyack, P. L. (1993). Animal language research needs a broader comparative framework. In H. L. Roitblat, L. M. Herman, & P. E. Nachtigall (Eds.), *Language and communication* (pp. 117–138). Hillsdale, NJ: Erlbaum.

Tyack, P. L. (1997). Development and social functions of signature whistles in bottlenose dolphins *Tursiops truncatus*. *Bioacoustics, 8*, 21–46.

Urian, K. W., Hofmann, S., Wells, R. S., & Read, A. J. (2009). Fine-scale population structure of bottlenose dolphins, *Tursiops truncatus*, in Tampa Bay, Florida. *Marine Mammal Science, 25*, 619–638.

Wahlberg, M., Jensen, F. H., Soto, N. A., Beedholm, K., Bejder, L., Oliveira, C., … , Madsen, P. T. (2011). Source parameters of echolocation clicks from wild bottlenose dolphins (*Tursiops aduncus* and *Tursiops truncatus*). *Journal of the Acoustical Society of America, 130*(4), 2263–2274.

Wartzok, D., Erbe, C., Getz, W. M., & Thomas, J. (2012). Marine mammal acoustics exposure analysis models used in U.S. Navy environmental impact statements. In A. N. Popper & A. Hawkins (Eds.), *The effects of noise on aquatic life* (pp. 551–557). New York: Springer Science+Business Media.

Wells, R. S. (2003). Dolphin social complexity: Lessons from long-term study and life history. In F. B. M. de Waal & P. L. Tyack (Eds.), *Animal social complexity: Intelligence, culture, and individualized societies* (pp. 32–56). Cambridge, MA: Harvard University Press.

Wells, R. S. (2010). Feeling the heat: Potential climate change impacts on bottlenose dolphins. *Whalewatcher Journal of the American Cetacean Society, 39*(2), 12–17.

Wells, R. S. (2014). Social structure and life history of common bottlenose dolphins near Sarasota Bay, Florida: Insights from four decades and five generations. In J. Yamagiwa & L. Karczmarski (Eds.), *Primates and cetaceans: Field research and conservation of complex mammalian societies* (pp. 149–172). Tokyo, Japan: Springer.

Wells, R. S., Allen, J. B., Hofmann, S., Bassos-Hull, K., Fauquier, D. A., Barros, N. B., et al. (2008). Consequences of injuries on survival and reproduction of common bottlenose dolphins (*Tursiops truncatus*) along the west coast of Florida. *Marine Mammal Science, 24*, 774–794.

Wells, R. S., McHugh, K. A., Douglas, D. C., Shippee, S., Berens-McCabe, E. J., Barros, N. B., et al. (2013). Evaluation of potential protective factors against metabolic syndrome in bottlenose dolphins: Feeding and activity patterns of dolphins in Sarasota Bay, Florida. *Frontiers in Endocrinology, 4*, 139. doi:10.3389/fendo.2013.00139.

Wells, R. S., & Scott, M. D. (1990). Estimating bottlenose dolphin population parameters from individual identification and capture–release techniques. In P. S. Hammond, S. A. Mizroch, & G. P. Donovan (Eds.), *Report of the International Whaling Commission, Special Issue 12* (pp. 407–415). Cambridge.

Wells, R. S., Scott, M. D., & Irvine, A. B. (1987). The social structure of free-ranging bottlenose dolphins. In H. Genoways (Ed.), *Current Mammalogy* (Vol. 1, pp. 247–305). New York: Plenum Press.

Wells, R. S., Tornero, V., Borrell, A., Aguilar, A., Rowles, T. K., Rhinehart, H. L., et al. (2005). Integrating life history and reproductive success data to examine potential relationships with organochlorine compounds for bottlenose dolphins (*Tursiops truncatus*) in Sarasota Bay, Florida. *Science of the Total Environment, 349*, 106–119.

Whitehead, H. (1996). Babysitting, dive synchrony, and indications of alloparental care in sperm whales. *Behavioral Ecology and Sociobiology, 38*, 237–244.

Whitehead, H. (1998). Male mating strategies: Models of roving and residence. *Ecological Modelling, 111*, 297–298.

Whitehead, H. (2003). *Sperm whales: Social evolution in the ocean.* Chicago: University of Chicago Press.

Whitehead, H. (2008a). *Analyzing animal societies: Quantitative methods for vertebrate social analysis.* Chicago: University of Chicago Press.

Whitehead, H. (2008b). Precision and power in the analysis of social structure using associations. *Animal Behaviour, 75,* 1093–1099.

Whitehead, H. (2009). SOCPROG programs: Analyzing animal social structures. *Behavioral Ecology and Sociobiology, 63,* 765–778.

Whitehead, H. (2010). Conserving and managing animals that learn socially and share cultures. *Learning and Behavior, 38,* 329–336.

Whitehead, H., Antunes, R., Gero, S., Wong, S. N. P., Engelhaupt, D., & Rendell, L. (2012). Multi-level societies of female sperm whales (*Physeter macrocephalus*) in the Atlantic and Pacific: Why are they so different? *International Journal of Primatology, 33,* 1142–1164.

Whitehead, H., & Lusseau, D. (2012). Animal social networks as substrate for cultural behavioural diversity. *Journal of Theoretical Biology, 294,* 19–28.

Whitehead, H. & Rendell, L. (2014). *The cultural lives of whales and dolphins.* Chicago: University of Chicago Press.

Whitehead, H., Richerson, P. J., & Boyd, R. (2002). Cultural selection and genetic diversity in humans. *Selection, 3,* 115–125.

Whitehead, H., & Van Parijs, S. (2010). Studying marine mammal social systems. In I. Boyd, D. Bowen, & S. Iverson (Eds.), *Marine mammal ecology and conservation: A handbook of techniques.* Oxford: Oxford University Press.

Whitehead, H., & Weilgart, L. S. (1991). Patterns of visually observable behaviour and vocalizations in groups of female sperm whales. *Behaviour, 118,* 275–296.

Williams, T. M., Kastelein, R. A., Davis, R. W., & Thomas, J. A. (1988). The effects of oil contamination and cleaning on sea otters (*Enhydra lutris*): I. Thermoregulatory implications based on pelt studies. *Canadian Journal of Zoology, 66*(12), 2776–2781.

Wood, F. G. (1953). Underwater sound production and concurrent behavior of captive porpoises, *Tursiops truncatus* and *Stenella plagiodon. Bulletin of Marine Science, 3,* 120–133.

Woods, D. L., Ridgway, S. H., Carder, D. A., & Bullock, T. H. (1986). Middle and long latency auditory event related potentials in dolphins. In R. Schusterman, J. Thomas, & F. G. Wood (Eds.), *Dolphin cognition and behavior: A comparative approach* (pp. 61–77). Hillsdale, NJ: Erlbaum.

Xitco, M. J., Jr., & Roitblat, H. L. (1996). Object recognition through eavesdropping: Passive echolocation in bottlenose dolphins. *Animal Learning and Behavior, 24,* 355–365.

Yurk, H., Barrett-Lennard, L. G., Ford, J. K. B., & Matkin, C. O. (2002). Cultural transmission within maternal lineages: Vocal clans in resident killer whales in southern Alaska. *Animal Behaviour, 63,* 1103–1119.

Contributors

Mats Amundin Linköping University; Kolmården Wildlife Park, Kolmården, Sweden

Whitlow Au Chief Scientist, Marine Mammal Research Program, University of Hawaii at Manoa

Ted W. Cranford Department of Biology, San Diego State University

Nicola Erdsack Department of Sensory and Cognitive Ecology Marine Science Center, University of Rostock

John Ford Department of Zoology, University of British Colombia

Wolf Hanke Department of Sensory and Cognitive Ecology Marine Science Center, University of Rostock

Louis M. Herman Professor Emeritus, Department of Psychology, University of Hawaii at Manoa

Denise L. Herzing Research Director, Wild Dolphin Project; Department of Biological Sciences, Florida Atlantic University

Christine M. Johnson Department of Cognitive Science, University of California, San Diego

Petr Krysl Jacob School of Engineering, University of California, San Diego

Stan A. Kuczaj II Director, Marine Mammal Behavior and Cognition Lab; Department of Psychology, University of Southern Mississippi

Marc O. Lammers Associate Researcher, Hawaii Institute of Marine Biology; President, Oceanwide Science Institute

Lori Marino Executive Director, the Kimmela Center for Animal Advocacy

Paul Nachtigall Director, Marine Mammal Research Program, University of Hawaii at Manoa

Julie N. Oswald Bio-Waves Inc., Encinitas, CA

Adam A. Pack Departments of Psychology and Biology, University of Hawai'i at Hilo

Heidi C. Pearson University of Alaska Southeast

Sam Ridgway Professor Emeritus, Department of Pathology, University of California, San Diego and President, National Marine Mammal Foundation

Jeanette Thomas Professor Emerita, Department of Biology, Western Illinois University

Randall Wells Director, Chicago Zoological Society's Sarasota Dolphin Research Program

Thomas I. White Center for Ethics and Business, Loyola Marymount University

Hal Whitehead Department of Biology, Dalhousie University, Nova Scotia

Kelley A. Winship Marine Mammal Behavior and Cognition Lab; Department of Psychology, University of Southern Mississippi

Bernd Würsig Departments of Marine Biology and Wildlife and Fisheries Sciences, Texas A&M University at Galveston and College Station

Index

Page numbers in italics indicate figures.

Printed in the United States
by Baker & Taylor Publisher Services